Corrosion of Ceramic Materials

Third Edition

T0300060

CORROSION TECHNOLOGY

Editor, Philip A. Schweitzer, P.E.,
Consultant, York, Pennsylvania

Corrosion of Ceramic Materials

Third Edition

Ronald A. McCauley

CRC Press
Taylor & Francis Group
Boca Raton London New York

CRC Press is an imprint of the
Taylor & Francis Group, an **informa** business

CRC Press
Taylor & Francis Group
6000 Broken Sound Parkway NW, Suite 300
Boca Raton, FL 33487-2742

First issued in paperback 2017

© 2013 by Taylor & Francis Group, LLC
CRC Press is an imprint of Taylor & Francis Group, an Informa business

No claim to original U.S. Government works

Version Date: 20121106

ISBN 13: 978-1-4398-2022-3 (hbk)
ISBN 13: 978-1-138-07364-7 (pbk)

Library of Congress Cataloging-in-Publication Data

McCauley, Ronald A., 1942-
    Corrosion of ceramic materials / Ronald A. McCauley. -- 3rd ed.
      p. cm. -- (Corrosion technology)
    Rev. ed. of: Corrosion of ceramic and composite materials / Ronald A. McCauley. 2004.
    Includes bibliographical references and index.
    ISBN 978-1-4398-2022-3 (hardback)
    1. Ceramic materials--Corrosion. I. McCauley, Ronald A., 1942- Corrosion of ceramic and composite materials. II. Title.

TA455.C43M4 2013
620.1'404223--dc23
                                                          2012032573

Visit the Taylor & Francis Web site at
http://www.taylorandfrancis.com

and the CRC Press Web site at
http://www.crcpress.com

To my parents

Harry Sylvester McCauley
1909–1966

and

Hazel Lorena (Rishel) McCauley
1912–2004

Contents

Know where to find the information and how to use it.

Albert Einstein

Section I   Fundamentals

Section II   Corrosion Analysis

Section III   Corrosion of Specific Materials

Section IV   Properties and Corrosion

Section V   Minimization of Corrosion

Preface to the Third Edition

It has been almost 10 years since the second edition was published. Since then, much has happened in the field of corrosion of ceramics. Much has occurred in the field of bioceramics including some information on nanomaterials. Although an attempt was made in the second edition to add more on these areas and the weathering of construction materials, insufficient material was incorporated to warrant whole chapters on these subjects. In addition, these topics were scattered throughout the book. These shortcomings have been remedied in this third edition by including new chapters on corrosion by biological sources (Chapters 4 and 14) and corrosion of architecture materials (Chapter 13), although no claim to completeness is made. Other areas have been added to existing chapters, especially the area of thermal and environmental barrier coatings.

The author has always thought that the best way to classify corrosion was by chemistry, although it could also be classified by application (i.e., bioceramics) or even type of material (i.e., composites). This third edition has been reorganized into four main sections—Fundamentals, Corrosion Analysis, Corrosion of Specific Materials, and Properties and Corrosion. A strong attempt has been made to follow the original plan of classification by chemistry. Because of the importance of the areas of bioceramics and architectural materials, it was believed that individual chapters on these subjects were warranted. Thus there will be by necessity some overlap and repetition; however, every attempt was made to minimize this. Also, the chapter on composites that was added to the second edition has been kept and expanded. New literature sources have been added to all chapters where appropriate.

The first edition was written to be primarily a reference edition, whereas the second edition with its included chapter-ending questions and selected readings was meant to address those readers who may want to use it as a textbook. This third edition continues that trend with more questions and examples. It is still probably most appropriate as a reference book, but it could suffice as a graduate-level text or even be used as a senior-level undergraduate text if all the background fundamental sciences have been mastered.

A special thank you is due Nan Gao, who offered to translate the first edition into Chinese, which was eventually published by the Metallurgical Industry Press in 2003.

Ronald A. McCauley

Preface to the Second Edition

Although a better understanding of the mechanisms of corrosion has occurred over the past 10 years since the publication of the first edition, corrosion still remains a major problem. One area where some advances have been made is the understanding of the weathering mechanisms of building materials, especially where related to monuments of the past. The weathering of building materials has been added as new sections in Chapters 2 (Fundamentals) and 5 (Corrosion of Specific Crystalline Materials). Although the work on weathering of building materials has been done predominantly by civil engineers, it should be of interest to the ceramic engineer. More information has become available concerning the corrosion of composite materials of all types. The strong interest in composites materials has been enhanced by the aerospace industry and the military. Because of this, a whole new chapter (Chapter 7) has been devoted to composite materials. In addition new sections have been added on bioceramics (Chapter 5). New literature sources have been added to all chapters where appropriate.

The first edition of this book was written to be used primarily as a reference book. Questions have been added at the end of most chapters and additional examples have been included along with recommended reading lists so that this second edition may be used also as a textbook for either a senior-level undergraduate or a graduate course on corrosion.

Preface to the First Edition

One of the most important problems confronting engineers today is the development of materials that are reliable under various environmental conditions. In some cases these conditions are considered extremely hostile—very high temperatures, mechanical loading, and/or aggressive chemical attack. Ambient temperature aqueous attack can also be extremely detrimental, especially over an extended period of time, as in the case of hazardous waste disposal. Engineers and scientists have been combating the attack upon ceramics of molten glass, molten metals and slags, and molten salts for hundreds of years with many improvements. Most of these improvements have occurred through experimentation, eventually finding the material that worked best. Only during the past 25 years has a true understanding of the complexities of corrosion of ceramics begun to develop. Major advances have been made in recent years; however, the details in many cases are still questionable or at least debatable.

The cost to industry due to corrosion is considerable and only a thorough understanding of all the complexities of the process will help to minimize that cost. There will undoubtedly be many applications of ceramics where the ceramic will be consumed during service, but maximizing service life will greatly reduce the overall cost.

While several books, mostly in the form of symposia proceedings, have been published on various aspects of corrosion of crystalline and glassy ceramics, generally on the newer, advanced materials, none has addressed the subject in a comprehensive manner. The most significant works have been reported in the technical literature; however, reading all the published articles is a formidable task. This book is an attempt to discuss all aspects of the corrosion of ceramics, but no attempt has been made to complete an exhaustive literature review. Although not all areas have been described in great detail, a summary of some of the most important work has been given with references for the interested reader.

This book is based upon a combination of lecture notes from the Advanced Refractories course that the author has taught at Rutgers during the past 34 years and the author's industrial and consulting experiences. It is intended predominantly as a reference work for practicing engineers and research scientists but could also be used as a text for a graduate-level course in corrosion of ceramics. Any comments or suggestions about the content of this book will be most welcome.

Acknowledgments

The author would like to thank the faculty and students of the Department of Ceramic and Materials Engineering (Materials Science & Engineering since 2005) at Rutgers, The State University of New Jersey, for many helpful and thoughtful discussions during the preparation of this book, and especially Drs. John Wachtman and M. John Matthewson for reviewing a portion of the manuscript (first edition) and for their valuable suggestions. A special thank you is due Major General Dr. George Weightman for reviewing the chapters related to biological sources of corrosion in this third edition.

The author would like to extend a very special thank you to Mr. William Englert of PPG Industries, who first introduced the author to the fascinating field of corrosion of ceramics.

Gratitude must also be extended to Mrs. Mary Guerin for her help in preparation of the first edition manuscript and to Paul Mort, Robert Sabia, John Martin, and Ryan McCuiston for their help in preparing the figures of the first edition, which have been used in the second and third editions.

The author would also like to extend a very special thank you to his wife, Eleanora, and his son, Matthew, for their understanding and encouragement during the many long hours required to complete this task.

Introduction

If we begin with certainties, we shall end in doubt; but if we begin with doubt, and are patient in them, we shall end in certainties.

Roger Bacon

Most engineers at one time or another will be confronted with corrosion whether it will be their sole endeavor or whether it will be a minor unexpected nuisance. The actual study of corrosion—its causes, effects, and means of elimination—is not as common in the field of ceramics as it is in the field of metallurgy. Actually, many engineers have the mistaken belief that ceramics do not corrode. Although many engineers study the corrosion of ceramics all their lives, they normally do not consider themselves as corrosion engineers, but as ceramic engineers or process engineers or possibly some other type of engineer. There are no corrosion engineering courses offered in the several undergraduate ceramic engineering curricula in the United States. Even at the graduate level, there are no courses dedicated to corrosion of ceramics, although several contain a large amount of information related to corrosion, such as those related to high-temperature materials, thermodynamics, etc. There is definitely no such thing as a bachelor's degree in corrosion engineering of ceramic materials. The University of Akron, however, has developed an undergraduate BS program in corrosion and reliability engineering with the help of National Association of Corrosion Engineers (NACE International) and the Department of Defense (DOD). This program was offered for the first time in the fall of 2010. This is the first degree program of this type in the United States. The American Ceramic Society does not have a specific division devoted to corrosion like some other major societies. For example, the Electrochemical Society has a Corrosion Division. They also publish a Corrosion Monograph Series. NACE International publishes a journal of corrosion science. But again these society divisions and publications are heavily weighted toward metals.

In 1998 the U.S. Congress approved an amendment to the Transportation Equity Act of the 21st Century for the cost of corrosion study due to *metallic* corrosion.[*] Total annual direct cost was estimated to be $276 billion per year for the period 1999–2001. Various strategies were given for the management and prevention of corrosion. Probably the most important strategy was the need for improved education and training. An outcome of that study was a report by the National Academy of Sciences titled *Assessment of Corrosion*

---

[*] Report No. FHWA-01-156 "Corrosion Cost and Preventive Strategies in the United States," September 30, 2001.

Education released in 2009.* This report stated that "first, there is an urgent
need to revitalize the corrosion education of our country's engineering
workforce and, second, that revitalization can be accomplished only in the
context of a healthy corrosion engineering research community." Although
this study addressed corrosion of metals it is appropriate also for ceramics.
As is the case for ceramics, this report found that "corrosion engineering is
not a required course in the curriculum of most bachelor's degree programs
in Materials Science and Engineering and related engineering fields in many
programs, corrosion is not only not a required subject, it is not even available." In addition, the FY2007 DOD Authorization Act created a corrosion
executive position and a policy and oversight office at the Pentagon.

Throughout the history of the ceramic industries, various material types
or compositions have been used because of some particular advantageous,
intrinsic property. High strength, low electrical conductivity, or some other
property may be the primary concern for a particular application. However,
excellent resistance to attack by the environment always plays a role and
may, in some cases, be the prime reason for the selection of a particular material. This is especially true for those materials selected for furnace construction in the metal and glass industries.

Almost all environments are corrosive to some extent. For practical applications it comes down to a matter of kinetics—how long will a material last
in a particular environment? In some cases corrosion may be beneficial, such
as in the preparation of samples by etching for microscopic evaluation; in
chemical polishing to obtain a flat, smooth surface; or in bioactive materials where reaction with bone or complete disappearance is required. The
selective leaching of the sodium- and boron-rich phase in phase separated
borosilicate glass to produce a high silica content glass (called Vycor™) is
an excellent example of how corrosion can be put to a beneficial use. Other
examples include dissolution and reprecipitation in liquid phase sintering
(also crystal growth studies) and the dissolution of various raw materials in
molten glass in the manufacture of glass products.

The proper selection of materials and good design practices can greatly
reduce the cost caused by corrosion. To make the proper selection, engineers
must be knowledgeable in the fields of thermodynamics, physical chemistry, electrochemistry, and even meteorology. In addition, engineers must be
familiar with the corrosion testing of materials, the nature of corrosive environments, and the manufacture and availability of materials and have a good
sense of the economics of the whole process. There is a growing need in many
ceramic applications to be able to predict the service life based upon laboratory tests. The limiting factors in making such predictions are more often
than not due to a lack of a thorough knowledge of the industrial operating

---

* "Assessment of Corrosion Education," National Research Council, National Academy of
Sciences, Washington, DC, 2009.

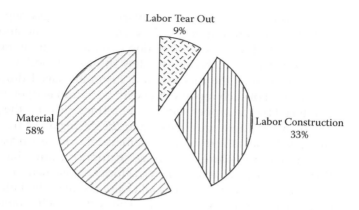

FIGURE I.1
Glass furnace repair estimated cost percentages.

conditions rather than to devising the proper laboratory test. A thorough knowledge of the microstructure and phase assemblage of the material, however, is critical to an understanding of the corrosion that may take place. The National Science Foundation has provided funding to several universities[*] to develop a Digital Library of Ceramic Microstructures (DLCM), which when completed will be a tremendous aid to the engineer involved in corrosion studies. This library will contain, in addition to selected micrographs, chemistry, phases, and some properties. Access to the database will be via the Internet. Even though a material may be listed in some handbook as having excellent resistance to some particular environment, it is important to know the form of the material. Were the data listed for a single-crystal, a powder, or a dense (or porous) sintered component? Were there any secondary phases present or was it a pure material? *The form and processing of a material will affect its corrosion.*

The cost of corrosion to ceramics in the United States is enormous; however, in many cases this corrosion is looked upon as a necessary expense in the production of some product. An excellent example is the corrosion of the refractories in a glass-melting furnace. These furnaces are shut down periodically to replace the worn-out materials. Not only is the cost of the refractories involved but also the cost of the tear out, the cost of the reconstruction, and the cost of any lost production. The total cost of such a repair can be in excess of $15 million for a single furnace. Figure I.1 shows an average estimate of the percentages for labor and materials for a typical furnace repair. Business interruption costs have not been included, since these will vary considerably depending upon the product being produced and the size

[*] University of Dayton Research Institute has the primary role. Georgia Institute of Technology, North Carolina A&T State University, and University of Missouri–Rolla along with Mechanical Test Instrumentation & Control are supplying the information.

of the furnace. The downtime for repairs also varies considerably depending upon the extent of the repair, but generally varies between one and three months. The cost due to business interruption can amount to as much as $1 million per month. There are, however, a few things that add a credit to the costs of a repair, such as the fuel and raw materials saved during the shutdown. The total cost of such repairs can never be eliminated; however, it can be greatly reduced by the proper selection of refractories and the proper operation of the furnace. There are times when the corrosion of the refractories goes unnoticed and failure occurs prematurely. Since furnaces are insured against premature failure, very large insurance claims have been filed. In addition to the costs related to the refractories, construction, and lost production are the costs related to additional cleanup due to the failure and the costs of insurance adjusters and lawyers. The cost of such a failure can exceed $20 million. Thus it should be obvious that the equation to determine the exact costs of a particular repair is quite complex.

Environmental problems can also add to the total cost of corrosion. For example, a trend toward the use of nonchrome-containing refractories for furnace construction has been ongoing for about the past 30 years. Used refractories have in the past been disposed of by burying them in landfills. Chrome-containing refractories have the potential of contaminating ground waters with hexa-valent chrome, a carcinogen. If chrome-containing refractories are used, upon disposal they must be hauled to toxic waste dumps with an added cost of disposal. To eliminate this problem some industries have been leaning toward the use of other materials for the construction of their furnaces. In some cases, the replacement material does not last as long as the chrome-containing material, thus shortening the time between repairs and adding to the cost.

The products of corrosion may enter the product being manufactured and lower the quality of the product or decrease the yields. Although this is a cost due to corrosion, it is one that is extremely difficult to quantify. This is often an indirect cost assigned to the consumer, since the effects of the corrosion may not be noticed until the product is being further processed or even used by the final consumer. Baboian 1986 listed the following areas, which all add to the cost of corrosion:

1. human safety,
2. plant shutdowns,
3. waste of resources,
4. contamination of products,
5. reduced efficiency,
6. expensive maintenance,
7. overdesign, and
8. inhibition of technological progress.

Although no accurate numbers are available for the annual cost to U.S. industries for the corrosion of ceramic materials,* an estimate of $3 billion does not seem unreasonable† and is probably quite conservative. Only through the intelligent selection of ceramic materials can the cost of corrosion be minimized. By using the best of current technology, Baboian estimated that 15 to 25% of the cost of corrosion could be avoided. This intelligent selection of materials can be obtained only through a thorough understanding of all the complexities of ceramic materials and the effect that the environment has upon them. These complexities involve the thermodynamics and kinetics of chemical reactions, which if understood thoroughly will allow the engineer to solve any corrosion problem in any field of study.

The corrosion of ceramic materials has not been as well categorized as it has been for metals. Similar terms do, however, appear in the literature. The more common types referred to in the literature are *diffusion corrosion,* which is very similar to concentration cell corrosion in metals; *galvanic cell corrosion; grain boundary corrosion;* and *stress corrosion.* A more common trend in ceramic materials is to group corrosion under a more general mechanism, such as *dissolution corrosion* (i.e., corrosion of a solid by a liquid). In this type of corrosion, diffusion, galvanic cell, grain boundary, and stress corrosion may all be present.

There are also many words used to describe corrosion, and if one is looking for information on the subject of corrosion of a particular material, a search including many of the following words should be performed: *consumption, decomposition, degradation, deterioration, dissolution, durability, instability, oxidation, reduction,* and *erosion.* Although erosion is technically not the same as corrosion, being a physical effect rather than a chemical one, erosion in many cases provides a means for continued corrosion.

Corrosion of ceramic materials and its relationship to various property degradation does not receive as wide a recognition as it probably should at the various technical and professional society meetings. For example, only about 2 1/2% of the approximately 1500 presentations at the 2002 Annual American Ceramic Society meeting were devoted to topics related to corrosion. As can be seen from Table I.1, the percentage of presentations on corrosion has been between 4 1/2 and 2 1/2% during the decade of the 1990s. The data are truly not sufficient to conclude that a trend exists. Even though the main topic of interest changes somewhat from year to year, hazardous waste materials appears to remain a major area of interest. Selecting a major topic is complicated by the fact that areas often overlap—for example, general glass durability versus hazardous waste glass leaching. At meetings of a less general nature, such as the Unified International Technical Conference

---

* Most studies evaluate the cost of metallic corrosion or combined materials cost.
† Although not all dental restorations (i.e., fillings) are ceramic many are composites containing substantial amounts of ceramic. The annual cost in the United States for replacement dentistry is $5 billion, Jokstad et al. 2001.

TABLE I.1

Presentations on Corrosion at American Ceramic Society
Annual Meetings

Year	Total Papers[a]	% on Corrosion[b]	Main Topic[c]
1990	1425	4 1/2	HWM[d]
1992	1450	3	Refractories/Composites
1994	1545	3 1/2	Composites/Basic Sci
1996	1578	2 1/2	Refractories
1998	1591	2 1/2	Refractories/HWM
2000	1455	3 3/4	Refractories/HWM
2002	1514	2 1/2	Nonoxide oxidation/HWM
2004	1620	2	Various

[a] Includes posters.
[b] Only presentations where corrosion was main topic.
[c] Does not include hydration of cements and fabrication reactions.
[d] HWM = hazardous waste materials.

on Refractories (UNITECR) in 1989 about 12 1/2% of the approximately 150 papers were on topics related to corrosion of refractories. In 2011 that number had dropped to about 9% of the 242 papers presented. This is a much better situation than the general field of ceramics, but historically the corrosion of refractories has received more attention than the corrosion of ceramics. At the First Ceramic Science & Technology Congress in 1989 an international symposium was held titled *Corrosion and Corrosive Degradation of Ceramics.* Of the 26 papers presented in the symposium more than 1/2 were devoted to silicon carbide and silicon nitride, indicative of the importance that is placed upon the corrosion resistance of these advanced ceramic materials. This symposium, however, was only a small portion of the parent congress where more than 625 papers were presented. At the 2003 Annual Conference & Exposition on Advanced Ceramics and Composites about 3 3/4% of the almost 500 presentations were on corrosion, a situation not much different than the Annual Ceramic Society meetings. Tracking the numbers of presentations on the corrosion of ceramics has become more difficult since the merger in 2006 of major conferences of several societies* to form the Materials Science and Technology (MS&T) Conferences.

---

* The co-sponsors for the MS&T conferences now include the American Ceramic Society, Association for Iron and Steel, ASM International, TMS—The Mineral, Metals, and Materials Society, and NACE International.

Exercises, Questions, and Problems

1. List the different fields that an engineer might want to study to be proficient in evaluating corrosion problems.

2. List all the various courses taught in a standard ceramic engineering curriculum that might contain information related to corrosion.

3. Discuss whether it is more important to know whether a material will corrode or how long it will last.

4. Discuss several applications where the corrosion is a beneficial factor rather than a problem. Find several applications not listed in the text.

5. Develop an equation that represents all the various factors involved in determining the total cost of a particular corrosion problem. Pick a particular product and try to find as much information as possible concerning the actual costs.

6. Discuss the relationship of erosion to corrosion and how it can be a major actor in the overall process of corrosion.

References

Baboian, R., 1986. Corrosion—A National Problem, *ASTM Standardization News*, Mar, 34–9.

Jokstad, A., Bayne, S., Blunck, U., Tyas, M. and Wilson, N. 2010. Quality of Dental Restorations, *Int. Dent. J., 51*, 117–58.

Section I

Fundamentals

1

Corrosion by Liquids

Before anything else, getting ready is the secret of success.

Henry Ford

Introduction

Corrosion of ceramics can take place by any one or a combination of mechanisms. Various models have been proposed to describe these mechanisms, several of which will be discussed below. In general the environment will attack a ceramic, forming a reaction product. The reaction product may be solid, liquid, gas, or any combination of these. This reaction product may remain attached to the ceramic or it may be fugitive, in the case when gaseous species make up the reaction product, or it may be a combination of both. When the reaction product remains as a solid, quite often it forms a protective layer toward further corrosion. At other times, for example if the reaction product is a combination of solid and liquid, this reaction layer may be removed through the process of erosion.* Thus to analyze corrosion one must have some idea of the type of processes that are in operation. When the reaction product remains as an intact interfacial layer, analysis is relatively easy. When gaseous species are formed, the consumption of the ceramic manifests itself as a weight loss. An understanding of the mechanism, however, requires analysis of the evolved gases. Many times the interface formed is very porous and/or friable requiring special care in preparing samples for analysis. Because of the various processes that may take place during corrosion, there is no one general model that can explain all cases of corrosion. In addition, a single ceramic material will react differently to different environments and thus there is no single explanation for the corrosion of a particular material for all environments. *It is also true that the manufacturing history of a ceramic material will affect its performance.* This may manifest itself, for example, as a low-corrosion-resistant grain boundary phase or a pore size distribution that greatly increases the exposed surface area to corrosion. Thus it should be obvious that a simple all-encompassing general theory of corrosion of ceramics does not exist and, because of the nature of corrosion

---

* The combined processes of erosion and corrosion are called *tribocorrosion* by the metallurgists.

and ceramics, will most likely never exist. There does, however, appear to be a common thread connecting all the various studies that have been reported. That is, *corrosion is dependent upon the structural characteristics of the material. The more compact or tightly bonded materials corrode less whether they are glasses or crystalline materials.* Thus it appears that if a general theory is to be developed, a comprehensive investigation of single crystals and some structurally well-characterized glasses should be investigated.

Corrosion, being an interfacial process, requires a thorough understanding of the surface structure of the material being corroded. Thus the study of single crystals is the best method to determine the fundamentals of corrosion mechanisms. It is not always possible, however, to obtain single crystals of sufficient size for appropriate measurements. Although the crystal surface characteristics determine short-term corrosion behavior, they may not be as important for long-term corrosion. Single crystals do lend themselves to evaluation of the effects that various dopants and defects (e.g., dislocations) have upon dissolution kinetics.

When attempting to understand the corrosion of a ceramic, it is a good idea to remember some of the fundamental concepts of chemistry that are too often forgotten. The following are just a few concepts that go a long way in helping one to understand corrosion:

1. A ceramic with acidic character tends to be attacked by an environment with a basic character and vice versa.

2. The vapor pressure of covalent materials is generally greater than that of ionic materials and therefore they tend to vaporize or sublime more quickly.

3. Ionic materials tend to be soluble in polar solvents[*] (e.g., salt in water) and covalent materials tend to be soluble in nonpolar solvents (e.g., SiC in hexane).

4. The solubility of solids in liquids generally increases with increasing temperature.

The solubility of materials in liquids can be obtained from phase diagrams, which give the saturation composition at a given temperature. Unfortunately, for many practical systems, phase diagrams are either very complex or nonexistent. Many data are available, however, for two- and three-component systems, and these should be consulted before attempting to evaluate the corrosion of a specific material.[†] The corrosion of a single pure compound by a liquid can be evaluated by use of the Gibbs Phase Rule. For example, the

---

[*] Polar solvents have a high dielectric constant, whereas nonpolar solvents have a low dielectric constant. At 20°, the dielectric constant of water is 80 and that of hexane is 1.874.

[†] *Phase Diagrams for Ceramists*, Vols. I–XIV, Am. Ceram. Soc., Westerville, Ohio. A software package called *Thermo-Calc* is available for calculating phase diagrams from the Dept. of Materials Science & Engineering, KTH, S-100 44 Stockholm, Sweden.

system of a binary oxide $A_xB_yO_z$ corroded by a liquid M_aO_b contains three components, where a solid and liquid are in equilibrium at some fixed temperature and pressure. This system has only one degree of freedom. Thus if the concentration of one dissolved component is changed, the concentrations of the others must also change. A good discussion of the use of phase diagrams in dissolution studies is that by Cooper 1970.

The corrosion of a solid crystalline material by a liquid can occur through the formation of an interface or reaction product formed between the solid crystalline material and the solvent. This reaction product, being less soluble than the bulk solid, may or may not form an attached surface layer. This type of mechanism has been called *indirect dissolution, incongruent dissolution,* or *heterogeneous dissolution* by various investigators. There are many examples of this reported in the literature. In another form, the solid crystalline material dissolves directly into the liquid either by dissociation or by reaction with the solvent. This type of mechanism is called *direct dissolution, congruent dissolution,* or *homogeneous dissolution.* The term *selective dissolution* is also found in the literature but is used to imply that only a portion of the species in the solid is dissolved whether or not an interface is formed. The saturation solution concentrations of the crystalline species in the liquid along with the diffusion coefficients of the species involved, all determine whether one mechanism will exist or the other. The most abundant species and their concentrations in the liquid must be known for one to determine the degree of saturation. This in turn will determine whether or not the solid will dissolve. The corrosion rate-limiting step in the indirect type may be the chemical reaction that forms the interfacial layer, diffusion through this interfacial layer, or diffusion through the solvent. When one is involved in the study of single crystals dissolution of the various crystallographic planes may be different. If the dissolution rate is the same for all planes, it is called *isotropic dissolution.* If the dissolution rate varies among the various planes, it is called *anisotropic dissolution.* It is easy to understand why dissolution may be different for different planes due to the differences in density of atoms, charges, and/or bonding; however, isotropic dissolution is not as easy to understand, even for a single cation-anion combination. In the case of a cubic crystal such as NaCl where the crystal faces exposed to dissolution are all of the same family, then isotropic dissolution is easy to understand.

Examination of the appropriate phase diagram will aid one in determining whether or not a particular combination of liquid and solid will form an interface. An example is given in Figure 1.1 that shows a portion of the CaO–Al_2O_3–SiO_2 diagram. If the 1265°C eutectic composition were selected as the liquid and melted in a crucible of Al_2O_3 at 1500°C, the dissolution would be indirect and an interface of $CaAl_{12}O_{19}$ (CA_6) would form. As the melt temperature increases, the dissolution type changes to direct above 1700°C, with the eutectic melt being in equilibrium with the alumina crucible. A slightly different situation exists if mullite were selected as the crucible material. At 1400°C, the dissolution is indirect with the interface being anorthite (CAS_2).

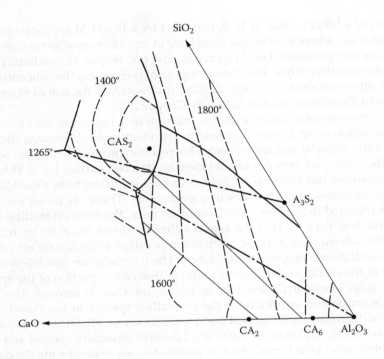

FIGURE 1.1
A portion of the CaO-Al$_2$O$_3$-SiO$_2$ phase diagram. The heavy lines between 1265°C eutectic and either alumina or mullite are construction lines. The dashed lines are isotherms. The solid lines are liquidus field boundary curves. (After Phase Diagrams for Ceramists.)

As the temperature increases, the dissolution remains indirect but the interface changes to alumina above 1500°C. As can be seen from these examples, the interface is determined by the intersection of the isotherm of interest and the construction line joining the liquid and the solid compositions.

The wetting characteristics between a ceramic and a liquid are very important in determining the degree of contact that is developed. Although the investigation of Kramer and Osborne 1983 was performed in an attempt to understand better the parameters involved in the processing of glass-ceramic-to-metal seals, their data exhibited the importance of atmospheric effects upon the degree of contact between a ceramic and a liquid. The better the contact the more aggressive the attack can be. Kramer and Osborne studied the atmospheric effects upon the wetting of a glass-ceramic by Ni-based alloys. They found that helium lowered the contact angle more than argon or an argon/hydrogen mixture. They also found that contact angles were inversely dependent upon the atmospheric dew point.

Noyes and Whitney 1897, in their classic work of the dissolution of lead chloride in boric acid and water, speculated that the rate of corrosion of a solute by a solvent was controlled by the diffusion rate of atoms away from the solute surface. Nernst 1904 postulated that a thin layer of solvent adjacent to the solute became rapidly saturated and remained saturated during the dissolution process and that beyond a certain distance, the concentration was that of the bulk solution. The following equation, now called the Noyes-Nernst equation, represents the flux density across the solute interface:

$$VdC_\infty/dt = jA = (D/\delta^*)(C_{sat} - C_\infty)A \tag{1.1}$$

where:
  V = volume of solution,
  C_∞ = concentration in the bulk,
  C_{sat} = saturation concentration,
  A = area of interface,
  D = diffusion coefficient,
  δ^* = boundary layer thickness, and
  t = time.

By including the surface reaction rate constant, K, Berthoud 1912 derived the following equation:

$$j = \frac{K}{\{1 + (K\delta^*/D)\}}(C_{sat} - C_\infty) \tag{1.2}$$

which indicates that the driving force for dissolution is a combination of both the interface chemical reaction and the interdiffusion of the products and reactants. The derivation from first principles of the empirical constant, δ^*, came after the development of boundary layer theory by Prandtl 1928 and Levich 1962. The most important consequence of these theories for the experimentalist was that the effective boundary layer thickness, δ^*, of a rotating disk was independent of its radius and proportional to the square root of the angular velocity, Levich 1947.

Crystalline Materials

Attack by Aqueous Media

Introduction

Probably some of the more significant work being done today concerning the understanding of ceramic-water interfaces is that being done in the area of *ab initio* calculations, Lasaga 1990. Large modern computers have made it possible to obtain accurate calculations that describe the potential surface of silicates through solutions to the Schrodinger equation. The major assumption in these calculations is that the local chemical forces of the first and second nearest neighbors determines the largest portion of the dynamics and energetics of the chemisorption process. Calculations of this type have shown that the adsorption of water onto silica terminal OH groups (silanols) is more stable than onto bridging OH groups. If these silanol groups are removed through heating, leaving a surface of essentially siloxane bonds (Si-O-Si), the surface becomes hydrophobic. In addition to the *ab initio* calculations, the use of molecular dynamics has allowed the description of the collective atomic motions on mineral surfaces and the surrounding fluids. One result of these studies has been the finding that for at least a few atomic layers the surface structure of solids can deviate considerably from that of the bulk. In general, all these studies indicate that the dissolution rate is controlled by the chemical species adsorbed onto the surface. In addition the amount and type of surface species is dependent upon the *p*H, which also determines the surface charge.

The structure of the water molecule is very important in the understanding of aqueous dissolution. The single oxygen atom has four sp^3 atomic hybrid orbitals (2s, $2p_x$, $2p_y$, and $2p_z$) that are directed toward the corners of a tetrahedron. Each of the hybrid orbitals contains two electrons. Two of these hybrid orbitals react with a 1s hydrogen atomic orbital, leaving the two remaining sp^3 hybrid orbitals uncommitted (called the *lone pairs*). This configuration thus produces an asymmetric electric charge across the water molecule making water a *polar molecule*. The small mass of the hydrogen allows the water molecule to rotate so that the negative charge of the lone pairs can interact with a positively charged ceramic surface or, alternatively, form hydrogen bonds with a negatively charged surface. The dissociation of water into ions of H^+ and OH^- most likely occurs as an excess or deficiency of protons on a water molecule creating H_3O^+ and HO^-, maintaining electrical neutrality. The relative concentration of these ions is the basis for the *p*H scale of acidity and alkalinity.

Minerals

A tremendous amount of information is available concerning the leaching or dissolution of minerals, especially silicates, in the soils literature and anyone interested in corrosion in aqueous systems should avail themselves of that literature. The Jackson weathering sequence discussed by Marshall 1977 exhibits a trend that may be applicable, to some degree, to the dissolution of silicate ceramics. *The greater the degree of bonding of the silica tetrahedra, the more difficult the weathering.* Some variation and overlap occurs due to the specific chemistry for an individual mineral, with the minerals containing alkalies and alkaline earths being less stable than those containing alumina and the transition metals. This is supported by Huang 1977 who in his studies of olivines, pyroxenes, and amphiboles reported that the relative stability of these materials appeared, among other things, to be related to the degree of polymerization of the tetrahedra, with more highly polymerized materials being more stable. A related phenomenon reported by Casey and Bunker 1990 was that minerals with a low density of cross-links tend to dissolve congruently and rapidly, while minerals with a high density of cross-links, such as the tektosilicates, dissolve incongruently producing a leached surface layer. Incongruent dissolution and selective leaching result from the three processes of hydration, hydrolysis, and ion-exchange. Hydration is more prevalent in materials with a low degree of covalent character to the cross-links and a structure that allows water penetration into the structure (i.e., those containing pore sizes > 2.8 Å).

This relationship between structure and leaching was described by Casey and Bunker in a comparison of quartz containing a small percentage of aluminum, forsterite, and albite. The quartz structure being a completely linked network of silica tetrahedra with very small pores exhibits only a near surface leaching of the aluminum. Forsterite, on the other hand, is a structure with a cross-link density of zero, containing independent silica tetrahedra bonded by magnesium ions. In acid solutions the silica tetrahedra are converted intact into silicic acid without hydrolysis. If a leached layer were to form, it would be very thin, since no bridging oxygen bonds would remain when the magnesium ions were removed. The albite structure contains exchangeable ions (sodium), hydrolyzable silica and alumina tetrahedra, and a cross-link density the same as quartz. One-third of the cross-links in albite are of the Al–O–Si type, a more reactive cross-link than the Si–O–Si bonds. The resultant hydrolysis of these Al–O–Si bonds opens the structure, allowing deep penetration of the solutes and water. As this occurs, structural integrity is maintained by the residual silicate framework, which allows a very thick leached layer to form.

Dissolution studies of smectite minerals by acids were shown by Borchardt 1977 to take place through the following steps:

1. Exchange of cations with H_3O^+,
2. Removal of the octahedral Al, Mg, and Fe, and
3. Removal of the tetrahedral Si and Al.

Humic substances were reported by Schnitzer and Kodama 1977 to exhibit strong solvent activity toward minerals, with silicate minerals generally being more resistant to attack by humic and fulvic acids than nonsilicate minerals. Fulvic acid and low molecular weight humic acid attack minerals by forming water soluble complexes with the di- and trivalent cations via CO_2H and phenolic OH groups. In general, any strong bonding to the metal cations promotes dissolution.

The model of dissolution of minerals is based upon the diffusion of leachable species into a thin film of water ~110 μm thick that is stationary. The movement of soil water also receives these soluble materials by diffusion from this thin film. The following equation represents this process:

$$\text{Mineral A} + n\text{H}^+ + m\text{H}_2\text{O} \leftrightarrow \text{Mineral B} + q\text{M}^+ \tag{1.3}$$

where M+ is the soluble species. The equilibrium constant is:

$$k = \left(\frac{[M^+]^q}{[H^+]^n [H_2O]^m} \right) \tag{1.4}$$

From equation 1.4 it should be obvious that the dissolution of minerals, and actually any ceramic, is dependent upon the pH of the water. The mineral B may actually not be crystalline but may form a *gel layer* with a variation in composition through its thickness. Mineral B may also be a metastable form that may vary its structure and composition depending upon the test conditions as reported by Jennings 1986 for the action of water. In dilute aqueous systems at one atmosphere the activity of H_2O can be assumed to be unity. At higher pressures the activity of water is approximately proportional to the pressure. Thus the dissolution of mineral A is directly affected by the pH of the system. In complex minerals where several ions may be released into solution, the theory of Donnan discussed by Marshall 1975 predicted that the activity ratio of monovalent cations to the square root of divalent cations in solution should tend to be constant.

Elmer 1985 has shown how the water activity of the leachant decreases with increasing acid strength. The ion exchange reaction between the various leachable ions in the ceramic with hydronium ions would be expected to increase with acid strength. He showed that the reaction rate of cordierite (and a borosilicate glass) decreased after reaching a maximum at some intermediate acid strength. This he attributed to the reduction in the water activity with increasing acid strength.

When a silicate is leached by an aqueous solution, an ion is removed from a site within the crystal structure and is placed into the aqueous phase. Most of the transition metal ions and many other ions occur in sixfold coordination in the crystalline structure and also in solution as hexahydrated ions. Since the crystal field stabilization energies of the transition metal ions in oxide crystal structures and in aqueous solutions is about the same, whether or not leaching occurs depends upon the ease with which the ions are removed from the crystal structure.

The mechanism reported by Burns 1970 requires a water molecule to approach the metal ion along a vacant t_{2g} orbital (see Burns also for a discussion of orbital theory) forming a seven-fold coordinated intermediate state, which is the rate-determining step. This intermediate state spontaneously disproportionates into a metal-hydroxysilicate and a hydroxysilicate residue. Continued repetition of the process ultimately produces a metal hydroxide or hydrated oxide and a hydroxysilicate residue.

Ions with d^3, d^8, and low-spin d^6 configurations are the most resistant to leaching, since these configurations contain electrons in the low energy t_{2g}

orbitals, producing a larger energy barrier for the formation of the interme-diate state. In those ions with at least one empty t_{2g} orbital (i.e., d^1 and d^2) the energy barrier is much lower and thus these ions exhibit less resistance to leaching. In those ions with more than three d electrons, additional energy is required to pair electrons in the t_{2g} orbital thus creating a vacant t_{2g} orbital. Thus ions with d^4, d^5, d^6, and d^7 hi spin should exhibit intermediate leaching characteristics. Data reported by Hawkins and Roy 1963 are in very good agreement with the predicted results.

Lasaga and Lüttge 2001 reported that the lateral movement of surface steps is the rate-controlling process in the dissolution of minerals. This lateral step movement causes the formation and evolution of etch pits. Lüttge 2005 gave the following equation for the overall dissolution rate as a first approach based on the volume change of the pits with time:

$$\text{Rate} = k = dV_{\text{pit}}/dt \tag{1.5}$$

where:

V_{pit} = volume loss of etch pits after time t.

This equation is valid only if the etch pits do not grow sufficiently to coalesce. Thus it is realistic only for the beginning of dissolution. The surface area of dissolution was incorporated into the calculation of the volume of the etch pits by using the number of pits per cm^2.

To understand the interaction of a material with an electrolyte, one must have a good understanding of the electrical double-layer characteristics of the immersed material. The structure of this electrical double-layer is dependent upon the decay of the potential from that at the solid surface to the zero potential of the bulk electrolyte. A more detailed discussion of the electrical double-layer concept can be found in the book by Shaw 1980. The methods that are used to study the interaction of ceramics with electrolytes are generally that of pH changes, such as potentiometric titra-tion. One result of these studies is the determination of the pH at which a net zero surface charge exists (called the *zero point of charge, ZPC*), which may or may not correspond to a zero zeta potential. The condition of zero zeta potential is called the *isoelectric point (IEP)*. For those materials that exhibit some solubility, it is more appropriate to use the *IEP* value since this relates to an equal number of positively and negatively charged dis-solution species present at the solid surface. Because of this, the pH of the *IEP* also represents the pH of minimum solubility. Above and below the pH of the *IEP* the dissolution increases due to the presence of species that weaken the metal-oxygen bonds. At low pH, dissolution studies of alkaline earth oxides and transition metal oxides have indicated that dissolution is controlled by the strength of the metal-oxygen bond, Brady and House 1996. Thus dissolution rates can be predicted from the oxide free energy of formation. The driving force for dissolution is the difference in potentials

TABLE 1.1

Probable *IEP* Values

Oxide Type	*IEP* pH Range
M_2O_5, MO_3	< 0.5
MO_2	0–7.5
M_2O_3	6.5–10.4
MO	8.5–12.5
M_2O	> 11.5

Source: Parks, G.A. 1965. The Isoelectric Points of Solid Oxides, Solid Hydroxides, and Aqueous Hydroxo Complex Systems, *Chem. Rev.*, 65(2), 177–198.

between the solid surface and the Stern plane or outer Holmholtz plane (i.e., the closest distance of approach of hydrated ions). According to Parks 1965 the probable *IEP* of a material falls into a range of values depending upon the cation oxidation state as shown in Table 1.1. Parks also lists the *IEP* of many materials.

In addition to the soils literature, a large amount of work has been reported on the dissolution of oxides that form as protective or semiprotective coatings on metals. Diggle 1973 has reported a good review, although dated, of this literature up to about 1971. Diggle has divided the dissolution of these oxide coatings into two major groupings: those cases where the rate-determining step involves electronic charge transfer, which are called *electrochemical dissolution,* and those where no charge transfer is involved in the rate-determining step, which are called *chemical dissolution.* Electronic conductivity, which is related to the oxide structure and bonding, is more important the greater the covalent character of the oxide. In chemical dissolution, crystallographic and metal-oxygen bond strengths play a very important role; whereas in electrochemical dissolution the electronic structure is of prime importance.

Dissolution of solids in solutions is sometimes dependent upon surface controlled reactions at the solid/solution interface. The exchange rate between the solid and the solution ligands decreases as the cation charge increases, forming stronger bonds between the cation and the ligands. An example reported by Bright and Readey 1987 of this dependence is the comparison between Ti^{4+} and Mg^{2+} where the dissolution of MgO is much faster than that of TiO_2.

The dissolution of minerals occurs by the formation of simple ions or complexes, the nature of which is determined by the pressure, temperature, and composition of the fluid. One must know what the most abundant species are and their concentration if one is to know whether the fluid is saturated, under saturated, or over saturated. This will determine if the mineral will dissolve or not.

The solvating ability of water is related to its dielectric constant, Eugster 1986. Raising the temperature lowers the dielectric constant due to the increased kinetic energy of the dipoles. Raising the pressure has the opposite effect. Increased temperature initially leads to greater dissociation of the solvent due mostly to a decrease in hydrogen bonding. Eventually the decrease in dielectric constant dominates increasing association of the solvent.

Ionic crystals dissolve reasonably well at any *p*H near room temperature. Covalent crystals, however, dissolve only at low or high *p*H. At higher temperature water is an excellent solvent for most minerals.

Architectural Materials

The corrosion of architectural materials is generally referred to as weathering, especially in the ceramics community, and is generally restricted to structural clay products, concrete, sandstone, limestone, marble, granite, and glass. Chapter 16 discusses all the various corrosion aspects of glassy materials including weathering; therefore there is no need to repeat them here. A large amount of literature is available on the subject of weathering of rocks and minerals; however, this vast source is generally not tapped when concerned with the weathering of ceramics, although many similarities exist. The weathering of natural building stones is essentially identical to the weathering of rocks and soils. Winkler 1973a states that the decay of engineering structures and monuments of stone and concrete is closely related to the geologic process of weathering.

The chemical aspects of weathering of architectural materials are mostly a problem related to acid rain. Pure distilled water has a *p*H of 7.0. Unpolluted rainwater is slightly acid at a *p*H of 5.6 due to combination of the rainwater with carbon dioxide in the air. Acid rain (i.e., polluted rainwater) has been reported by the U.S. Geological Survey 2007 to have an average *p*H between 4.2 and 4.4 in the Washington, DC, area due to combination of the rainwater with vehicular exhaust pollutants. In addition to moisture effects, faulty handling, design errors, and inappropriate restoration methods may all enhance deterioration. Pollutants from the atmosphere settle out onto stone surfaces as aerosols, which range in size from molecules to raindrops. The most important aerosols in the decay of architectural stone are sizes in the range 0.1 to 0.001 μm, Winkler 1973a.

The effects of moisture on building materials in addition to causing deterioration can also cause human health problems due to microbial growth. Although the health problems associated with this type of corrosion are not the subject of this book, they are a closely related topic and thus cannot be completely ignored. Many building materials contain man-made mineral or glass fibers, such as ceiling tiles, wall panels, and insulation. These materials provide a huge surface area for moisture adherence and microbial growth. Most fungus spores require at least a level of 25% moisture to germinate and grow. Temperature is also important since optimum temperatures for growth are between 27 and 32°C.* Lower temperatures can halt growth but do not cause death. Higher temperatures may cause death. Diseases such as hay fever, asthma, and pneumonitis can be caused by allergens (i.e., fungi spores). Some bacteria can cause pneumonitis. Bacteria generally require more moisture than fungi to survive. Moisture is probably the greatest cause of deterioration in historic buildings. It is not merely a matter of replacing damaged parts but the source of the moisture must be eliminated, Burge et al. 1994. The National Park Service's policy on the preservation of historic buildings was first established in 1966. Standards have been developed for the treatment of historic properties. Most of the materials used in the construction of historic buildings are still available. Using sealers, coatings, etc., should be avoided since these will trap in moisture and cause continued deterioration. The Preservation Assistance Division of the National Park Service should be contacted for a list of the documents available for help in preserving historic properties, Park 1994.

Historically it has been a general practice to bury the used refractories from industrial furnace rebuilds in landfills. When chromium-containing basic refractories are buried a serious potential problem may arise. This problem relates to what happens to the chromium. Most references about basic refractories consider the chromium to exist as Cr^{3+} and reside on the octahedral sites of the mineral phases present. In the fusion cast type of chromium-containing refractories the chromium is located either in the spinel or the sesqueoxide phase. In any of the various unused refractories the chromium is thus generally accepted to be in the trivalent state. It is the hexavalent state of chromium that has been identified as a carcinogen, Joiner et al. 1983. Thus it is a question of whether or not the chromium will be oxidized while in the landfill and whether or not this oxidized ion will be leached into the surrounding ground waters. At present there still has not been a definitive study performed to determine accurately the fate of the chromium in these materials.

Today an extremely important area of corrosion problems related to weathering is that of historic buildings, monuments, and landmarks. The civil engineering community has made great strides in recent years in understanding the mechanisms of weathering of ancient building materials. The weakening of intergranular bonds is the major cause of deterioration of architectural materials. One of the more common building materials was a rock called *sandstone*† which is a rather porous, weak material. Other materials used were *limestone*‡ and *marble*.§ The acid-base character of the various

---

* King 2009 gives the optimal range as 35 to 40°C.
† Sandstone is a sedimentary rock composed mainly of quartz, feldspar, and mica bonded by lime.
‡ Limestone is a sedimentary rock composed of calcium carbonate and is often from marine sources.
§ Marble is a crystalline metamorphic limestone, often containing substantial magnesium in which case it may be called dolomitic marble.

TABLE 1.2

Volume Change during Hydration of Some Minerals

Mineral	Specific Vol. (cm³)	Hydrated Phase	Hydrated Vol. (cm³)	$\Delta V\%$
$CaSO_4$	45.94	$CaSO_4 \cdot 2H_2O$	74.31	61.8
CaO	16.75	$Ca(OH)_2$	33.06	97.3
MgO	11.25	$Mg(OH)_2$	40.90	263.6
Na_2SO_4	53.33	$Na_2SO_4 \cdot 10H_2O$	290.8	312.2

Source: Kühnel, R.A. 2004. Cause and Consequence: Volume Changes behind Building Material Deterioration, *Materials Characterization, 53*, 171–180.

materials is determined by the silica content. If the silica content is greater than 66 wt% the material is considered to be acidic, if less than 52 wt% it is considered to be basic. The porosity provides a route through capillary action; which is large for pore diameters less than 1 μm; for water, moisture, etc. to enter the sandstone. While inside some dissolution can take place depending upon the chemical makeup of the liquid. The most predominant species are CO_2 and SO_2, which when mixed with water form weak acids. After dissolution, the liquid migrates to the surface by capillary action and evaporates. When evaporation takes place at the surface various salts (e.g., gypsum) are left behind either on the surface or within the pores that can disrupt the mechanical integrity of the sandstone depending upon their expansion and growth characteristics upon precipitation from the liquid (see Table 1.2 for the volume changes that occur during hydration of some minerals). When salts are precipitated and left behind on the surface of the material the term *efflorescence* is used. This appears as a whitish-to-gray fluffy bloom on the surface. When precipitation occurs within the pores the terms *subflorescence* or *crytoflorescence* are used. Efflorescence may visually look bad, but generally is not a problem, whereas subflorescence or crytoflorescence can be very damaging. These soluble salts may originate from the material itself, from decomposition of the material, or from external sources like jointing and mortar materials, backing materials, etc. It should be remembered that with many salt solutions considerable supersaturation may take place before precipitation occurs and for mixed salts the individual solubilities may be increased due to various interactions. Supersaturation may occur by decreasing temperature, increasing concentration, or a combination of both decreasing temperature and increasing concentration (the more common situation). Supersaturated solutions generally produce larger crystals that in turn produce greater pressures for disruption. For one-directional crystal growth a pressure within small pores is developed that can be represented by the following equation Correns (1949):

$$P = (RT/V_s) \ln (C/C_s) \tag{1.6}$$

where:
  P = pressure (atm) produced by crystal growth,
  R = gas constant,
  T = temperature, degrees K,
  V_s = solid salt molecular volume,
  C = solute concentration, and
  C_s = solute saturation concentration.

If hydrated salts occur during the growth process, volume changes will occur if humidity changes take place causing the salts to adjust their water of hydration. These volume changes will cause pressure changes that have been represented by the following equation, Winkler 1973b:

$$P = (nRT/V)\ln(P_o/P_h) \tag{1.7}$$

where:
  P = hydration pressure,
  n = moles of water gained during hydration,
  R = gas constant,
  T = temperature,
  V = change in volume (volume of hydrate minus volume of original salt),
  P_o = partial pressure of water vapor, and
  P_h = vapor pressure of hydrated salt.

The use of equation 1.7 demonstrates that low temperatures and high humidities produce the greatest pressures. The reverse is also true.

In a discussion of the deterioration of architectural materials it is difficult to avoid the problems caused by mechanical forces. Although these processes do not fall under a strict definition of corrosion they most often occur simultaneously with the chemical aspects of deterioration. A phenomenon called *salt scaling* has been described by Valenza and Scherer 2006 as "superficial damage caused by freezing a saline solution on the surface of a cementitious body." The maximum damage is caused when the solute concentration is about 3 wt%. A survey of the literature over a 60-year period by Valenza and Scherer revealed the following characteristics of salt scaling:

  1. progressive removal of small flakes or chips of binder,
  2. maximum damage at a solute concentration of ~ 3 wt%,
  3. if solution pool is missing from surface, no scaling occurs,
  4. no damage occurs above –10°C, damage increases as temperature decreases below –10°C and with increased time at temperature,
  5. scaling resistance increases with air entrainment,
  6. salt concentration more important in pool solution than in pore solution,

7. salt scaling susceptibility not correlated with internal frost action susceptibility, and

8. surface strength governs scaling resistance.

Valenza and Scherer explain that pore chemistry and/or internal frost action cannot account for all the characteristics listed above but that a mechanism called *glue-spalling* can. The glue-spall mechanism causes the salt scaling when the salt solution pool freezes on the surface leading to a mismatch in thermal expansion and subsequent chipping of the surface. This process is very similar to the *crazing** of glazes on a ceramic substrate.

The porosity of limestone can increase if temperature fluctuations (on the order of –15 to +35°C) are present, Amoroso and Fassina 1983. It has been reported that some marbles have exhibited as much as 50% increase in porosity when subjected to temperature fluctuations. The porosity increase is a function of the differential thermal expansion of the calcite crystals present in limestone and marbles. Calcite actually expands along the *c*-axis while contracting in the perpendicular direction ($\alpha_c = +26.5 \times 10^{-6}/°C$ vs. $\alpha_a = -5.7 \times 10^{-6}/°C$). This expansion/contraction of the calcite crystals causes them to separate forming small pores. Thus the circulation of corrosive liquids increases causing increased deterioration of the limestones.

In all the processes discussed above it should be recognized that water is an essential ingredient. If there were no water present there would be no dissolution or chemical reactions, no transport of corrosive species, or no disruptive forces. The chemical makeup of the water depends on its origin: rainwater, groundwater, or condensation (similar to rain). Condensation is generally thought of as dew, but one must not forget that frost and fog are also forms of condensation that may lead to corrosion problems. The corrosion that results from contact with fog can be more serious than that from rainwater, since as reported by Hoffmann 1986 pollutants are more concentrated and the *p*H is generally much lower. Although somewhat dated, values for the *p*H of fog in the Los Angeles basin were reported to range from 2 to 4. The reason for these differences from rainwater was attributed to the droplet size being much smaller for fog (2–50 µm for fog and 0.1–3 mm for rain). Higher concentrations of pollutants have also been reported for dew by Mulawa et al. 1986. In addition to the higher pollutant concentrations in dew Mulawa et al. reported that the acidity on downward-facing surfaces was higher (i.e., lower *p*H) than on upward-facing surfaces. The chemical makeup of all these sources has changed over the years and thus the exact mechanisms of weathering including the kinetics have also changed. The sad thing is that the deterioration is becoming more rapid for some very famous landmarks.

---

* Crazing manifests as a network of small cracks in a glaze on a ceramic when the thermal expansion mismatch between glaze and substrate causes tension in the glaze upon cooling from the high processing temperature.

TABLE 1.3

Common Atmospheric Pollutants

CO_2 (combustion processes)
Cl (mostly marine atmosphere but also coal burning)
SO_2/SO_3 (combustion processes, especially those of coal)
NH_3 (waste treatment)
NO_x (combustion processes)

Rainwater chemistry is affected by local pollutants (see Table 1.3), which can have their source many hundreds of miles away and also change over time depending upon the socioeconomic progress of the region. Weather patterns are also of major importance since the wind direction plays a role in where various pollutants end up. The acidity of rainwater is governed mostly by the content of sulfate, nitrate, and ammonium ions. The acidity is generally represented by the pH, which has been reported to be less than 5 over almost the entire eastern United States, Semonin 1986. Neal et al. 1986 reported that the leaching rate of limestone doubled for each $pH = 1$ decrease. Thus one can see that to be able to understand the problems of weathering of ancient buildings one must, in addition to being an engineer, also be a historian and meteorologist.

Groundwater-related weathering implies that one knows a little geology, since dissolved species play a role in attacking the sandstone. Local water tables and flooding history are important aspects of the problem. To complicate the whole process a combination of the above mechanisms can take place.

A form of building stone weathering located mostly near marine coasts, called *salt weathering*, appears very much like what geologists call *cavernous or alveolar weathering* of natural rock. An example of salt weathering is shown in Figure 1.2. Compare this with Figures 1.8 and 1.9. Cavernous weathering is a process that leaves the surface of the rock full of pits in a honeycomb-like structure.* This structure is believed to be the result of repeated salt solution penetration, reaction and/or dissolution of various mineral species, drying associated with transport to the surface, and crystallization. This process is similar to if not identical to what is called *efflorescence* described above. In efflorescence the surface crystallization remains as a whitish deposit rather than being eroded or dissolved away. Even though the outward appearance of the various corrosion processes of cavernous weathering, salt weathering, and upward drilling (described below) looks very similar their actual mechanisms may vary considerably. The major difference is that with upward drilling the material is completely submerged into the liquid during corrosion.

Corrosion by condensation, which can be another form of rainwater, is actually called *dew point corrosion*. But again an understanding of meteorol-

---
* These pits have been called *tafoni* by some. See http://www.kuriositas.com/2011/08/tafoni-natures-rock-art.html for some amazing photos.

FIGURE 1.2
Salt weathering of building stone on the island of Gozo, Malta. (Reprinted with permission of Dr. S. MacLeod, 2005, http://en.wikipedia.org/wiki/File:Salt_weathering_in_gozo.jpg.)

ogy is helpful since one must know what species are contained within the atmosphere and at what temperatures they will condense.

Dew point corrosion is the deterioration of a solid ceramic material caused by the condensation of a corrosive liquid from a saturated gas when the temperature is lowered below the point (the dew point) where the liquid will condense. Although a considerable amount of dew point corrosion is from man-made industrial gases, a portion of dew point corrosion could actually be a subgroup of *atmospheric corrosion*, which is the degradation of materials by natural environments. *Atmospheric corrosion* is a term used by the metallurgist, whereas an equivalent term used by ceramists is *weathering*.

Some industrial gases may contain many species that will condense at different temperatures and thus exhibit more than one dew point. Since the dew point is the temperature at which evaporation and condensation rates are equal for a saturated vapor, it should be obvious that the dew point will vary depending upon the concentration (equal to the vapor pressure) of the species in the vapor. Stated in another way, *the dew point is the highest temperature at which a condensate can exist in equilibrium with the vapor from which it condensed.* Many of the condensable species condense in combination with water vapor forming acid condensates. For example, for a specific concentration of SO_3 the dew point temperature will increase as the water vapor concentration increases (see Figure 1.3). The actual corrosion rate reaches a maximum about 30°C below the dew point, Kobussen 1985. Atmospheric SO_2 as it diffuses upward may react with hydroxide radicals produced by sunlight forming H_2SO_4, Penkett 1979 (see equation 13.8 in Chapter 13). Just as with rainwater, these species may cause dissolution of the sandstone and subsequent mechanical problems. Some common constituents of industrial flue gases and their dew points are given in Table 1.4.

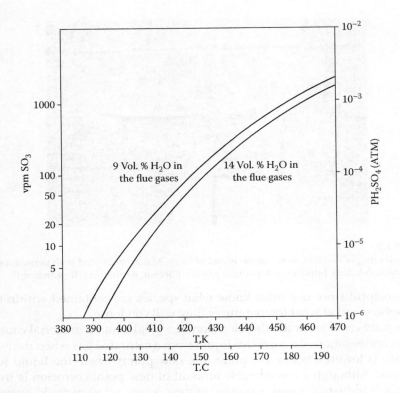

FIGURE 1.3
Sulfuric acid dew point curves. (Lees and Moore 1985, reprinted with permission of The Institute of Corrosion, United Kingdom.)

TABLE 1.4

Dew Points of Common
Constituents of Industrial
Flue Gases

Species	Dew Point, °C
SO_3/H_2O	100–140
HCl	40–50

The relationship between excess air and SO_3/SO_2 levels has been used by glass furnace operators for many years to adjust the fining* operation. Above about 10% excess air there is little additional conversion of SO_2 to SO_3.

Determination of the dew point in gaseous environments dates back to 1929, Cox 1985. Commercial equipment is available from several suppliers.

---

* The process of fining is the removal of small bubbles and general homogenization. The furnace atmospheric SO_3/SO_2 level is related to the residual sulfate level that remains within the glass and whether or not it exists as bubbles.

Cussler and Featherstone 1981 reported the action of acids upon porous ionic solids and concluded that within the porous solid, dissolution would take place only if the valence of the solid cation were between zero and one and that material would precipitate if the valence were greater than one. They verified this conclusion with experiments on $Ca(OH)_2$ that showed $Ca(OH)_2$ precipitates within the pores of the $Ca(OH)_2$ being dissolved. The assumptions of Cussler and Featherstone were that all reactions in the solid were much faster than diffusion so that the reactions reached equilibrium, the diffusion coefficients of all species were equal, and the porous solid was present in excess. Although these assumptions may yield reasonable first approximations for simple systems, they generally do not hold true, especially for the more complex type systems often encountered.

One of the more practical problems associated with service life of ceramics is the often observed degradation of mechanical properties attributed to attack by atmospheric water vapor. This is commonly called *stress corrosion*, is time dependent, and is capable of decreasing both Young's Modulus and fracture strength, Janowski and Rossi 1968. For more information concerning property degradation caused by corrosion see Chapters 18 and 19.

Electrochemical Corrosion

Very few studies have been reported over the last 30 years; however, much work was performed in the 1950s and 1960s on what has been called *galvanic corrosion* of refractories by glasses. Galvanic corrosion as defined by the physical chemist must occur between two materials in contact with one another and both must be in contact with the same electrolyte. Much of what has been reported should more appropriately be called *electrochemical corrosion*. One of the first reports of the existence of an electrical potential between refractory and glass was that in 1953. The setup is schematically represented in Figure 1.4. In such a case, the molten glass acts as the electrolyte and the platinum wire as a reference electrode (i.e., standard oxygen electrode). The use of platinum as a reference electrode requires that the atmosphere above the melt contain a reasonable oxygen partial pressure, since the reaction

$$O_2^- \leftrightarrow 1/2\, O_2 + 2e^- \tag{1.8}$$

must be present at the site where the platinum comes in contact with the melt. The overall accuracy of such potential measurements is critically dependent upon obtaining excellent electrical contact among the various components of the galvanic cell. An additional problem that undoubtedly causes variation or drift in the measurements is the formation of a reaction interface layer between the refractory and the molten glass.

Godrin 1975 has published a review of the literature on electrochemical corrosion of refractories by glasses. It has been shown that a potential difference does exist in such systems; however, no quantitative relationship

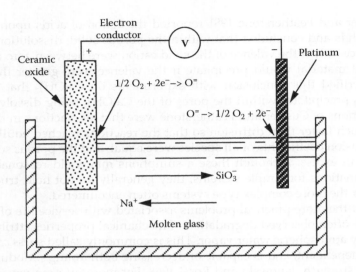

FIGURE 1.4
Electrochemical cell to determine potential difference between a ceramic and a molten glass.

between corrosion and potential has been reported. Since a potential difference exists in corroding systems, it has been tempting to assume that the potential is at least partly responsible for the corrosion; however, the application of a bias potential has been unsuccessful in eliminating corrosion. Even though not totally reliable, Godrin concluded that refractories that have an electrical potential with respect to glass that is positive 0.4 to 0.7 V are fairly resistant to corrosion, that refractories with a potential greater than 1.0 V have rather poor resistance, and that refractories that have a negative potential with respect to glass should not be used.

Pons and Parent 1969 have concluded that the oxygen ion activity is a very important parameter in corrosion and that its role is determined by the difference in oxygen potential between the molten glass and the refractory oxide. An additional interesting case is that of two different oxide materials (i.e., a multiphase polycrystalline material) in contact with the same glass, that have oxygen potentials on either side of that of the glass. In such a case, it is assumed that oxygen migrates from the oxide of higher potential toward that of lower potential. If the conduction mechanism of the two oxides is different (ionic versus electronic) the situation becomes more complex. When the oxygen potentials of the oxides are greater than the glass, oxygen ions are assumed to be transported from the ionic conductive oxide to the electronic conductive one, which may ultimately result in pitting caused by the release of oxygen. If the oxygen potential of the oxides is lower than the glass, alkali ions of the glass are transported to the electronic conductive oxide with oxygen release at the interface between the two oxides.

Although in theory the application of a bias potential to minimize or eliminate corrosion, which implies that the corrosion process is one that involves

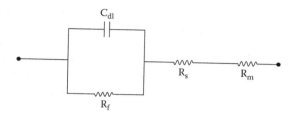

FIGURE 1.5
Equivalent circuit model for electrochemical interface. (After Wall et al. 1993.)

charge transfer, should produce noticeable results, a major practical problem has been that of making the electrical connection to the ceramic. The other problems relating to the success of a bias potential in eliminating corrosion are the other factors in corrosion—chemical reaction, diffusion, viscosity, solubility, etc. This topic is one of considerable importance and should receive a lot more attention than it has in recent years. A standard text that discusses electrode effects in liquid electrolytes should be consulted by the interested reader, Vetter 1967.

Wall et al. 1993 in their studies of graphite fiber/BMI composites in contact with various metals immersed into an electrolyte developed an electrical equivalent circuit of the electrochemical interface. The classic "electrical double layer" is established at the surface of the graphite fibers where the electrons at the surface are separated from the ionic charges in solution. This forms a capacitor, called the "double layer capacitance," C_{DL}. This capacitance is dependent upon the electrode surface area, which allows one to monitor surface roughening, surface adsorbed species, and ingress of solution between fiber and matrix by changes in capacitance. Charge leakage rate across the interface can be represented by a resistance, R_F, in parallel with C_{DL} that is inversely proportional to the reaction rate. Changes in exposed surface area of the fibers or changes in surface chemistry will affect this rate. The electrical equivalent circuit is completed by the addition of resistors in series representing the solution resistance, R_S, and the material resistance, R_M (circuit depicted in Figure 1.5). Imposing a high frequency AC signal to this circuit will allow one to determine the resistance $R_S + R_M$, since C_{DL} will short out R_F. If the solution resistance, R_S, is assumed to remain constant, changes in the sum, $R_S + R_M$, can be related to composite integrity. Wall et al. used electrochemical impedance spectroscopy to evaluate all the circuit elements under steady state conditions.

Attack by Molten Glasses

The use of a single diffusion coefficient, as was done in equations 1.1 and 1.2, even in multicomponent systems was verified by Cooper and Kingery 1964. They, along with Samaddar et al. 1964 and Oishi et al. 1965, described in detail the theory of corrosion by liquids in ceramic systems (i.e., alumina, mullite, fused silica, and anorthite in Al-Ca-silicate liquid). Diffusion through the boundary layer was determined to be the rate-limiting step during dissolution. The composition of the boundary layer may vary depending upon whether diffusion is more or less rapid than the boundary reaction. The basic equation describing the rate of solution under free convection with density being the driving force is:

$$j = \frac{-dR}{dt} = 0.505 \left(\frac{g\Delta\rho}{v_i x} \right)^{1/4} D_i^{3/4} C^* \exp\left(\frac{\delta^*}{R + \delta^*/4} \right) \tag{1.9}$$

where:

g = acceleration due to gravity,
$\Delta\rho$ = $(\rho_i - \rho_\infty)/\rho_\infty$ (ρ_i = saturated liquid density and ρ_∞ = original),
v = kinematic viscosity,
x = distance from surface of liquid,
D_i = interface diffusion coefficient,
C^* = a concentration parameter,
δ^* = effective boundary layer thickness, and
R = solute radius.

The exponential term was introduced as a correction for cylindrical surfaces. Since experimental tests often involve cylindrical specimens, these equations have been developed for that geometry. In practical applications, the condition relating to the corrosion of slabs is most predominant. However, if the sample diameter is large compared to the boundary layer thickness, the two geometries give almost identical results.

After a short induction period (which is of no consequence in practical applications) in which molecular diffusion predominates, the rate of corrosion becomes nearly independent of time. As a surface corrodes, the interface, if denser than the corroding medium, will be eroded away due to free convection caused by density variation. Use of this equation implies that one has at his disposal data relating to the variation of density and viscosity with temperature. In cases where these data are not available the investigator will need to determine them prior to any calculation of corrosion rates.

Hrma 1970 has used the work of Cooper and Kingery to discuss further the rates of corrosion of refractories in contact with glass. The following equation given by Hrma describes the corrosion under the condition of free convection due to density difference:

$$j_c = k\Delta c \left(\frac{D^3 \Delta\rho g}{vL} \right)^{1/4} \tag{1.10}$$

where:

j_c = rate of corrosion,
c = solubility of material in liquid,
D = coefficient of binary diffusion,
g = acceleration due to gravity,
v = kinematic viscosity,
L = distance from surface of liquid,
ρ = relative variation of density, and
k = constant = 0.482.

This is essentially the same equation as that of Cooper and Kingery, without the exponential term.

Many corrosive environments associated with ceramic materials involve diffusion into the corroding medium, and thus increased velocity of the medium increases corrosion. Thus, if transport in the liquid is important, the corrosion rate must be evaluated under forced convection conditions. In such cases, the rate will depend on the velocity of forced convection:

$$j = 0.61 \, D^{*2/3} v^{*-1/6} \omega^{1/2} C^* \qquad (1.11)$$

The terms D^* and v^* were introduced, since diffusivity and viscosity may be composition dependent. The important point of this equation is that the rate of corrosion depends on the square root of the angular velocity ω.

In the majority of practical cases, the solubility of the material in the liquid and the density of the liquid change much more slowly than the viscosity of the liquid. Under isothermal conditions the viscosity change is due to compositional changes. Thus, the predominant factor in the corrosion of a material by a liquid is the viscosity of the liquid, Lakatos and Simmingskold 1967 and Lakatos and Simmingskold 1967b. This, however, doesn't hold for every case, since liquid composition does affect the solubility of the solid, Lakatos and Simmingskold 1971. These relationships hold quite well for the corrosion of a solid below the liquid surface. At the surface, where three states of matter are present, the corrosion mechanism is different and much more severe.

At the liquid surface a sharp cut normally develops in the vertical face of the solid material being corroded as shown in Figure 1.6. This region has been called *flux-line, metal-line,* or *glass-line* corrosion (also called the *Marangoni* effect, Chung and Schlesinger 1994). Pons and Parent 1969 reported that the flux-line corrosion rate was a nonlinear function of the oxygen potential difference between the surface and the interior of a molten sodium silicate. Cooper and Kingery 1964 reported that flux-line corrosion was the result of natural convection in the liquid caused by changes in surface forces due to an increase in surface energy of the liquid as solid is dissolved. They also reported that if the surface energy of the liquid were independent of the amount of solid dissolved, no such excessive flux-line corrosion would occur. Hrma 1970 reported that the additional corrosion at the flux-line depended

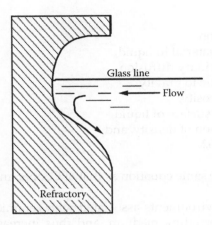

FIGURE 1.6
Corrosion of a vertical face by a liquid.

only upon the variation in surface tension and density, with surface tension being the more important factor. Although this is a well-known phenomenon, no one has investigated it thoroughly to determine a definitive mechanism. In actual practice, quite often a thermal gradient exists such that the highest temperature exists at the flux-line. This temperature difference, however, cannot be the sole driving force for excessive corrosion at the flux-line since the same phenomenon is observed in laboratory isothermal studies. This same excessive corrosion occurs at any location where three substantially different materials come in contact with one another. In the above case, it was ceramic, liquid, and atmosphere. It may also occur where two liquids come in contact with a ceramic – a well-known phenomenon in metallurgy. The two liquids in that case are molten metal and an oxide slag.

The temperature dependence of corrosion can be represented by the Arrhenius equation:

$$j = A \exp(E/RT) \tag{1.12}$$

Excellent fit of some experimental data to this equation reported by Samaddar et al. 1964 has indicated that corrosion corresponds to an activated process. Blau and Smith 1950 have attempted such an interpretation. However, the fact that variations of liquidus compositions, diffusion coefficients and liquid structure change with temperature suggests that interpreting corrosion as an activated process may be very misleading and at least ineffective. The Arrhenius dependence should be used only for cases where the liquid is far from being saturated with components from the solid, which according to Woolley 1989 is the case for practical glassmaking applications.

The corrosion of a flat vertical slab under a thermal gradient is depicted in Figure 1.7. As the convective flow of the liquid, caused by either forced

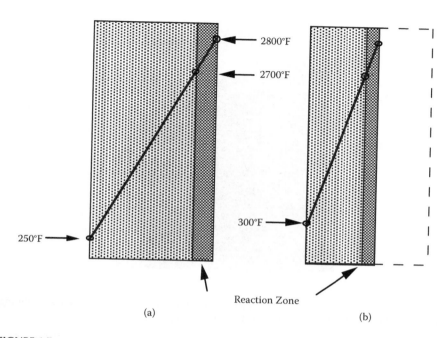

FIGURE 1.7
Effect of thermal gradient upon corrosion interface: (a) short time and (b) extended time.

convection or density changes, removes some of the reaction product interface, the total thickness of the slab decreases and the thermal gradient
becomes steeper, assuming that the hot face temperature remains constant,
which is very close to actual furnace operations. The actual cold face temperature will rise slightly but the overall result is a steeper thermal gradient. The
thermal gradient through a wall as depicted in Figure 1.7 is more complex
than presented here (actually it is not linear but a complex 3-D shape) but
the overall effect is the same. If the reaction product layer can form between
certain temperature limits (2800 and 2700°F in Figure 1.7), it is obvious that
the layer thickness must become smaller as corrosion proceeds. Thus the corrosion rate decreases with time. It is not uncommon for the flux-line of the
basin wall of a glass furnace to corrode away approximately one-half of its
thickness in less than one year, where the remaining half may take four or
five times as long to exhibit the same amount of corrosion.

A downward-facing horizontal surface also exhibits greater corrosion than
does a vertical or upward-facing horizontal surface. A downward-facing
surface can exhibit excessive corrosion if bubbles are trapped beneath the
horizontal surface. This is known as *upward-drilling,* since it results in ver-
tically corroded shafts (see Figure 1.8). Surface tension changes around the
bubble cause circulatory currents in the liquid that cause excessive corrosion
very similar to flux-line corrosion. Although no scientific comparisons have
been made to geological corrosion, examples of something similar to upward

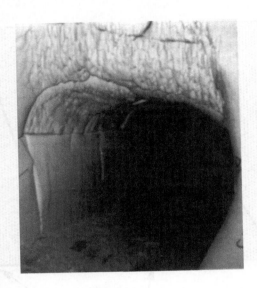

FIGURE 1.8
A fusion cast alumina-zirconia-silica refractory throat of a TV panel glass furnace exhibiting upward-drilling of the throat cover. (Courtesy of Corning, Inc.)

drilling can be found. Figure 1.9 shows a comparison of upward drilling of an AZS fusion cast refractory paver from a glass furnace and a dolomite boulder (compare also with Figure 1.8). The surface of the boulder shown in Figure 1.9 was not horizontal like the paver but sloped approximately 45 degrees.

Attack by Molten Metals

The potential reaction of molten metals with oxide ceramics can be easily obtained from an Ellingham type diagram, which is a compilation of free energy of formation of an oxide from its metal and oxygen at different temperatures. The simple redox mechanism is

$$x M + y/2\ O_2 \rightarrow M_x O_y \tag{1.13}$$

Thus one can easily determine compatibility between a metal and an oxide, since any metal will reduce any oxide that has a less negative free energy of formation for the oxide. The reaction of aluminum metal with silica is a good example:

$$3 SiO_2 + 4 Al \rightarrow 2 Al_2 O_3 + 3 Si \tag{1.14}$$

Since metals such as aluminum and magnesium have very large negative free energies of oxide formation, determining what to use as a container when melting these metals becomes a serious problem.

(a)

(b)

FIGURE 1.9
Comparison of upward drilling between a dolomite boulder (a) and an AZS fusion cast glass furnace paver (b).

Another mechanism for metal attack on ceramics is by the formation of a new compound by a reaction of the type:

$$A_xO_y + zM \rightarrow A_{x-w}M_zO_y + wA \tag{1.15}$$

An example of this is the formation of spinel:

$$4Al_2O_3 + 3Mg \rightarrow 3MgAl_2O_4 + 2Al \tag{1.16}$$

This reaction yields a lower free energy of reaction ($\Delta G°_{1000}$ = -52 kcal/mol) than the simple redox reaction ($\Delta G°_{1000}$ = -28kcal/mol) shown below:

$$Al_2O_3 + 3Mg \rightarrow 3MgO + 2Al \tag{1.17}$$

Thus a compound is formed that would appear to be a good container for molten Al/Mg alloys. However, Lindsay et al. 1964 reported that reaction 1.17 was preferred to reaction 1.16 due to the magnesium activity being sufficiently high to form MgO and the reaction of Al_2O_3 and MgO to form spinel being kinetically slow.

Another possible reaction of ceramics with metals is that of reduction to the metal and solution into the attacking molten metal as shown below:

$$A_xO_y \rightarrow xA + yO \tag{1.18}$$

The metal that forms may be in the gaseous state depending upon the environmental conditions, which may also be true for the oxide that forms in reaction 1.17.

It has been found that in the operation of commercial glass (especially lead-containing glasses) furnaces that metals cause a unique corrosion pattern on upward-facing horizontal surfaces by drilling vertical shafts into the bottom paving refractories. This is called *downward drilling* and is very similar to the upward drilling described in the section on Attack by Molten Glasses. According to Busby 1992 the excessive corrosion caused by molten metal droplets is due to a surface tension gradient on the surface of the droplet. The corrosion drilling rates are not dependent upon the type of metal but are dependent upon the quality of the corroded material. Smaller droplets are more corrosive than larger ones.

Attack by Molten Salts

The corrosion of ceramic components in gas turbine engines generally occurs through the action of condensed salts formed from impurities in the fuel and/or combustion air. Similar corrosion mechanisms occur in glass furnace regenerators and on glass furnace crowns. The condensation of molten salts occurs below their dew point and is thus dependent upon the temperature and pressure of operation, along with the concentration of the impurities in the fuel or air. Fox et al. 1990 listed the dew points for sodium sulfate deposition, a few of which are given in Table 1.5. Not only did a higher pressure raise the dew point for condensation but it also increased the deposition rate, which generally led to more severe corrosion, Jacobson et al. 1986. The effects of molten Na_2SO_4 upon the dissolution of silica and the importance of Na_2O activity and the partial pressure of oxygen is discussed in more detail in Chapter 15, equations 15.14 through 15.18.

TABLE 1.5

Dew Points (°C) for Na_2SO_4 Condensation

		Sodium (ppm)		
Pressure (atm)	Sulfur (ppm)	0.1	1.0	10
1	500	876*	937	991
10	500	969	1045	1111
1	5000	887	961	1025
10	5000	984	1075	1155

* Solid, since melting point is 884°C.

Source: Fox, D.S., Jacobson, N.S and Smialek, J.L. 1990. Hot Corrosion of Silicon Carbide and Nitride at 1000°C, in *Ceramic Transactions Vol. 10: Corrosion and Corrosive Degradation of Ceramics*, Tressler, R.E. and McNallan, M. (eds.), Am. Ceram. Soc., Westerville, OH, pp. 227–49.

A model developed by Cook et al. 1990 in their study of hot corrosion of ceramic (alumina) barrier coatings by sodium, sulfur, and vanadium molten salts gave the rate of solution of a ceramic when a steady state condition prevailed for the rate of salt removal equal to the rate of salt deposition. This provides a salt layer of constant thickness. The ceramic solution rate is then dependent upon the rate of salt deposition:

$$d(M_c/A)/dt = \left(\frac{C}{1-C}\right)d(M_s/A)/dt \tag{1.19}$$

where:
M_c = mass of ceramic dissolved,
A = surface area,
M_s = mass of salt deposited, and
C = concentration of ceramic in layer.

At low deposition rates when salts become saturated the solubility in the salt becomes important. Use of this model requires the calculation of the gas phase and condensed solution equilibria using a computer program such as that developed by the NASA-Lewis Research Center, Gordon and McBride 1971. In addition to the steady state assumption for salt deposition and removal, other assumptions included the parabolic rate law, known equilibrium solubilities, and congruent dissolution.

Corrosion by molten salts has several beneficial applications. One very important application where dissolution of a ceramic is desired is in the removal of the ceramic cores from metal castings manufactured by the investment casting technique. The solvent used for core removal must be highly reactive to the ceramic at rather low temperatures while not damaging the metal. The ceramic must be stable toward molten metal attack at high temperatures and highly reactive toward solvent attack at low temperatures. In a study of the leaching rates of Al_2O_3, Y_2O_3, La_2O_3, ZrO_2, ThO_2, and MgO by molten Li_3AlF_6, Borom et al. 1981 found that the corrosion appeared to involve a solid reaction layer and a boundary layer in the liquid. Vigorous solvent circulation was required to overcome the diffusion-controlled process. Thus it appeared that congruent dissolution was required for optimum core removal, since incongruent dissolution may form reaction layers that require forced convection for removal.

Glassy Materials

Bioactive Glass

Bioactive glasses were first discovered by Hench in 1969. The special chemistry of these glasses allowed them to bond to living bone. These Na_2O–CaO–P_2O_5–SiO_2 glasses have been trademarked as Bioglass® and marketed under several other names depending upon the application. The beneficial effect of these glasses is their controlled release of soluble silicon and calcium ions. In this way the glass acts as a substrate for the growth of new cells. Newer forms of these glasses have been prepared via sol-gel routes that contain numerous very fine interconnected pores. Dissolution kinetics are a function of the following variables, Hench 2002/03:

1. composition,
2. particle size,
3. pore size distribution, average size, and volume%,
4. surface area,
5. thermal stabilization temperature, and
6. chemical stabilization temperature.

The alumina content of bioactive glasses is very important in controlling the durability of the glass surface. The bioactivity, although dependent upon the bulk composition of the glass, decreases beyond acceptable levels once the alumina content rises above 1.0 to 1.5 wt%, Hench and Wilson 1993. This same phenomenon is present for glass compositions containing cations such as Ta_2O_5 except higher levels are tolerable (1.5–3.0 wt%).

Rare earth aluminosilicate (REAS) glasses have been developed for applications as delivery agents for radiation in the treatment of various cancerous tumors, White and Day 1994. In these cases the glass must be sufficiently durable to allow the release of beta-radiation over a specified period of time (about two weeks) while being lodged within the malignant tumor. Once the radiation treatment has been completed, the REAS can be resorbed into the body. It is important that these glasses not dissolve while being radioactive, which would release radioactive species into the other parts of the body damaging healthy tissue. These glasses are generally incorporated in the body as microspheres about 30 μm in diameter. A ^{90}Y-containing radiotherapeutic REAS is sold under the trade name TheraSphere™. White and Day 1994 reported no detectable weight loss of a $1 \times 1 \times 0.2$ cm glass sample before six weeks in 100 ml of distilled water ($pH = 7$) or saline ($pH = 7.4$) at 37, 50, or 70°C. Dissolution rates of $\leq 3 \times 10^{-9}$ g/cm²·min were determined after six weeks. In a comparison study of fused silica, a Corning glass (CGW-1723™) and yttria aluminosilicate (YAS), Oda and Yoshino 1991 showed that YAS was significantly more durable than fused silica in saturated steam at 300°C and 8.6 MPa. The dissolution mechanism is very important for applications in the human body; however, it is very difficult to determine whether these glasses exhibit congruent or incongruent dissolution. Surface analyses of microspheres and bulk glasses indicated that the mechanism was congruent, White and Day 1994. Using inductively coupled plasma and atomic adsorption spectroscopy it has been determined that the yttrium release from YAS microspheres in distilled water or saline at 37 or 50°C was below detectable limits, Erbe and Day 1991.

More recently Conzone et al. 2002 have reported the development of borate glasses for use in treatment of rheumatoid arthritis, since these glasses are potentially more reactive with physiological liquids. Borate glasses containing only alkali ions dissolve uniformly (i.e., congruently) in simulated physiological liquids at temperatures ranging from 22 to 75°C. When the borate glasses contain other cations (such as Ca, Mg, Fe, Dy, Ho, Sm, and Y) in amounts ranging from 2 to 30 wt% dissolution is non-uniform (i.e., incongruent) with the formation of new compounds. Day 2002/03 gave an example of Dy_2O_3-containing borate solid glass microspheres that reacted to form hollow spheres, shells of concentric layers, or microspheres filled with homogeneous gel-like material depending upon the Dy_2O_3 content. The dissolution mechanism involved the selective leaching of lithium and boron allowing the rare earth (i.e., Dy) to react and form an insoluble phosphate (phosphorus is from a phosphate-buffered saline simulated physiological liquid). When calcium-containing borate glasses were reacted a semicrystalline or gel calcium phosphate formed that had a composition very similar to hydroxyapatite. Although early work by Hench and colleagues has indicated the need for the formation of a silica gel surface layer for silicate glasses to be bioactive, the work of Day and colleagues has indicated that a silica gel is not always necessary for bioactivity.

In addition to the beneficial bioactive glasses discussed above, there is the extremely important area of hazardous health effects from glasses. One such case is that of inhalation of glass fibers. The dissolution of these fibers is very critical in determining their health risk. Bauer 2000 reported the work of Eastes and Hadley that glass fibers greater than 20 μm, if inhaled, have been correlated to respiratory disease in laboratory animals. The dissolution is dependent upon the fiber surface chemistry and physical nature. The continuous movement of fluids in the human lung increases the dissolution rate and also transports the dissolved species to other parts of the body via the blood stream. Aluminosilicate fibers are the most durable, while the dissolution rate of borosilicate fibers (e.g., home insulation) can be 1000 times greater. The biopersistence of 1 μm diameter fibers varies from several days to as long as 14 years depending upon their chemistry. Annealing fibers at temperatures below the transition temperature can decrease the dissolution rate in simulated extracellular fluid (pH = 7.4) by 2 to 3 times. The fact that they have not shown any major adverse reaction in human lungs was attributed by Bauer to the high dissolution rate of glass fibers.

Bulk Glasses

Probably the most abundant examples of glass corrosion are those caused by a liquid. Release of toxic species (such as PbO or radioactive waste) from various glass compositions has received worldwide interest during the past 20 to 30 years. Although glass is assumed by many to be inert to most liquids, it does slowly dissolve. In many cases, however, the species released are not harmful.

The corrosion resistance of glasses is predominately a function of structure, which is determined by the composition. Although some have related glass durability to the number of nonbridging oxygens, a function of composition, White 1988 has suggested that glass durability is more closely related to the presence of specific depolymerized units. He arrived at this conclusion through the correlation of vibration spectra with the effective charge on bridging and nonbridging oxygens. In a study of the leaching behavior of some oxynitride glasses, Wald et al. 1986 reported that the nitrogen-containing glasses exhibited a greater durability (i.e., silicon release) by at least a factor of two than either fused silica or quartz tested under identical conditions at 200°C in deionized water for 28 days. This they attributed to the increased amount of cross-linking of the silica network and the resultant reduction in hydrolysis.

Glasses can be soluble under a wide range of pH values from acids to bases, including water. Water-soluble sodium silicates form the basis of the soluble silicate industry that supplies products for the manufacture of cements, adhesives, cleansers, and flocculants. At the other extreme are glasses designed for maximum resistance to corrosion.

The mechanism of silicate glass corrosion by water involves competition between ion exchange and matrix dissolution, Douglas and El-Shamy 1967, that are affected by glass composition and the possible formation of a protective interfacial layer. The characteristics of this interfacial layer controls subsequent dissolution. Dealkalization of this layer, which generally causes further matrix dealkalization and dissolution, is dependent upon the ease of alkali

TABLE 1.6

Leached Glass Surface Types

Type I	Thin surface hydrated layer; < 5 Å thick; high durability
Type II	Surface layer depleted in alkali; medium durability
Type III	Silica-rich layer adjacent to bulk and cation-rich (leached from bulk) adjacent to solution; medium durability
Type IV	Silica-rich nonprotective layer; low durability
Type V	No layer formation; lowest durability

Source: Hench, L.L. and Clark, D.E. 1978. Physical Chemistry of Glass Surfaces, *J. Noncryst. Solids, 28*, 83–105.

diffusion through this layer, the physical properties of the layer (e.g., porosity, thickness), and the pH of the solution. The increase in pH of the solution caused by dealkalization causes increased silica dissolution. High initial reaction rates are quite often observed and are generally caused by an excessively large exposed surface area due to microcracks or generally rough surfaces. This excessive surface area can be eliminated by proper cleaning procedures.

Jantzen 1992 has used a thermodynamic approach to the corrosion of glasses, especially applied to nuclear waste glass leachability. The earlier work of Newton and Paul 1980 on a wide variety of glasses was expanded and then combined with that of Pourbaix 1974 and Garrels and Christ 1965 to describe the effects of natural aqueous environments. Using thermodynamic hydration equations, Newton and Paul predicted glass durability from composition. Jantzen showed that the kinetic contribution was primarily a function of the test conditions (SA/V^* ratio, time, and temperature). The major assumptions in Jantzen's approach were that the total free energy of hydration of the glass was the sum of the free energies of hydration of the components and that the glass structure was a primary function of glass composition. The activity-pH diagrams of Pourbaix provided the needed correlation between free energy of hydration and ion concentration in solution. Thus Jantzen was able to determine glass durability from glass composition by use of a pH-adjusted free energy of hydration term for several hundred compositions of nuclear waste glasses, man-made glasses, and natural glasses. The more negative the pH-adjusted free energy of hydration term the less durable the glass.

Species may be leached from a glass as a result of ion exchange with protons from solution or silica may be leached as the siloxane bonds of the matrix are attacked by hydroxyl ions from the solution. The former mechanism is predominant at low pH whereas the latter is predominant at high pH. Hench and Clark 1978 categorized leached glass surfaces into five groups. These groupings are listed in Table 1.6. In Types I, II, and III the surface reaction layer that forms has a low solubility if composed of metal hydrates or hydrated silicates. Quite often these layers are protective, essentially stopping further attack.

---

[*] SA = surface area; V = volume.

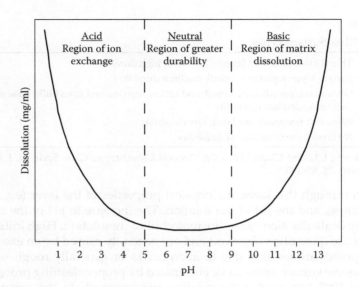

FIGURE 1.10
Effect of *p*H upon glass dissolution.

Many consider the *p*H of the solvent to be the most important parameter that affects glass durability, McVay and Peterson 1981. At *p*H < 5, ion exchange is the predominant mechanism and at *p*H > 9, matrix dissolution is predominant. Between *p*H 5 and 9, the corrosion is a minimum. This is represented schematically in Figure 1.10. Thus dissolution is rapid when the metal-oxygen bond is extensively coordinated to hydrogen or hydroxyl ions, and is a minimum under neutral conditions.

Since the driving force for dissolution of silicate glasses is the hydrogen ion activity, the loss of hydrogen ions during dissolution causes a continuous decrease in the dissolution rate. The change of the *p*H of the solution can drift into the basic region causing breakdown of the silica matrix with subsequent increasing dissolution rates. To determine accurate rate constants, the experimenter must hold the *p*H constant by use of buffer solutions or an automatic titration system. Silica solubility increases significantly above about *p*H = 9.

In systems containing ions of variable valence one must consider the redox potential, Eh, of the system, since the solubility of ions is dependent upon their valence state. A general rule of thumb is that higher valence states are more soluble.

Hogenson and Healy 1962 developed the following equation:

$$W = a\varnothing^{b_2} \exp(-b_1/T) \tag{1.20}$$

where:

W = weight loss,
a = experimentally determined coefficient,
b_1 = experimentally determined coefficient,
b_2 = experimentally determined coefficient,
$\emptyset$ = time, and
T = temperature.

for describing the effects of time and temperature upon the acid (10% HCl) corrosion of silicate glasses. This equation, since it relates total multicomponent weight loss to time and temperature assuming a uniform surface corrosion, does not take into account the mechanism of dissolution but instead determines the total overall corrosion. This is probably sufficient for practical problems but does not allow one to study mechanisms.

Budd 1961 has described the corrosion of glass by either an electrophilic or a nucleophilic mechanism, or both. The surface of the glass has electron-rich and electron-deficient regions exposed. Various agents attack these regions at different rates. Exposed negatively charged nonbridging oxygens are attacked by H^+ (or H_3O^+), whereas exposed network silicon atoms are attacked by O^2, OH^-, and F^-.

Budd and Frackiewicz 1961 found that by crushing glass under various solutions, an equilibrium pH value was reached after sufficient surface area was exposed. The value of this equilibrium pH was a function of the glass composition, and it was suggested that it was related to the oxygen ion activity of the glass. When foreign ions were present, the amount of surface required to reach an equilibrium pH was greater.

The rate of hydrolysis of a glass surface is one of the major factors that delineates the field of commercial glasses. The rate of hydrolysis is of great importance because it determines the service life of a glass with respect to weathering or corrosion and also because it influences the mechanical properties. Glass fracture is aided by hydrolysis. The rate of hydrolysis of alkali-silicate glasses of the same molar ratios proceeds in the order Rb > Cs > K > Na > Li.

The mechanism of corrosion of fluorozirconate glasses is substantially different from that of silicate-based glasses, Simmons and Simmons 1986. The fluorozirconate glass corrodes by matrix dissolution, with the components going into solution as fluorides, without first hydrolyzing as in the silicates. These glasses are also characterized by the formation of a nonprotective porous hydrated interfacial layer. Compounds highly insoluble in water remain in the porous layer. The formation of a hydroxylated zirconia fluoride complex in solution causes the pH of the solution to decrease considerably increasing the solubility of zirconia fluoride, thus increasing the overall dissolution rate by orders of magnitude.

The properties of the leached layers that build up can dramatically affect the dissolution rate, since the silanol groups present can polymerize, various solutes and colloids present can react with the leached layer, and stress buildup can cause cracking and spalling. If the flow rate is sufficiently low the dissolution fluid may approach saturation values thus slowing the dissolution rate and/or changing its nature. In addition the dissolution of alkalis and alkaline earths from the glass consumes hydrogen ions in solution. If this exceeds the buffer capacity of the solution, the *p*H will increase.

Fiberglass

A discussion of glass would not be complete if some mention of glass fibers were not made. The corrosion of fibers is inherently greater than bulk glass simply because of the larger surface-to-volume ratio. Since one of the major applications of fibers is as a reinforcement to some other material, the main property of interest is that of strength. Thus, any corrosion reactions that would lower the strength are of interest. This effect is important both when the fiber is being manufactured and after it has been embedded in another material. For example, the strength of E-glass (borosilicate) fibers in dry and humid environments was studied by Thomas 1960 with the observation that humid environments lower strength. The mechanisms of environmentally enhanced stress corrosion of glass fiber are discussed in more detail in Chapter 19.

Wojnarovits 1983 reported that multicomponent glass fibers exhibited a variation in dissolution in acid and alkaline environments due to the existence of a layered structure, each having a different dissolution rate, with the core generally having the highest rate. Single component fibers (i.e., silica) did not show this layering effect and thus no variation in dissolution rate.

Additional Recommended Reading

Amoroso, G.G. and Fassina, V. 1983. *Stone Decay and Conservation*, p. 12, Elsevier, Amsterdam, 453pp.

Bockris, O.M. and Reddy, A.K.N. 1970 *Modern Electrochemistry*, Vol. 2, Plenum Press, NY.

Burns, R.G. 1970 *Mineralogical Applications of Crystal Field Theory*, Cambridge University Press, Cambridge.

Goudie, A.S. and Viles, H.A. 1997. *Salt Weathering Hazards*, Wiley, Chichester, 256pp.

Lasaga, A. and Lüttge, A. 2004. Mineralogical Approaches to Fundamental Crystal Dissolution Kinetics, *Amer. Mineral.*, 89, 527–40.

Exercises, Questions, and Problems

1. Using equation 1.6, calculate the pressures produced by crystallization of NaCl and calcium sulfate. V_s for NaCl = 28 and for CaSO$_4$

= 46 cm^3/mol. $T = 20°C$. Assume $C/C_s = 10$. Which material would produce the worst decay?

2. Using equation 1.7, calculate the hydration pressure produced by the hydration of $CaSO_4 \cdot 1/2H_2O$ to $CaSO_4 \cdot 2H_2O$ at 20°C and relative humidity of 60 and 80%. Vapor pressures must be looked up in any appropriate handbook. Molar volumes will also have to be calculated.

3. Using the results of problems 1 and 2 above discuss which is more important, crystallization pressure or hydration pressure.

4. Discuss the reaction products that may form and how they may relate to any interfacial reaction layer formed.

5. If a "unified theory of corrosion of ceramics" were to be developed, what structural characteristic would be included and why?

6. Why does the corrosion rate decrease when a thermal gradient is present?

7. Discuss the difference between direct and indirect dissolution. What other terms are used to describe these types of dissolution?

8. What is the most predominant parameter in the equation for corrosion rate under free convection? Why is this parameter more predominant than the others?

9. Discuss the various problems relating to the experimental verification of the galvanic corrosion of ceramics.

10. Describe how the cross-linking of silica tetrahedra affect corrosion in silicates by aqueous solutions.

11. How does pH affect the corrosion of crystalline ceramics and how does this relate to *IEP* (isoelectric point)?

12. Discuss the difference between electrochemical and chemical dissolution. What material parameters are important in each type?

13. The Arrhenius equation has been used to represent the temperature dependence of corrosion. Discuss when this equation is most appropriate and why.

References

Amoroso, G.G. and Fassina, V. 1983. *Stone Decay and Conservation*, p. 12, Elsevier, Amsterdam, 453pp.

Appleton, J.D., Fuge, R. and McCall, G.J.H. (eds.) 1996. *Environmental Geochemistry and Health, with Special Reference to Developing Countries*, Geological Society Special Pub No. 113, The Geological Society, London, 264pp.

Bauer, J.F. 2000. Corrosion and Surface Effects of Glass Fiber in Biological Fluids, *The Glass Researcher*, 9(2), 4–5.

Berthoud, A. 1912. Formation of Crystal Faces, *J. Chem. Phys.*, 10, 624–35.

Blau, H.H. and Smith, C.D. 1950. Refractory Problems in Glass Manufacture, *Bull. Am. Ceram. Soc.*, 29 (1), 6–9.

Borchardt, C.A. 1977. Montmorillonite and Other Smectite Minerals, Ch. 9 in *Minerals in Soil Environments*, Dinauer, R.C. (ed.), Soil Sci. Soc. Am., Madison, WI, pp. 293–330.

Borom, M.P., Arendt, R.H. and Cook, N.C. 1981. Dissolution of Oxides of Y, Al, Mg, and La by Molten Fluorides, *Ceram. Bull.*, 60(11), 1168–74.

Brady, P.V. and House, W.A. 1996. Surface-Controlled Dissolution and Growth of Minerals, Ch. 4, pp. 225–305 in *Physics and Chemistry of Mineral Surfaces*, Brady, P.V. (ed.), CRC Press, New York.

Bright, E. and Readey, D.W. 1987. Dissolution Kinetics of TiO_2 in HF-HCl Solutions, *J. Am. Ceram. Soc.*, 70(12), 900–6.

Budd, S. M. 1961. The Mechanism of Chemical Reaction between Silicate Glass and Attacking Agents; Part 1. Electrophilic and Nucleophilic Mechanism of Attack, *Phys. Chem. Glasses*, 2(4), 111–4.

Budd S. M. and Frackiewicz, J. 1961. The Mechanism of Chemical Reaction between Silicate Glass and Attacking Agents; Part 2. Chemical Equilibria at Glass-Solution Interfaces, *Phys Chem. Glasses*, 2(4), 115–8.

Burge, H.A., Su, H.J. and Spengler, J.D. 1994. Moisture, Organisms, and Health Effect, pp. 84–90 in *Moisture Control in Buildings*, Trechsel, H.R. (ed.), ASTM Manual Series: MNL 18, Philadelphia, PA.

Burns, R.G. 1970. *Mineralogical Applications of Crystal Field Theory*, Cambridge University Press, pp. 162–7.

Busby, T. 1992. Hotter Refractories Increase the Risk of Downward Drilling, *Glass Ind.*, 73(1), 20, 24.

Casey, W.H. and Bunker, B. 1990. Leaching of Mineral and Glass Surfaces during Dissolution, Ch. 10 in *Reviews in Mineralogy, Vol. 23: Mineral-Water Interface Geochemistry*, Hochella Jr., M.F. and White A.F. (eds.), Mineral. Soc. Am., Washington, DC, pp. 397–426.

Chung, Y-D. and Schlesinger, M.E. 1994. Interaction of $CaO-FeO-SiO_2$ Slags with Partially Stabilized Zirconia, *J. Am. Ceram. Soc.*, 77 (3), 612.

Conzone, C.D., Brown, R.F., Day, D.E. and Ehrhardt, G.J. 2002. *In Vitro* and *In Vivo* Dissolution Behavior of a Dysprosium Lithium Borate Glass Designed for the Radiation Synovectomy Treatment of Rheumatoid Arthritis, *J. Biomedical Materials Research*, 60(2), 260–68.

Cook, L.P., Bonnell, D.W. and Rathnamma, D. 1990. Model for Molten Salt Corrosion of Ceramics, in *Ceramic Transactions Vol. 10: Corrosion and Corrosive Degradation of Ceramics*, Tressler, R.E. and McNallan, M. (eds.), Am. Ceram. Soc., Westerville, OH, pp. 251–75.

Cooper, A.R. 1970. The Use of Phase Diagrams in Dissolution Studies, in *Refractory Materials*, Vol. 6-III (Alper, A.M. ed.), Academic Press, New York, pp. 237–50.

Cooper Jr., A.R. and Kingery, W.D. 1964. Dissolution in Ceramic Systems: I, Molecular Diffusion, Natural Convection, and Forced Convection Studies of Sapphire Dissolution in Calcium Aluminum Silicate, *J. Am. Ceram. Soc.*, 47(1), 37–43.

Correns, C.W. 1949. Growth and Dissolution of Crystals under Linear Pressure, *Discussions of the Faraday Society*, No. 5, 267–71.

Cox, W.M., Farrell, D.M. and Dawson, J.L. 1985. Corrosion Monitoring for Process Control, pp. 191–217 in *Dewpoint Corrosion*, Holmes, D.R. (ed.), Ellis Horwood, Chichester, UK.

Cussler, E.L. and Featherstone, J.D.B. 1981. Demineralization of Porous Solids, *Science*, 213, 28 Aug, 1018–9.

Day, D.E. 2002/03. Reactions of Bioactive Borate Glasses with Physiological Liquids, *Glass Researcher*, 12(1&2), 21–2.

Diggle, J.W. 1973. Dissolution of Oxide Phases, Ch. 4 in *Oxides and Oxide Films, Vol. 2*, Diggle, J.W. (ed.), Marcel Dekker, New York, pp. 281–386.

Douglas, R.W. and El-Shamy, T.M.M. 1967. Reaction of Glass with Aqueous Solutions, *J. Am. Ceram. Soc.*, 50(1), 1–8.

Elmer, T.H. 1985. Role of Acid Concentration in Leaching of Cordierite and Alkali Borosilicate Glass, *J. Am. Ceram. Soc.*, 68(10), C273–4.

Erbe, E.M. and Day, D.E. 1991. p. 105 in *Proceedings: Science and Technology of New Glasses*, Sakka, S. and Soga, N. (eds.), Ceram. Soc. Tokyo, Japan.

Eugster, H.P. 1986. Minerals in Hot Water, *Amer. Mineral.*, 71, 655–673.

Finkelman, R.B., Skinner, C.W., Plumlee, G.S. and Bunnell, J.E. 2001. Medical Geology, *Geotimes, 46*, 20–23.

Fox, D.S., Jacobson, N.S and Smialek, J.L. 1990. Hot Corrosion of Silicon Carbide and Nitride at 1000°C, in *Ceramic Transactions Vol. 10: Corrosion and Corrosive Degradation of Ceramics*, Tressler, R.E. and McNallan, M. (eds.), Am. Ceram. Soc., Westerville, OH, pp. 227–49.

Gordon, S. and McBride, B.J. 1971. Computer Program for Calculation of Complex Chemical Equilibrium Compositions, Rocket Performance, Incident & Reflected Shocks, and Chapman-Jongnet Detonations, NASA SP-273, U.S. Printing Office, Washington, DC.

Grodrin, Y. 1975. Review of the Literature on Electrochemical Phenomena, International Commission on Glass, Paris.

Hawkins, D.B. and Roy, R. 1963. Distribution of Trace Elements between Clays and Zeolites Formed by Hydrothermal Alteration of Synthetic Basalts, *Geochim. Cosmochim. Acta*, 27(165), 785–95.

Hench, L.L. 2002/03. Bioactive Glasses Help Heal, Repair and Build Human Tissue, *Glass Researcher*, 12(1&2), 18.

Hench, L.L. and Clark, D.E. 1978. Physical Chemistry of Glass Surfaces, *J. Noncryst. Solids, 28*, 83–105.

Hoffmann, M.R. 1986. Fog and Cloud Water Deposition, pp. 64–91 in *Materials Degradation Caused by Acid Rain*, ACS Symposium Series 318, Baboian, R. (ed.), Am. Chem. Soc., Washington, DC.

Hogenson, D.K. and Healy, J.H. 1962. Mathematical Treatment of Glass Corrosion Data, *J. Am. Ceram. Soc.*, 45(4), 178–81.

Hrma, P. 1970. Contribution to the Study of the Function between the Rate of Isothermal Corrosion and Glass Composition (Fr), *Verres Refract.*, 24 (4–5), 166–8.

Huang, P.M. 1977. Feldspars, Olivines, Pyroxenes, and Amphiboles, Ch. 15 in *Minerals in Soil Environments*, Dinauer, R.C. (ed.), Soil Sci. Soc. Am., Madison, WI, pp. 553–602.

Jacobson, N.S., Stearns, C.A. and Smialek, J.L. 1986. Burner Rig Corrosion of SiC at 1000°C, *Adv. Ceram. Mat.*, 1(2), 154–61.

Janowski, K.R. and Rossi, R.C. 1968. Mechanical Degradation of MgO by Water Vapor, *J. Am. Ceram. Soc.*, 51(8), 453–5.

Jantzen, C.M. 1992. Thermodynamic Approach to Glass Corrosion, Ch. 6 in *Corrosion of Glass, Ceramics, & Ceramic Superconductors*, Clark, D.E. and Zoitos, B.K. (eds.), Noyes Publications, Park Ridge, NJ, pp. 153–217.

Jennings, H.M. 1986. Aqueous Solubility Relationships for Two Types of Calcium Silicate Hydrate, *J. Am. Ceram. Soc.*, 69 (8), 614–18.

Joiner, R.L., Rench, J.D., Zanetos, M.A. and Brauning, S.E. 1983. Report to The Refractories Institute, Pittsburgh, PA, Feb.

Kobussen, A.G. 1985. Corrosion in Condensing Gas-Fired Central Heating Boilers, pp. 179–90 in *Dewpoint Corrosion*, Holmes, D.R. (ed.), Ellis Horwood, Chichester, UK.

Kramer, D.P. and Osborne, N.R. 1983. Effects of Atmosphere and Dew Point on the Wetting Characteristics of a Glass-Ceramic on Two Nickel-Based Superalloys, pp. 740–50 in *Ceramic Engineering and Science Proceedings*, Smothers, W.J. (ed.), Vol. 4, (9–10).

Kühnel, R.A. 2004. Cause and Consequence: Volume Changes behind Building Material Deterioration, *Materials Characterization, 53*, 171–180.

Lasaga, A.C. 1990. Atomic Treatment of Mineral-Water Surface Reactions, Ch. 2 in *Reviews in Mineralogy, Vol. 23: Mineral-Water Interface Geochemistry*, Hochella, M.F., Jr. and White, A.F. (eds.), Mineral. Soc. Am., Washington, DC, pp. 17–85.

Lasaga, A.C. and Lüttge, A. 2001. Variation of Crystal Dissolution Rate Based on a Dissolution Stepwave Model, *Science, 291*, 2400–04.

Lakatos, T. and Simmingskold, B. 1967. Influence of Constituents on the Corrosion of Pot Clays by Molten Glass, *Glass Technol., 8*(2), 43–7.

Lakatos, T. and Simmingskold, B. 1967b. Corrosion Effect of Glasses Containing Na_2O-CaO-MgO-Al_2O_3-SiO_2 on Tank Blocks Corhart ZAC and Sillimanite, *Glastek. Tidskr., 22*(5), 107–13.

Lakatos, T. and Simmingskold, B. 1971. Influence of Viscosity and Chemical Composition of Glass on Its Corrosion of Sintered Alumina and Silica Glass, *Glastek. Tidskr., 26*(4), 58–68.

Leake, B.E. et al. 1997. Nomenclature of Amphiboles: Report of the Subcommittee on Amphiboles of the International Mineralogical Association, Commission on New Minerals and Mineral Names, *Amer. Mineral., 82*, 1019–1037.

Le Clerc, P. and Peyches, I. 1953. Polarization of Refractory Oxides Immersed in Molten Glass (Fr), *Verres Refract., 7* (6), 339–45.

Lees, D.J. and Moore, W. 1985. A Review of Factors Which Exacerbate Acid Dewpoint Corrosion in Coal-Fired Power Stations, pp. 50–78 in *Dewpoint Corrosion*, Holmes, D.R. (ed.), Ellis Horwood, Chichester, UK.

Levich, B.G. 1947. Theory of Concentration Polarization, *Discussions Faraday Soc., 1*, 37–43.

Levich, B.G. 1962. *Physicochemical Hydrodynamics*, Prentice Hall, Englewood Cliffs, NJ.

Lindsay, J.G., Bakker, W.T., and Dewing, E.W. 1964. Chemical Resistance of Refractories to Al and Al-Mg Alloys, *J. Am. Ceram. Soc., 47*(2), 90–4.

Lüttge, A. 2005. Etch Pit Coalescence, Surface Area, and Overall Mineral Dissolution Rates, *Amer. Mineral., 90*, 1776–83.

Marshall, C.E. 1977. *The Physical Chemistry and Mineralogy of Soils, Vol. II: Soils in Place*, Wiley, New York, p. 39.

Marshall, C.E. 1975. *The Physical Chemistry and Mineralogy of Soils, Vol. I: Soil Materials*, Krieger, Huntington, NY.

McVay, G.L. and Peterson, L.R. 1981. Effect of Gamma Radiation on Glass Leaching, *J. Am. Ceram. Soc., 64*(3), 154–8.

Mulawa, P.A., Cadle, S.H., Lipari, F., Ang, C.C. and Vandervennet, R.T. 1986. Urban Dew: Composition and Influence on Dry Deposition Rates, pp. 64–91 in *Materials Degradation Caused by Acid Rain*, ACS Symposium Series 318, Baboian, R. (ed.), Am. Chem. Soc., Washington, DC.

Neal, K.M., Newnam, S.H., Pokorney, L.M. and Rybarczyk, J.P. 1986. Elemental Analysis of Simulated Acid Rain Stripping of Indiana Limestone, Marble, and Bronze, pp. 285–300 in *Materials Degradation Caused by Acid Rain*, ACS Symposium Series 318, Baboian, R. (ed.), Am. Chem. Soc., Washington, DC.

Nernst, W. 1904. Theory of Reaction Velocities in Heterogeneous Systems (Ger), *Z. Physik. Chem., 47*, 52–55.

Newton, R.G. and Paul, A. 1980. A New Approach to Predicting the Durability of Glasses from Their Chemical Compositions, *Glass Tech., 21*(6), 307–9.

Noyes, A.A. and Whitney, W.R. 1897. Rate of Solution of Solid Materials in Their Own Solutions (Ger), *Z. Physik. Chem., 23*, 689–92.

Oda, K. and Yoshino, T. 1991. Properties of Y_2O_3-Al_2O_3-SiO_2 Glasses as a Model System of Grain Boundary Phase of Si_3N_4 Ceramics, Part 2: Leaching Characteristics, *J. Ceram. Soc. Japan Int. Ed., 99*(11), 1150–1152.

Oishi, Y., Copper Jr., A.R. and Kingery, W.D. 1965. Dissolution in Ceramic Systems: III, Boundary Layer Concentration Gradients, *J. Am. Ceram. Soc., 48*(2), 88–95.

Park, S.P. 1994. Moisture in Historic Buildings and Preservation Guidance, pp. 402–419 in *Moisture Control in Buildings*, Trechsel, H.R. (ed.), ASTM Manual Series: MNL 18, Philadelphia, PA.

Parks, G.A. 1965. The Isoelectric Points of Solid Oxides, Solid Hydroxides, and Aqueous Hydroxo Complex Systems, *Chem. Rev., 65*(2), 177–98.

Penkett, S.A. 1979. Chemical Changes in the Air, *SCI Sulfur Symposium*, 109–22, May.

Pons, A. and Parent, A. 1969. The Activity of the Oxygen Ion in Glasses and Its Effect on the Corrosion of Refractories (Fr), *Verres Refract., 23* (3), 324–33.

Prandtl, L. 1928. NACE Tech. Memo, No. 452.

Reddy, M.M., Sherwood, S.I. and Doe, B.R. 1986. Limestone and Marble Dissolution by Acid Rain: An Onsite Weathering Experiment, pp. 226–38 in *Materials Degradation Caused by Acid Rain*, ACS Symposium Series 318, Baboian, R. (ed.), Am. Chem. Soc., Washington, DC.

Samaddar, B.N., Kingery, W.D. and Cooper Jr., A.R. 1964. Dissolution in Ceramic Systems: II, Dissolution of Alumina, Mullite, Anorthite, and Silica in a Calcium-Aluminum-Silicate Slag, *J. Am. Ceram. Soc., 47*(5), 249–54.

Sato, T., Ohtaki, S. and Shimada, M. 1985. Transformation of Yttria Partially Stabilized Zirconia by Low Temperature Annealing in Air, *J. Mater. Sci., 20*(4), 1466–70.

Schnitzer, M. and Kodama, H. 1977. Reactions of Minerals with Soil Humic Substances, Ch. 21 in *Minerals in Soil Environments*, Dinauer, R.C. (ed.), Soil Sci. Soc. Am., Madison, WI, pp. 741–70.

Semonin, R.G. 1986. Wet Deposition Chemistry, pp. 23–41 in *Materials Degradation Caused by Acid Rain*, ACS Symposium Series 318, Baboian, R. (ed.), Am. Chem. Soc., Washington, DC.

Shaw, D.J. 1980. Charged Interfaces, Ch. 7 in *Introduction to Colloid and Surface Chemistry*, 3rd edition, Butterworths, London, pp. 148–82.

Shelby, J.E. (ed.), 1994. *Key Engineering Materials*, Vols. 94–95, Trans Tech Pub., Switzerland.

Simmons, C.J. and Simmons, J.H. 1986. Chemical Durability of Fluoride Glasses: I, Reaction of Fluorozirconate Glasses with Water, *J. Am. Ceram. Soc.*, 69(9), 661–9.

Smith, K.S. and Huyck, H.L. 1999. An Overview of the Abundance, Relative Mobility, Bioavailability, and Human Toxicity of Metals, in *The Environmental Geochemistry of Mineral Deposits: Part A. Processes, Techniques and Health Issues*, Plumlee, G.S. and Logsdon, M.J. (eds.), Soc. Econ. Geol. Rev. Econ. Geol. 6A, Society of Economic Geologists, Littleton, CO, pp. 29–70.

Thomas, W. F. 1960. An Investigation of the Factors Likely to Affect the Strength and Properties of Glass Fibers, *Phys. Chem. Glasses*, 1(1), 4–18.

U.S. EPA, http://www.epa.gov/ceampubl/mmedia/minteq/index.htm.

U.S. Geological Survey. 2007. Building Stones of Our Nation's Capital: Acid Rain in Washington, http:/pubs.usgs.gov/gip/stones/acid-rain.html (accessed on 1/11/2011).

Valenza II, J.J. and Scherer, G.W. 2006. Mechanism for Salt Scaling, *J. Am. Ceram. Soc.*, 89(4), 1161–79.

Vetter, K.J. 1967. *Electrochemical Kinetics*, Academic Press, NY.

Wald, J.W., Messier, D.R. and DeGuire, E.J. 1986. Leaching Behavior of Si-Y-Al-O-N Glasses, *Int. J. High Tech. Ceram.*, 2 (1), 65–72.

Wall, F.D., Taylor, S.R. and Cahen, G.L. 1993. The Simulation and Detection of Electrochemical Damage in BMI/Graphite Fiber Composites Using Electrochemical Impedance Spectroscopy, pp. 95–113 in *High Temperature and Environmental Effects on Polymeric Composites*, STP 1174, Harris, C.E. and Gates, T.S. (eds.), ASTM, Philadelphia, PA.

White, W.B. 1988. Glass Structure and Glass Durability, in *Materials Stability and Environmental Degradation*, Barkatt, A., Verink Jr, E.D. and Smith, L.R. (eds.), Vol. 125 of *Materials Research Society Symposium Proceedings*, Mater. Res. Soc., Pittsburgh, PA, pp. 109–14.

Winkler, E.M. 1973a. *Stone: Properties, Durability in Man's Environment*, Springer-Verlag, New York, 230 pp.

Winkler, E.M. 1973b. Salt Action on Stone in Urban Buildings in *Application of Science in Examination of Works of Art*, Joung, W.J. (ed.), Museum of Fine Arts, Boston, MA.

Wojnarovits, I. 1983. Behavior of Glass Fibers in Strong Acidic and Alkaline Media, *J. Am. Ceram. Soc.*, 66(12), 896–8.

Woolley, F.E. 1989. Prediction of Refractory Corrosion Rate from Glass Viscosity and Composition, *UNITECR '89 Proceedings*, Trostel Jr., L.J. (ed.), Am. Ceram. Soc., Westerville, OH, pp. 768–79.

Yoshimura, M., Hiuga, T. and Somiya, S. 1986. Dissolution and Reaction of Yttria-Stabilized Zirconia Single Crystals in Hydrothermal Solutions, *J. Am. Ceram. Soc.*, 69(7), 583–4.

2

Corrosion by Gas

Wisdom lies not in possessing knowledge—but in perpetually seeking it.

Paul Johnson, author

Crystalline Materials

The corrosion of a polycrystalline ceramic by vapor attack can be very serious, much more so than attack by either liquids or solids. One of the most important material properties related to vapor attack is that of porosity or permeability. If the vapor can penetrate into the material, the surface area exposed to attack is greatly increased and corrosion proceeds rapidly. *It is the total surface area exposed to attack that is important.* Thus not only is the volume of porosity important, but the pore size distribution is also important. See Chapter 12 for a discussion on porosity determination.

Vapor attack can proceed by producing a reaction product that may be either solid, liquid, or gas, as in the following equation:

$$A_s + B_g \rightarrow C_{s,l,g} \tag{2.1}$$

As an example, the attack of SiO_2 by Na_2O vapors can produce a liquid sodium silicate.

In another type of vapor attack, which is really a combined sequential effect of vapor and liquid attack, the vapor may penetrate a material under thermal gradient to a lower temperature, condense, and then dissolve material by liquid solution. The liquid solution can then penetrate further along temperature gradients until it freezes. If the thermal gradient of the material is changed, it is possible for the solid reaction products to melt, causing excessive corrosion and spalling at the point of melting.

The driving force for ionic diffusion through a surface reaction layer and for continued growth is thermal energy. If sufficient thermal energy is not provided, layer growth falls off rapidly. Across very thin (< 5 nm) films at low temperatures, strong electric fields may exist that act to pull cations through the film, much like that which occurs in the room temperature oxidation of metals, Kubaschewski and Hopkins 1962. The growth of the reaction layer generally can be represented by one of the following equations for thin films:

$$y = K_1 \log t \text{ (logarithmic)} \tag{2.2}$$

$$1/y = K_2 - K_3 \log t \text{ (inverse log)} \tag{2.3}$$

and

$$y = K_4(1 - \exp[-K_5 t]) \text{ (asymptotic)} \tag{2.4}$$

and for thick films:

$$y^2 = K_6 t \text{ (parabolic)} \tag{2.5}$$

and

$$y = K_7 \text{ (rectilinear)} \tag{2.6}$$

where:
  y = film thickness;
  t = time;
  K_i = rate constant.

Equation 2.5 is the simplest form of the parabolic law and is quite often used to describe corrosion by oxidation. Oxidation processes are generally more complex than the simple mechanism of a single species diffusing through an oxide layer. Preferential diffusion along grain boundaries can alter the oxide layer growth substantially. This is especially important if the grain boundary phase is glassy providing a continuous medium for transport of cations outward and oxygen inward. Grain boundary diffusion is a lower energy process than bulk diffusion and thus will be more important at lower temperatures. Quite often a higher reaction rate will be observed at lower temperatures than expected if one were to extrapolate from high temperature reaction rates. Thus the microstructure of the layer, especially grain size, is particularly important. In addition fully stoichiometric reaction layers provide more resistance to diffusion than anion and/or cation deficient layers, which provide easy paths for diffusion.

　　Readey 1990 has listed the possible steps that might be rate controlling in the kinetics of gas-solid reactions. These are given below:

  1. diffusion of the gas to the solid,
  2. adsorption of the gas molecule onto the solid surface,
  3. surface diffusion of the adsorbed gas,
  4. decomposition of reactants at surface specific sites,
  5. reaction at the surface,

6. removal of products from reaction site,

7. surface diffusion of products,

8. desorption of gas molecules from the surface, and

9. diffusion away from solid.

Any one of these may control the rate of corrosion.

Much attention has been given recently to the oxidation of nonoxide ceramics, especially silicon carbide and nitride. In general, the stability of nonoxides toward oxidation is related to the relative free energy of formation between the oxide and nonoxide phases. When studying the oxidation of nitrides, one must not overlook the possibility of the formation of an oxynitride, either as the final product or as an intermediate. The stability of the oxide versus the nitride, for example, can be represented by the following equation:

$$2M_xN_y + O_2 \leftrightarrow 2M_xO + yN_2 \tag{2.7}$$

As the difference in free energy of formation between the oxide and the nitride becomes more negative, the greater is the tendency for the reaction to proceed toward the right. Expressing the free energy change of the reaction in terms of the partial pressures of oxygen and nitrogen one obtains:

$$\Delta G^\circ = -RT \ln \frac{(pN_2)^y}{pO_2} \tag{2.8}$$

One can then calculate the partial pressure ratio required for the oxide or nitride to remain stable at any temperature of interest. For example, the oxidation of silicon nitride to silica at 1800 K yields a partial pressure ratio of nitrogen to oxygen of about 10^7. Thus very high nitrogen pressures are required to stabilize the nitride. Anytime the permeability of the product gas through the reaction layer is less than that of the reactant gas, the product gas pressure can build at the interface to very high levels with the result being bubbles and/or cracks in the reaction interface layer. This subsequently leads to continued reaction.

The reduction of oxide ceramics at various partial pressures of oxygen may also be of interest and can be obtained from the examination of Ellingham plots of $\Delta G^\circ = -RT \ln pO_2$ versus temperature. If one is interested in the reduction of a binary compound, such as mullite, the presence of a second more stable oxide that forms the compound increases the stability of the less stable oxide by decreasing $RT \ln pO_2$. Although increasing the stability of the less stable oxide, the magnitude of this change is not large enough to increase the stability of the more stable oxide. Thus the free energy of formation of mullite will be between that of silica and alumina but closer to that of silica.

A reaction that is very common when transition metals are present is exhibited by the reduction of one of the constituent oxides of binary compounds with decreasing oxygen partial pressure. An example is that of lanthanum cobalt oxide.

$$4LaCo^{3+}O_3 \rightarrow La_4Co_2^{3+}Co^{2+}O_{10} + CoO + 1/2O_2 \qquad (2.9)$$

These reactions become very important when applications of double oxides (or multicomponent oxides) require placement in an environment containing an oxygen potential gradient. In more general terms, this is true for any gaseous potential gradient if the gas phase is one of the constituents of the solid.

As reported by Yokokawa et al. 1989, a double oxide may decompose kinetically even if the oxygen potential gradient is within the stability region of the double oxide. This kinetic decomposition is due to cation diffusivity differences along the oxygen potential gradient.

Another factor that might enhance the reduction of an oxide is the formation of a more stable lower oxide and the vaporization of the reaction products. An example of this is the reduction of silica by hydrogen at elevated temperature to the monoxide, which is highly volatile above 300°C.

A loss of weight by oxidation to a higher oxide that is volatile can also occur. A good example of this is the assumed vaporization of Cr_2O_3 that actually occurs through oxidation to CrO_3 gas by the following equation:

$$Cr_2O_3 + 3/2O_2 \leftrightarrow 2CrO_3 \text{ (g)}^* \qquad (2.10)$$

This reaction is one that is not easily proven experimentally since CrO_3 upon deposition/condensation dissociates to Cr_2O_3 and O_2. CrO_3 gas, however, has been identified by mass spectrometry, Grimley et al. 1961. Diffusion of CrO_3 gas through a stagnant gaseous boundary layer was determined to be rate controlling as opposed to the surface reaction for the reaction above, Graham and Davis 1971.

A gas that is often encountered in practical applications is water vapor. Increase in corrosion rates when moisture is present has been reported by many investigators. This is apparently related to the ease with which gaseous hydroxide species can form. It is well known that a silica protective scale forms on materials such as SiC that is very resistant to oxidation in dry air. This scale, however, reacts with water vapor to form volatile silicon hydroxide phases.

A possible rate-controlling step in vapor attack is the rate of arrival of a gaseous reactant and also possibly the rate of removal of a gaseous product. One should realize that many intermediate steps (i.e., diffusion through a

---

* This reaction occurs above 1000°C according to Chatterjee et al. 2001.

gaseous boundary layer) are possible in the overall reaction, and any one of these may also be rate controlling. It is obvious that a reaction cannot proceed any faster than the rate at which reactants are added, but it may proceed much more slowly. The maximum rate of arrival of a gas can be calculated from the Hertz-Langmuir equation:

$$Z = \frac{P}{(2\pi MRT)^{1/2}} \qquad (2.11)$$

where:
  Z = moles of gas that arrive at surface in unit time and over unit area,
  P = partial pressure of reactant gas,
  M = molecular weight of gas,
  R = gas constant, and
  T = absolute temperature.

Using P and M of the product gas, the rate of removal of gas product can be calculated using the same equation. To determine if service life is acceptable, these rates may be all that is needed. Actual observed rates of removal may not agree with those calculated if some surface reaction must take place to produce the species that vaporizes. The actual difference between observed and calculated rates depends upon the activation energy of the surface reaction. If the gaseous reactant is at a lower temperature than the solid material, an additional factor of heat transfer to the gas must also be considered and may limit the overall reaction.

According to Readey 1990 in the corrosion of spheres, the rate of corrosion is proportional to the square root of the gas velocity. If the gas vapor pressure and velocity are held constant, the corrosion rate is then proportional to the square root of the temperature. At low gas vapor pressures, transport of the gas to the surface controls the corrosion rate. At high vapor pressures, the reaction at the surface is controlling. The gaseous reaction products many times cause formation of pits and/or intergranular cracking. This can be very important for materials containing second phases (composites) that produce gaseous reaction products.

Pilling and Bedworth 1923 have reported the importance of knowing the relative volumes occupied by the reaction products and reactants. Knowing these volumes can aid in determining the mechanism of the reaction. *When the corrosion of a solid by a gas produces another solid, the reaction proceeds only by diffusion of a reactant through the boundary layer when the volume of the solid reactant is less than the volume of the solid reaction product.* In such a case the reaction rate decreases with time. If the volume of the reactant is greater than the product, the reaction rate is usually linear with time. These rates are only guidelines, since other factors can keep a tight layer from forming (i.e., thermal expansion mismatch).

When a surface layer is formed by the reaction through which a gas must diffuse for the reaction to continue, the reaction can generally be represented by the parabolic rate law, which is discussed in more detail in Chapter 7 on kinetics. Jorgensen et al. 1960 have shown that the theory put forth by Engell and Hauffe 1952 that describes the formation of a thin oxide film on metals is applicable to the oxidation of nonoxide ceramics. In this case the rate constant being dependent upon oxygen partial pressure has the form:

$$k = A \ln pO_2 + B \tag{2.12}$$

where A and B are constants. The driving force for diffusion was reported to be mainly an electric field across the thin film (100 to 200 nm thick) in addition to the concentration gradient.

Glassy Materials

The corrosion of glasses by atmospheric conditions, referred to as *weathering*, is essentially attack by water vapor. Weathering occurs by one of two mechanisms. In both types, condensation occurs on the glass surface; however, in one type it evaporates, whereas in the other it collects to the point where it flows from the surface, carrying any reaction products with it. The latter type is very similar to corrosion by aqueous solutions. The former type is characterized by the formation of soda-rich films, according to Tichane and Carrier 1961. This soda-rich film has been shown to react with atmospheric gases such as CO_2 to form Na_2CO_3, according to the work of Simpson 1958 and Tichane 1966.

The electronics industry is another area where vapor attack of glasses may be of importance. Sealing glasses and glass envelopes have been developed that resist attack by alkali vapors and mercury vapors. In their study of some CaO- and Al_2O_3-containing glasses, Burggraaf and van Velzen 1969 reported that alkali vapor attack increased greatly above a temperature that coincided approximately with the transformation range (T_g) of the glass, indicating that one should use a glass with the highest possible T_g.

In the manufacture of flat glass by the Pilkington or PPG processes, glass is floated onto a bed of molten tin in a chamber containing a reducing atmosphere ($N_2 + \sim 10\% H_2$). The hydrogen present in the atmosphere above the glass can act upon the top surface of the glass causing reduction of the most reducible species present. All commercial flat glass contains some iron and that present near the top surface is predominantly in the reduced ferrous state. This is generally not a problem; however, those glasses containing NiO can exhibit small metallic droplets on the top surface which is cause for rejection. Based upon the modified Ellingham diagram shown in Figure 9.2, this should not occur if the pO_2 is maintained greater than 10^{-9} atmosphere, assuming a maximum temperature no greater than 1100°C.

Johnston and Chelko 1970 proposed the mechanism of reduction of ions in glass by hydrogen diffusion through the glass to the reducible ions that act as immobile traps reacting with the hydrogen and stopping further diffusion.

Vacuum

It is generally believed that all materials vaporize; however, several modes of vaporization are possible. Some materials will vaporize congruently to a gas of the same composition as the solid, which is also called *sublimation*. Others will vaporize incongruently to a gas and a different condensed phase. It is also possible for more than one stable gas molecule to form. Decomposition to the elements may also occur, which is called *direct vaporization*. In multicomponent materials where the various components exhibit greatly different heats of vaporization, selective vaporization may occur.

The deterioration of ceramics in a vacuum in many cases is the equilibration of the material with a low partial pressure of oxygen. In such a case a lower oxide of the metal forms along with some oxygen represented by the following equation:

$$MO_2 \text{ (s)} \rightarrow MO \text{ (g)} + O_2 \text{ (g)} \tag{2.13}$$

Sublimation of solid spheres controlled by gaseous diffusion through a boundary layer was first suggested by Langmuir in 1918. The reduction in size was given by the equation:

$$r_o^2 - r^2 = \frac{KDV_oP}{RT}t \tag{2.14}$$

where:

r_o = initial radius,
r = radius at time t,
K = geometrical constant (~ 2),
D = diffusion coefficient of gas through boundary layer,
V_o = molar volume of evaporating species,
P = equilibrium partial pressure of gas,
R = gas constant,
T = temperature, and
t = time.

Exercises, Questions, and Problems

1. What is the most important material parameter when exposed to gaseous attack?
2. Discuss the use of equation 2.5 in explaining corrosion by oxidation.
3. Of the nine possible rate-controlling steps listed by Readey, discuss one and its relationship to the others.
4. Explain the problems associated with placing a material containing transition metals into an environment containing an oxygen potential gradient.
5. Although the silica scale that forms on SiC in dry air is protective, explain why it is not in moist air.

References

Chatterjee, U.K., Bose, S.K. and Roy, S.K. 2001. *Environmental Degradation of Metals*, Dekker., New York, Ch. 6, p. 283.

Burggraaf, A.J. and van Velzen, H.C. 1969. Glasses Resistant to Sodium Vapor at Temperatures to 700°C, *J. Am. Ceram. Soc.*, *52*(5), 238–42.

Engell, H.J. and Hauffe, K. 1952. Influence of Adsorption Phenomena on Oxidation of Metals at High Temperatures, *Metallurgy*, *6*, 285–91.

Graham, H.C. and Davis, H.H. 1971. Oxidation/Vaporization Kinetics of Cr_2O_3, *J. Am. Ceram. Soc.*, *54*(2), 89–93.

Grimley, R.T., Burns, R.P. and Inghram, M.G. 1961. Thermodynamics of Vaporization of Cr_2O_3: Dissociation Energies of CrO, CrO_2 and CrO_3, *J. Chem. Phys.*, *34*(2), 664–7.

Johnston, W.D. and Chelko, A.J. 1970. Reduction of Ions in Glass by Hydrogen, *J. Am. Ceram. Soc.*, *53*(6), 295–301.

Jorgensen, P.J., Wadsworth, M.E. and Cutler, I.B. 1960. Effects of Oxygen Partial Pressure on the Oxidation of Silicon Carbide, *J. Am. Ceram. Soc.*, *43*(4), 209–12.

Kubaschewski, O. and Hopkins, B.E. 1962. *Oxidation of Metals and Alloys*, Butterworths, London.

Langmuir, I. 1918. Evaporation of Small Spheres, *Phys. Rev.*, *12*(5), 368–70.

Pilling, N.B. and Bedworth, R.E. 1923. The Oxidation of Metals at High Temperature, *J. Inst. Met.*, *29*, 529–91.

Readey, D.W. 1990. Gaseous Corrosion of Ceramics, in *Ceramic Transactions Vol. 10, Corrosion and Corrosive Degradation of Ceramics*, Tressler, R.E. and McNallan, M. (eds.), Am. Ceram. Soc., Westerville, OH, pp. 53–80.

Simpson, H.E. 1958. Study of Surface Structure of Glass as Related to Its Durability, *J. Am. Ceram. Soc.*, *41*(2), 43–9.

Tichane, R.M. 1966. Initial Stages of the Weathering Process on a Soda-Lime Glass Surface, *Glass Technol.*, *7*(1), 26–9.

Tichane, R.M. and Carrier, G. B. 1961. The Microstructure of a Soda-Lime Glass Surface, *J. Am. Ceram. Soc.*, *44*(12), 606–10.

Yokokawa, H., Kawada, T. and Dokiya, M. 1989. Construction of Chemical Potential Diagrams for Metal-Metal-Nonmetal Systems: Applications to the Decomposition of Double Oxides, *J. Am. Ceram. Soc.*, *72*(11), 2104–10.

3

Corrosion by Solid

The important thing is not to stop questioning. Curiosity has its own reason for existing.

Albert Einstein

Many applications of materials involve two dissimilar (e.g., an acid and a base) solid materials in contact. Corrosion can occur if these materials react with one another. Common types of reactions involve the formation of a third phase at the boundary, which can be a solid, a liquid, or a gas. An example of the formation of a solid interface forming between an acid and a base is the formation of mullite between alumina and silica. In some cases the boundary phase may be a solid solution of the original two phases (e.g., the formation of red ruby between white alumina and green chromia). Again, phase diagrams will give an indication of the type of reaction and the temperature where it occurs.

When the reaction that takes place is one of diffusion as a movement of atoms within a chemically uniform material, it is called *self-diffusion*. When a permanent displacement of chemical species occurs, causing local composition change, it is called *interdiffusion* or *chemical diffusion*. The driving force for chemical diffusion is a chemical potential gradient (concentration gradient). When two dissimilar materials are in contact, chemical diffusion of the two materials in opposite directions forms an interface reaction layer. Once this layer has been formed, additional reaction can take place only by the diffusion of chemical species through this layer.

Solid-solid reactions are predominantly reactions involving diffusion. Diffusion reactions are really a special case of the general theory of kinetics (discussed in Chapter 7) since the diffusion coefficient, D, is a measure of the diffusion reaction rate. Thus diffusion can be represented by an equation of the Arrhenius form:

$$D = D_o \exp\left(-Q/RT\right) \tag{3.1}$$

where:
  D = diffusion coefficient,
  D_o = constant,
  Q = activation energy,
  R = gas constant, and
  T = absolute temperature.

The larger the value of Q, the activation energy, the more strongly the diffusion coefficient depends upon temperature.

The diffusion in polycrystalline materials can be divided into *bulk diffusion, grain boundary diffusion,* and *surface diffusion.* Diffusion along grain boundaries is greater than bulk diffusion because of the greater degree of disorder along grain boundaries. Similarly, surface diffusion is greater than bulk diffusion. When grain boundary diffusion predominates, the log concentration decreases linearly with the distance from the surface. When bulk diffusion predominates, however, the log concentration of the diffusion species decreases with the square of the distance from the surface. Thus by determining the concentration gradient from the surface (at constant surface concentration) one can determine which type of diffusion predominates.

Since grain boundary diffusion is greater than bulk diffusion, it would be expected that the activation energy for boundary diffusion would be lower than that for bulk diffusion. The boundary diffusion is more important at lower temperatures, and bulk diffusion is more important at high temperatures.

Chemical reactions wholly within the solid state are less abundant than those that involve a gas or liquid, owing predominately to the limitation of reaction rates imposed by slower material transport. The solid-solid contact of two different bulk materials also imposes a limitation on the intimacy of contact—much less than that between a solid and a liquid or gas.

Applications of ceramic materials commonly involve thermal gradients. Under such conditions it is possible for one component of a multicomponent material to diffuse selectively along the thermal gradient. This phenomenon is called *thermal diffusion* or the *Sorét effect,* Platten and Costesèque 2004. This diffusion along thermal gradients is not well understood, especially for ceramic materials. See Chapter 6 for a discussion of diffusion.

An example of a solid-solid corrosion problem when a third medium is the originator of the corrosion is the intimate contact of limestone and sandstone exposed to atmospheric conditions. In this case the third medium, water containing dissolved sulfur, reacts with the limestone forming calcium sulfate. This sulfate when coming in contact with the sandstone reacts with the silica causing the sandstone to crumble. This type of corrosion is termed *associative decay,* Schaffer 2004.

References

Platten, J.K. and Costesèque, P. 2004. Charles Soret. A Short Biography. On the Occasion of the Hundredth Anniversary of His Death. *Eur. Physical J. E.,* 15, 235–9.

Schaffer, R.J. 2004. *The Weathering of Natural Building Stones,* Donhead Publishing, Shaftesbury, UK, p. 149 (first published in 1932).

4

Corrosion by Biological Sources

Science depends not on speculation but on conclusions verified through experiment.

James Schlesinger

Introduction

Biological sources of corrosion are becoming an increasingly more important area of study. Although ceramics have been used in the human body for many years, it is only recently that their reaction with body fluids has become more understandable. Probably the oldest use of ceramics in the human body was that of artificial teeth and teeth fillings. Ceramic teeth prostheses[*] date back as far as the Egyptians, Ironside and Swain 1998. Restorative dental ceramics date back to about 1840. But today ceramic prostheses are used in many locations within the body. Knee and hip replacements are prevalent as are many bone repair surgeries. Materials can be inert to body fluids, which is the case for the bone replacement materials; they can react with the body fluids (i.e., bioreactive),[†] as is the case for the bone repair materials; or they can be soluble, which are also used for repair surgeries where the ceramic dissolves after the original bone heals.[‡] Bioceramics must not dissolve faster than the rate of bone formation otherwise the bone will not form properly. Another use of soluble ceramics is in the therapeutic materials used for drug delivery. In these cases controlled dissolution is of extreme importance. Inhaled, ingested, or adsorbed mineral dusts have been studied for over 50 years but again it is only recently that the interactions with body fluids are becoming understandable. Those minerals that are soluble are generally not a problem unless they contain a toxic chemical species (e.g., Pb, Cr). Those that are inert or nearly so can cause various forms of cancer. Particle size and morphology play an important role in where the particles ultimately reside and subsequently with what fluids they come in contact. The importance of

---

[*] Ox bone was one material used by the Egyptians as a tooth prosthesis.
[†] Chemical bonding of a ceramic, tricalcium phosphate, to bone was first observed by Driskell et al. 1973.
[‡] The field of bone regeneration through the use of resorbable ceramic scaffolds is a subfield of the new field of tissue engineering.

mineral or ceramic solubility to human health is the potential toxic response that may occur.

The processes that control metal mobilities include not only dissolution/precipitation but also redox reactions, complexation with ligands, sorption/desorption by solids, and biotransformation, Reeder et al. 2006. The water saturation states of various metals with respect to various solids can be calculated by use of programs like MINTEQA2 (US EPA), PHREEQC, Parkhurst and Appelo 1999, or the Geochemists Workbench, Bethke 2002. The phases of interest, however, may lack the necessary thermodynamic stability data or they may be complicated by being a solid solution. Metal solubility can be significantly enhanced through complex formation with various organic or inorganic ligands. An example is that of the affinity of UO_2^{2+} for dissolved $CO_3^=$. The solubility of UO_2^{2+} in equilibrium with *schoepite**[*] at a pH of 7.0 is 3.4 µM. With 1 µM dissolved CO_2 the solubility increases to 56 µM.

Sorption includes the accumulation of ions at the surface (i.e., adsorption), formation of a distinct surface phase (i.e., precipitation), or incorporation of ions into a phase (i.e., co-precipitation). Many factors affect sorption including pH, presence of complexing ligands, ionic strength, surface charge, and site coordination. Quite often sorption processes occur simultaneously with redox and biotransformation processes.

The area of microbial enhanced corrosion is related almost entirely to architectural materials. Many of these are ceramics. The action of microbes, although biological, doesn't generally fall under the category of bioceramics. This important area will be discussed in greater detail in Chapter 13 on architectural materials.

Although one may think that the interaction of ceramics and the human body is of extreme importance, very few interdisciplinary studies have been done. However, the field of Medical Mineralogy and Geochemistry has started to emerge as evidenced by the short course offered by the Mineralogical Society of America and the subsequent published Volume 64 of *Reviews in Mineralogy and Geochemistry* titled *Medical Mineralogy and Geochemistry* in 2006. Although many topics were covered in this volume, several are more closely related to corrosion of ceramics than others. These include the development of ceramics for use as orthopedic and dental implants, oxide-encapsulated living cells for the development of biosensors, and inhaled mineral dusts. See Table 4.1 for other names used by various researchers to describe this field of mineral effects upon the human body. Although some of the scientific fields combined in these interdisciplinary approaches appear to be very different, they all use the same physicochemical principals to solve their problems. Thus it is important that these fields communicate horizontally with one another to share their wealth of expertise. The fundamental sciences of thermodynamics, kinetics, phase

---

[*] $UO_3 \cdot 2H_2O$.

TABLE 4.1

Names Used to Describe the Interaction of Minerals with the Human Body

Name	Ref.
Medical Mineralogy and Geochemistry	Plumlee et al. 2006
Medical Geology	Finkelman 2001
Medical Geochemistry	Smith & Huyck 1999
Environmental Geochemistry and Health	Appleton et al. 1996

equilibria, and physical chemistry are used by all involved. In addition, the emerging field of nanomaterials is playing an ever important role.

Several different but similar terms have been used to describe how minerals react with body fluids. Some of these are essentially the same as used in other areas of ceramic corrosion. Terms such as *biosolubility, bioreactivity,* and *biodurability* (without the prefix *bio*) are found in all areas of ceramic corrosion. *Biotransformation, bioaccessibility, bioavailability,* and *biopersistence* are terms that are used by the medical mineralogy and toxicology fields. Although these terms are in general self-explanatory, some confusion does exist between bioaccessibility and bioavailability, Plumlee et al. 2006. Bioaccessibility is that fraction of the mineral that can be dissolved by body fluids and is therefore available for absorption. Bioavailability is the fraction that is absorbed. Thus the amount that is accessible is always greater than the amount that is absorbable. Biotransformation is any change that may occur to dimensions, composition, or surface morphology during residence in the human body. Biopersistence is the ability to persist after being inhaled, ingested, etc., into the human body. In addition to dissolution rates, biopersistence is dependent upon the effects of dimensions and surface morphology.

Interaction is physiological but influenced by properties of minerals. The mineral properties of importance are surface properties. Jolicoeur and Poisson 1987 indicated that surface chemical composition, surface functionality, and surface specific sites are the ones of relevance. The various body fluids that are involved include respiratory, gastrointestinal, perspiration, blood serum, and inter- and intracellular fluids. Saliva, tears, and gastric fluids all contain digestive enzymes and the intestines additionally contain bacteria that are essential to the digestive process. Of importance are the fluid-mineral interactions. These minerals can encounter a wide range of compositions, pH, Eh, electrolyte content, and concentration of organics plus all of these vary by their location within the body. For example, the overall Eh of plasma is on the order of -100 millivolts due to the combined action of the many redox couples present, whereas it is in the range of -200 to -240 millivolts for intracellular fluids due to a greater abundance of the glutathione redox couple. Gastric fluids are quite reduced with an Eh of about -200 millivolts at a pH of 1.5–2.0, Davis et al. 1992.

Minerals that are dissolved by the body fluids can then be absorbed through various membranes and eventually end up in the bloodstream. Dissolved chemical species with net neutral charge or ones that can mimic physiologically important species such as sulfate of phosphate species, Taylor and Williams 2005, can cross membranes very easily and then be absorbed by the body.

Human Body Fluids

Biogenic Corrosion

Several terms have been used to describe the effects of biological organisms upon materials. The two most prevalent are *biodeterioration* and *biodegradation*. Biodeterioration was first defined by Hueck in 1965 as *"any undesirable change in the properties of a material caused by the vital activities of organisms,"* Hueck 1965. Biodegradation has been used to describe the beneficial effects caused by the activities of organisms upon materials, although the actual degradation could be essentially the same as any deterioration. Although the biodeterioration of materials in general can be very complex, if limited to the effects upon ceramics the situation becomes much more tolerable. *Biocorrosion, bacterial corrosion, microbial (or microbiologically) influenced corrosion* (MIC) *and biologically influenced corrosion* have also been used. These latter three have been mostly associated with the corrosion of metals. Thierry 1988 has stated that corrosion related to microbes is not a new form of corrosion but merely an enhanced form of electrochemical corrosion.

In some cases the biological effect is purely physical—for example, the growth of an organism within the cracks, joints, etc., of masonry products. In these cases the growth may cause spalling at the surface. Common examples include the cracking of terra cotta sewer pipes by the growth of tree roots. Many other physical or mechanical effects can be envisioned (e.g., bird impact of airplane windows); however, these will not be discussed in this book. Instead only the chemical effects of corrosion will be discussed.

In most cases the corrosion that takes place is the result of a combination of factors and/or microorganisms. Nonbiological factors always play a role and can enhance or decrease the effects of the microorganisms. Atmospheric effects can enhance the corrosive conditions. For example windblown fertilizers can provide the nutrients for growth of algae, especially in humid regions.

There are several different types of mechanisms related to biogenic corrosion of materials, several of which are discussed below, Allsopp et al. 2004.

The corrosion of teeth enamel (a calcium-deficient carbonate hydroxyapatite), for example, is caused by the formation of a biofilm on the surface of the tooth, which eventually leads to decay.

Corrosion of stonework can be caused by the production of acids, the chelation of ions, and the uptake of H^+ ions by fungi, bacteria, cyanobacteria, and lichens.

The solubilization of silicates is caused by the production of citric and oxalic acids by fungi such as Botrytis, Mucor, Penicillium, and Trichoderma. Certain fungi and bacteria produce acids through the metabolism of hydrocarbons (from pollution) that cover masonry surfaces.

Although little data appears in the literature concerning the effects of bacteria and the corrosion of concretes and stonework several mechanisms have been proposed. Sulfate-reducing bacteria (SRB) through a reaction with calcium sulfate in stonework produce hydrogen sulfide. This is then carried in solution throughout the stonework by capillary action. These bacteria require anaerobic conditions and a reduced environment.

One proposed mechanism, although doubted, includes the following steps:

1. SRB form H_2S through their normal anaerobic metabolism,
2. the H_2S is dissolved by water and carried throughout by capillary action,
3. H_2S is oxidized by autotropic sulfur-oxidizing bacteria,
4. this then creates sulfuric acid, and
5. then, the sulfuric acid attacks the stone, forming calcium sulfate.

Concrete has a pH of approximately 12 to 13 after installation. This must be reduced before bacteria activity can commence. This reduction in pH is accomplished through the action of atmospheric carbon dioxide.

Another mechanism although probably of minor importance is through the action of nitrifying bacteria. The steps are listed below:

1. ammonia in the air is oxidized by these bacteria, forming nitrates,
2. this nitrate then reacts with calcium carbonate of the stonework,
3. this reaction forms a more soluble calcium nitrate,
4. the calcium nitrate is then leached from the stonework by rain, and
5. the leaching process leaves behind a loose mass of silica particles.

Heterotropic bacteria, which produce organic acids, have been reported to cause the release of insoluble silicates and phosphates from rocks.

Lichens, which are a complex association of alga or cyanobacterium and fungus, excrete organic acids that can chelate various cations (e.g., Ca and Si) contained within stonework thus removing them. In addition to this chemical attack, lichens also degrade stonework via their root structures that cause mechanical degradation.

Biocidal washes have been used to remove lichens from stonework, but care must be exercised that the wash is slightly alkaline or neutral and does not form salts.

Biological sources that cause corrosion of ceramics are various organisms such as algae, bacteria, fungi, etc. Their natural functions produce organic acids that can leach sodium ions from glass surfaces through ion exchange. Etching occurs as a result. Enhanced attack can occur through the water-soluble complexes formed by some fungi with metals present in the glass. Lichens did not grow on stained glass church windows where silver salts and iron oxides containing borax were used. Other metals used as stains did not prevent growth, Allsopp et al. 2004.

The Environmental Scanning Electron Microscope has made it possible to examine biofilms more accurately, McNeil and Odom 1994. This in turn has allowed more appropriate models to be constructed and tested to evaluate corrosion processes. Some information can be gained that is useful for ceramics by examining what has been reported for metals. Quite often during the corrosion of a metal an oxide layer forms. These oxides are then transformed by the action of SRB to sulfides. The thermodynamics of the reaction will aid in determining if the reaction is spontaneous or not. If spontaneous, then the oxide will be consumed. These reactions are also influenced by local conditions of pH, dissolved oxygen, and sulfide concentrations. The general rule of thumb from all this is that if the sulfide has a higher free energy than the oxide, no reaction will occur. If the reverse is true, then reaction is possible.

Several mechanisms are possible that relate to the way microorganisms enhance or control corrosion reactions, Wagner and Ray 1994. These are as follows:

1. formation of concentration cells,
2. formation of aggressive metabolites, such as sulfides and various acids,
3. redox reactions, and
4. deactivation of corrosion inhibitors.

Bacteria have been found in both fresh and sea water at temperatures from 1 to 320°C. They have been also found where oxygen concentrations vary from 0 to 100%. The presence of terminal electron acceptors is important, such as organic carbon nutrients, CO_2, NO_3, O_2, and SO_4. In anaerobic environments SRB have been found to be responsible for MIC through the production of H_2S. The presence of filamentous material (i.e., slime) and debris suggests the presence of algae, fungi, and bacteria. Algae produce oxygen in the presence of light (i.e., photosynthesis) and consume oxygen when light is absent.

Ceramic Implants

Four types of interface reaction can occur between a ceramic implant and human tissue. These are as follows (Hench and Wilson 1993):

1. formation of a fibrous capsule around the implant,
2. formation of an interfacial bond between the implant and tissue (called a bioactive interface),
3. dissolution of the implant that may be inert, and
4. dissolution of the implant that may be toxic.

In the first type of interface reaction, the implant (essentially inert) does not form a bond with the bone.[*] In the second type, *bioactive,*[†] the implant forms a bond with the bone through a chemical reaction at the interface. Bioactive implants characteristically form a layer of hydroxyl-carbonate apatite on their surface along with collagen fibrils that bind the ceramic implant to the organic tissues. This reaction forms an interface that is essentially identical to the naturally occurring interfaces between bone and ligaments or tendons. If the reaction at the interface is purely mechanical where the bone grows into a porous implant it is not termed *bioactive*. In the last two types of reaction, *dissolution,*[‡] the implant is replaced by bone. Very small variations in chemistry can change a material from being inert to one that is bioactive or resorbable. Although alumina and zirconia tend to be inert they do form thin layers of fibrous tissue at the interface. The thickness of the fibrous layer is dependent upon the chemical, physical, and mechanical properties of both the implant and the tissue. These various types of implants, along with their different reactions, were developed for specific applications in the human body.

Bioapatites have the general formula $A_{10}(BO_4)_6X_2$, where A is mostly Ca^{2+}, B is mostly P^{5+}, and X is mostly OH^- but also F^-. The most common crystal symmetry is hexagonal, although when the X anions become ordered, the symmetry lowers to monoclinic, Mathew and Takagi 2001. Bioapatites are uncommonly stoichiometric with a variety of substitutions on the A, B and X sites. Bioapatites contained in bone and teeth are generally carbonate-containing apatites. The carbonate ions $(CO_3^=)$ are substituted on either the B or X sites, the B site being most common. B site substitution is greater by a factor of about five, Leventouri et al. 2009. Due to the widespread use of fluoridated drinking water and fluoride-containing toothpastes and mouthwashes to prevent tooth decay, dental apatites contain fluorine on the X site

---

[*] The human body considers the implant as a foreign material and isolates it by encapsulation in scar tissue.

[†] Bioactivity is the bonding to bone along with stimulated bone growth without the addition of drugs or other biological agents.

[‡] Dissolution of the implant is also termed bioresorbable.

along with the hydroxyl ion. Kwon et al. 2011 reported that too much fluorine (no limits were given) can cause staining, pitting and tooth formation defects.

Although at first thought the applications of ceramic implants into the human body appear to be very different than the conventional applications of ceramics (e.g., in furnaces) there are many similarities. For example, mechanical loading at joints in the human body and their movement can be related to convective flow of metal at the bottom of a blast furnace—both environments exhibit a mechanical load and an erosive effect. These two environmental factors produce similar effects upon the interface reactions that take place. The initial dissolution reactions of ceramic implants also are very similar to the weathering reactions of silicate minerals, Sahai 2007. *One must know the environment and the ceramic thoroughly to understand and explain any reactions or corrosion that takes place.*

Orthopedic implants generally of type one (e.g., hip or knee replacements) do not have the critical characteristics of living tissue and therefore have limited life spans. These critical characteristics are as follows (Jones et al. 2007):

1. self-repair capability,
2. necessity for a blood supply, and
3. modification in response to stimuli.

It has been long known that the surface of a crystal is greatly distorted due to the asymmetry of the electronic force field. This is the result of broken bonds at the surface. Thus the surface structure is quite different from that of the bulk material. Surface reactivity is dependent upon the following parameters:

1. characteristics of the surface structure including porosity, composition, and possibly crystallographic orientation,
2. characterization of the solution composition including pH, and structure,
3. characteristics of the adsorbent ion or molecular structure and hydrolysis of surface structural groups, and
4. temperature.

Mineral Dusts

Although not an area that one would normally classify as bioceramic, the whole area of the effects of mineral dusts upon the human body are closely related to and actually very similar to bioceramics. The literature on the health effects of inhaled mineral dusts dates back more than 65 years. Fubini and Areán 1999 cite references that date back as far as 61–113 AD. A tremendous amount of data has been reported, but today the mechanisms whereby minerals cause various health issues are still not fully understood. The area

TABLE 4.2

Amphibole Chemistry

General Formula	$AB_2C_5T_8O_{22}W_2$	Regulated
Anthophyllite	$\square Mg_7Si_8O_{22}(OH)_2$	Yes
Edenite	$NaCa_2Mg_5(Si_7Al)O_{22}(OH)_2$	No
Ferroactinolite	$\square Ca_2Fe^{2+}_5Si_8O_{22}(OH)_2$	Yes
Grunerite[a]	$\square Fe^{2+}_7Si_8O_{22}(OH)_2$	Yes
Riebeckite[b]	$\square Na_2(Fe^{2+}_3Fe^{3+}_2)Si_8O_{22}(OH)_2$	Yes
Richterite	$Na(Ca,Na)Mg_5Si_8O_{22}(OH)_2$	Yes
Tremolite	$\square Ca_2Mg_5Si_8O_{22}(OH)_2$	Yes
Winchite	$\square(Ca,Na)Mg_4(Al,Fe^{3+})Si_8O_{22}(OH)_2$	No

[a] Asbestiform grunerite is termed *amosite*.
[b] Asbestiform riebeckite is termed *crocidolite*.
Source: Hawthorne, F.C. and Oberti, R. 2007. Classification of the Amphiboles, *Reviews in Mineralogy and Geochemistry*, 67, pp. 55–88.

of the relationship between the surface activity of minerals and their toxicity to humans recently has been receiving more and more attention. Most of the biological studies of toxicity examine only the effects of particle size and shape, and the mass concentration. Anyone interested in this topic should consult the review by Guthrie and Mossman 1993 who pointed out the need for collaborative studies between the health and mineral scientists.

Whenever researchers from different fields begin to overlap their studies, there is bound to be some misunderstandings. These are often due to the meaning of various terms—either a different term is used for the same meaning or the same term has a different meaning in the various fields of study. This has been the case for asbestos-related studies, which have also included the regulatory and legal fields. The term *asbestos* as used by the mineralogist means that the mineral is fibrous. The asbestiform minerals that are the ones of concern by the medical field are mostly amphiboles, a double-chain silicate (see Table 4.2). The term *asbestos* has been used by the general public (and regulatory and legal fields) to define a mineral that is 95% chrysotile, a mineral that is not an amphibole but a sheet silicate that rolls up into a tube (about 25 nm in diameter) forming a fibrous type (i.e., asbestiform) material. Since the inhaled minerals will undergo some form of alteration during the time they remain within the human body, it is of interest to study the bio-durability of these minerals as a factor in mineral dust-related diseases. The diseases that have received the most attention are asbestos-related: asbestosis, mesothelioma, and lung cancer. Mesothelioma is a rare cancer that affects the external lining of the lungs. *Evidence indicates that mesothelioma is an amphibole asbestos related disease and not caused by chrysotile, which is the main actor in asbestosis, a noncancerous chronic restrictive lung ailment,* Gunter et al. 2007.

The composition of mineral dusts varies depending upon the particular source of the dusts (e.g., asbestos or coal). Mining operations will produce

TABLE 4.3

Sources and Constituents of Various Dusts

Mineral mining	Clay minerals, carbonate minerals, asbestos, asbestiform minerals, silica, coal
Atmospheric dusts	Quartz, feldspars, micas, clay minerals, carbonates, oxides
Volcanic ash	Erionite, silica
Construction materials[a]	Cement, concrete, mineral and glass fiber insulation, gypsum wallboard

[a] Can also be from demolition or collapse of buildings.

dusts related to the particular mineral being mined. Dusts also occur from the weathering of rocks. These dusts generally are composed of a mix of clay-based-iron oxide minerals, carbonate-containing minerals and quartz* in the size range of <1 to about 30 μm. These dusts are often inhaled on a daily basis by people in certain parts of the world and often cause no serious health problems. However, these dusts can cause climatic changes. Mineral dusts (see Table 4.3 for the various constituents of mineral dusts) that are inhaled may end up lodged in and affect the tissues of the lungs. Since lung fluids are undersaturated with respect to most chemical species contained in minerals, these minerals will dissolve in the lung. The health aspect of all this is a kinetics problem—how long will it take to dissolve completely? Minerals that dissolve within a matter of days are thought to have no health effect, although this may not be true. Health effects are dependent upon whether or not the released chemical species are toxic. Many studies have been done to determine whether the body's response to these minerals lies with the chemistry of the minerals or with their morphology (i.e., shape and/ or size). Stanton (Nolan and Langer 1993) and his colleagues have proposed a hypothesis that mesothelioma is caused by the morphology of ceramic fibers. The general result was that decreasing fiber diameter and increasing fiber length were related to increased activity. Morphology affects the aerodynamics as the fibers pass through the respiratory tract and determines the depth to which they will penetrate. If their size is too large for the macrophages[†] to digest then various toxic effects may result.

Incongruent dissolution or selective leaching leads to transverse breakage of fibers which in turn alters the bioreactivity. As more breakage occurs the fibers become smaller allowing for easier consumption by macrophages of the lung. This is apparently the case for chrysotile.

Amphiboles, Hawthorne and Oberti 2007, (more than 80 species) are double chains of Si/Al tetrahedra cross-linked with various cations that hold the chains together and also balance the charge. The Si/Al ratio and the cross-linking cations cause a large variety of chemically different minerals. Some

---

* The largest source of atmospheric mineral dusts is North Africa contributing 58wt% of the total dust emissions, Tanaka and Chiba 2006.
† Macrophages are white blood cells that engulf and then digest cellular debris and pathogens.

of these are very prevalent, others are much rarer. The general chemical formula for an amphibole is $AB_2C_5T_8O_{22}W_2$, where $A = Li^+$, Na^+, K^+, Ca^{2+}, or $\square$; B $= Li^+$, Na^+, Ca^{2+}, Mg^{2+}, Mn^{2+}, or Fe^{2+}; $C = Li^+$, Mg^{2+}, Fe^{2+}, Mn^{2+}, Al^{3+}, Fe^{3+}, Mn^{3+}, or Ti^{4+}; $T = Si^{4+}$, Al^{3+}, or Ti^{4+}; and $W = (OH)^-$, F^-, Cl^-, or O^{2-}. C can also contain minor amounts of Zn^{2+}, Ni^{2+}, Co^{2+}, V^{3+}, Sc^{3+}, Cr^{3+}, or Zr^{4+}. The C cations are located on octahedrally coordinated sites labeled M_1, M_2, and M_3. Sites M_1 and M_3 are coordinated with two W cations, whereas M_2 is coordinated with only oxygens. The B cations are located on an eight-coordinated site labeled M_4 with square anti-prism symmetry. There are two tetrahedrally coordinated sites* containing the T cations, labeled T_1 and T_2. The tetrahedra form a double chain sharing corners with each other and sharing edges with the octahedra. The M_4 site is at the junction of the tetrahedra and the octahedra. Below the hexagonal ring of tetrahedra is the A site. Both the M_1 and M_3 sites are coordinated to oxygen and anions of the W site. The M_2 site is coordinated to only oxygens. The A cations are surrounded by 12 oxygens but not all always bond to the A cation. For a more detailed description of amphiboles see Hawthorne and Oberti 2007. All these various structural features cause a difference in bond lengths, and thus a difference in bond strengths. This in turn will contribute to a variation in the dissolution rates for the various amphiboles. *Stronger bonds generally mean that a mineral is more durable.* It is known that as the aluminum content increases the T–O bond distances increase. The T_1–O bond length increases strongly with increasing Al content, whereas the T_2–O bond length increases less strongly and with considerable scatter. Many of the cations located on the M sites can be selectively leached in acidic conditions and therefore may be more important than the T site cations for mineral human body fluid interactions. T site dissolution is greater than M site dissolution only at $pH > 9$. Since the amphibole structure is rather complex and the chemistry is even more complex, determining a ranking based on bond strengths for which one is more durable is a monumental task. One factor controlling the dissolution of these minerals is most likely the number of various bond types—for example, Si–O–Si, Si–O–Mg, and Mg–O and their combinations. The hydrolysis of Mg–O and Si–O–Mg bonds is much faster than Si–O–Si. Thus minerals containing a mixture of these bonds will invariably exhibit selective leaching of magnesium.

Guthrie 1992 has suggested a way to predict potential toxicity based on structural-activity relationships. These relationships are based on various active sites within the amphibole structure. Naturally occurring structural defects and chemical substitutional defects can provide additional surface sites for enhanced chemical activity. Four possible sites were suggested as shown in Table 4.4. The toxicity mechanisms associated with these sites, as stated by Guthrie, are too poorly understood to allow an accurate prediction of a mineral's toxicity. Jurinski and Rimstidt 2001 have stated that, "...

---

* Some amphiboles have four distinct tetrahedral sites, T_{1A}, T_{1B}, T_{2A}, and T_{2B}, all four coordinated but with slightly different symmetry.

TABLE 4.4

Possible Active Sites within the Amphibole
Structure

Site 1	Between two octahedral layers—A cation sites
Site 2	Protons associated with tetrahedral Al substitution
Site 3	Octahedral polyvalent cations
Site 4	Protons associated with underbonded oxygen

Source: Guthrie Jr., G.D. 1992. Biological Effects of Inhaled
Minerals, *Am. Miner., 77*, 225–243.

dissolution rates of the magnesium silicates … are easily explained in terms
of a combined chemical and geometrical model," indicating the importance
they placed on the structural characteristics of dissolution.

The mineral chrysotile that has been confused as an amphibole is a sheet
silicate; however, the sheet rolls up, forming fibrils. The sheet contains two layers, one of magnesium-containing octahedra and the other silica-containing
tetrahedra. The octahedra are larger than the tetrahedra and therefore the
two-layered sheet curls with the magnesium-containing octahedra exposed
on the exterior. The average diameter of these curled sheets (i.e., fibrils) is
about 25 nm, Fubini and Areán 1999. The sheets themselves are about 0.8
nm thick, Anon. 2009. These fibrils are what have caused chrysotile to be
confused with the asbestiform amphiboles. The cross-linking, if somewhat
loose, allows the chains to part, forming elongated fibers. The acid environment of the macrophages within the lung causes the sheet structure to break
up, forming small pieces that can be readily cleared by normal mechanisms.
In contrast, the exterior surface of an amphibole is silica-rich and therefore
has durability similar to quartz. At the pH levels normally encountered their
solubility is negligible, Anon. 2009.

Inhaled particles are exposed to water vapor since the airways are either
close to or fully saturated with water. Fluid lining the respiratory tract has
a pH = 7–7.4, Plumlee et al. 2006. Most minerals are more likely to be dissolved in the gastric fluids of the stomach than in the near-neutral fluids of
the respiratory tract.

Although the term *speciation*[*] has been defined in many different ways by
various groups, it relates to the oxidation state among other things. Even
though a particular chemical species may enter the body as a nontoxic species, interaction with body fluids may alter the oxidation state (i.e., speciation) causing it to become toxic. The reverse may also occur. For example,
trivalent chrome is rather inert, whereas hexavalent chrome is a carcinogen.
Hexavalent chrome, if present, can be reduced by ascorbate or other organic
reductants to trivalent chrome that can precipitate out in the intestines and
thus not be absorbed into the bloodstream. Hexavalent chrome, on the other

---

[*] The DOE has cited the following aspects relating to speciation: element identity, physical
state, oxidation state, chemical formula, and molecular structure [DOE 1995].

hand, is relatively soluble in the fluids of the lungs and is therefore absorbed into the bloodstream, Plumlee et al. 2006.

Chemical species with intermediate oxidation states (called *reactive oxygen species*[*]), such as hydrogen peroxide and hydroxyl radicals, can lead to cancer. It is suspected that coal-workers *pneumoconiosis*[†] is caused by sulfides (Fe and Ni) in coal that can produce reactive oxygen species. Silica is well known for forming oxygen reaction species, which are caused by grinding-induced structural defects, Plumlee et al. 2006.

Fibers less than about 0.5–1.5 μm in diameter can penetrate deep into the alveoli. Size and density determine depth of penetration into the respiratory tract. Shape can affect also the degree of penetration. Aspect ratio is of importance since fibers as they travel through the various passageways tumble or wobble presenting a larger equivalent diameter than their actual diameter. Thus an equiaxed particle of the same diameter as a fiber will penetrate deeper into the respiratory tract.

The respiratory tract includes the nose and mouth, the larynx, the trachea, and the two main bronchi that supply each lung. These bronchi divide into smaller and smaller ducts, the smallest being called bronchioles that end up forming alveolar ducts with a diameter of about 0.2 mm that end in the alveoli. Gas exchange in the lungs is performed by the alveoli that have a total surface area of about 80 m². Thus larger particles (larger than 10 μm) are caught in the upper portions of the respiratory tract, generally engulfed with mucus and expelled by sneezing, coughing, and/or swallowing. Smaller particles (generally less than 5 μm) can penetrate deep into the lungs and end up in the alveoli. Specialized cells in the lungs secrete a phospholipid-rich pulmonary surfactant (lowers surface tension and maintains a wet surface) needed to prevent alveolar collapse that also contains antioxidants (ascorbate and glutathione) and proteins (immunoglobulins). This surfactant can interact with inhaled minerals. Mineral particles can also pass into the lymphatic system through the action of macrophages. Once in the lymphatic system particles can end up in the bloodstream and eventually other body organs, Fubini and Areán 1999. Particles that penetrate deep into the lung subsequently are in proximity to the lung/pleura[‡] interface.

Various mechanisms are operative that rid the body of foreign particles. Macrophages in the airway fluids contain lysosomes, which are acidic (on the order of pH 4.5), and digestive enzymes (i.e., acid hydrolases) that can dissolve mineral particles. The acidic lysosomal fluids are rich in cytotoxic acid proteases[§] and reactive oxygen species. Due to their size (i.e., 5–10 μm) macrophages can engulf particles only less than 2–5 μm. Thus shorter fibers can be completely engulfed by the macrophages and subsequently cleared

---

[*] Reactive oxygen species are formed by incomplete oxidation of molecular oxygen and redox sensitive organic species present in the body, Plumlee et al. 2006.
[†] Colloquially known as black lung disease.
[‡] The pleura is the two-layered lining surrounding the lungs.
[§] Enzymes that digest proteins.

TABLE 4.5

Body Fluid pH Ranges

Interstitial blood plasma	~7.5
Extracellular lung fluid	~7.0
Saliva	~6.5–8.0
Macrophage lysosomal fluid	~4.5
Sweat	~4.0–6.5
Gastric fluid (absence to presence of food)	~1.5–6.5

Source: Sahai, N. 2007. Medical Mineralogy and Geochemistry: An Interface Science, *Elements*, 3(6), 381–384.

from the body just as nonfibrous particles would be. Longer fibers that are not completely engulfed are persistent and eventually lead to chronic inflammation and may result in fibrosis* and pulmonary tumors. The body fluid-mineral interaction, which may involve dissolution, chelation, precipitation, and/or redox changes, can strongly influence mineral absorption.

Ingested particles are exposed to several different environments as they pass from the mouth, pharynx, esophagus, stomach, and the intestines. Not only are ingested particles affected by the fluids they encounter but the presence of other materials (i.e., food) and reactions involving resident microbes can also affect the dissolution. Saliva in the mouth contains near neutral enzyme-rich fluids, the stomach contains very acidic gastric fluids, and the intestines contain near neutral pancreatic and bile fluids. See Table 4.5 for the pH ranges of various body fluids. Ingested particles are generally larger than inhaled ones and can be as large as 1 mm but most often are on the order of several hundred μm, Plumlee et al. 2006. Once in the intestines dissolved minerals, if in a nonionized, lipid-soluble form, can diffuse across the intestinal wall. These dissolved mineral species, after passing through the intestinal wall, enter the bloodstream and may end up in the liver. In addition, minerals that have been dissolved in the acidic gastric fluids of the stomach may re-precipitate in the lower pH fluids of the intestines.

Nanoparticles[†]

Nanoparticles, those with a diameter less than 100 nm[‡], have been shown to have greater chemical reactivity predominantly due to their greater surface area to volume ratio. Solubility of nanoparticles increases sharply as

---

* Fibrosis is the accumulation of collagen during prolonged tissue repair, i.e., scarring.
† The prefix *nano* has been used in conjunction with many other words. This is unfortunate since the meaning in some cases may be confusing. In most cases this prefix implies that the substance is composed of nanoparticles. For example, *nanocomposite* does not relate to a very small composite but to a composite made of nanoparticles. *Nano* comes from the Greek or Latin word meaning *dwarf*. The modern scientific definition means one-billionth or 10^{-9}.
‡ More precisely, nanoparticles are those with one or more dimensions less than 100 nm.

their size decreases below about 10 nm, Banfield and Zhang 2001. The excess energy associated with the broken bonds and unsatisfied coordination requirements at exposed surface sites along with the associated roughness caused by edges, corners, etc. relates to the increased reactivity. This greater chemical reactivity relates to a greater biological activity. This enhanced reactivity, when interacting with body fluids, results in the production of reactive oxygen species and free radicals, Nel et al. 2006. Adsorption of macromolecules within the human body onto their surfaces becomes much easier with the increased surface area, thus affecting the regulation of enzymes and proteins.

Being extremely small, nanoparticles can gain entry into the human body very easily and eventually end up in the bloodstream. Once in the bloodstream they can be transported around the body and end up in the liver, kidneys, spleen, heart, brain, bone marrow, and the nervous system, Oberdörster et al. 2005. Just as in larger particles, other factors such as chemical composition, shape, surface properties, and solubility, affect the toxicity of nanoparticles.

Burello and Worth 2010 developed a model that predicts available electronic energy levels of nanoparticle structures and then matches them with oxidation potentials of reactions that remove antioxidants from cells or creates reactive oxygen species. Where the two values overlap equates with the creation of oxidative stress in cells. Titanium dioxide, which causes oxidative stress, was used as a test for the model with excellent results.

Dermal absorption of mineral particles can be either directly through the skin or through wounds or pores in the skin. Reaction with perspiration can result in allergic reactions, irritation, or even burns. Recently there has been reported evidence that nanoparticles can be absorbed directly through the skin, Berger 2007.

Not only can nanoparticles have toxic effects, but they have been used recently for drug delivery systems, Yang et al. 2010. The high surface area-to-volume ratio of nanoparticles offers the medical practitioner the ability to deliver high drug payloads and also a prolonged drug release profile. The longer degradation times of ceramic nanoparticles, as compared to other materials, is crucial to the kinetics of diffusion-controlled drug release. Drugs have also been encapsulated into hollow nanospheres. The drug release is then controlled by collapsing various amounts of nanospheres by the use of ultrasound.

Synthetic Glass Fibers

Bauer et al. evaluated the earlier work of Leineweber 1982 and Scholze 1988 who described the dissolution of synthetic glass fibers in extracellular fluid. Under sufficiently high fluid flow rates (where the products of dissolution do not interfere with subsequent dissolution) the mass loss rate of the fibers was proportional to the fiber surface area. The dissolution rate was assumed to be independent of time. Two other assumptions were uniform fiber diameter and congruent dissolution. A consequence of these conditions was that the fiber diameter decreased at a constant rate. *The real problem with actual samples is that individual fibers do not necessarily have uniform diameters and all fibers of an aerosol sample do not have equal diameters.* Bauer et al. suggested that the mass loss should be calculated at each experimental time for each fiber diameter and that the dissolution constant should be varied until agreement with the experimental data is obtained. Bauer et al. also suggested that the constant velocity dissolution rate constant was probably the best single parameter for estimating synthetic glass fiber endurance in the lung. However, they warned that this method does not adequately reproduce the dissolution of rock wools that contain more than 20 wt% Al_2O_3 and less than 42 wt% SiO_2 and also that the manufacturing process may affect the results. The method of Potter and Mattson 1991, where dissolution rate constant was determined by the decrease in fiber diameter, being less sensitive and more time consuming, was not recommended by Bauer et al. Not knowing the role that simulated lung fluid constituents like proteins, enzymes, and surfactants play in dissolution, Bauer et al. left these out. Thus Bauer et al. used the following equations to evaluate the dissolution of uniform diameter synthetic glass fibers:

$$dM/dt = -k_{dis}A \qquad (4.1)$$

where dM/dt is the rate of mass loss, k_{dis} is the dissolution constant (ng/cm^2·hr), and A is the fiber surface area. Solving equation 4.1 for uniform diameter fibers dissolving congruently with an initial mass M_o, initial diameter D_o, and initial density ρ one obtains

$$1 - (M/M_o)^{1/2} = 2k_{dis}t/D_o\rho \qquad (4.2)$$

Equations 4.1 and 4.2 led to the fact that the fiber diameter decreases at a constant rate according to

$$D = D_o - 2k_{dis}t/\rho \qquad (4.3)$$

Other fiber situations (i.e., nonuniform diameters) were also discussed by Bauer et al. Equation 4.3 can be used to determine the lifetimes of various fibers by setting $D = 0$. Doing this, Eastes and Hadley 1996 compared the lifetimes of glass, rock, and slag wools with chrysotile. The estimate for chrysotile was done by using a glass fiber of similar composition. They obtained approximately 50 dys, 328 dys, 15 dys, and 7 yrs, respectively, for glass, rock, slag wools, and chrysotile.

Eastes et al. 2000 developed a method to determine glass fiber (borosilicate) dissolution rates from chemical composition. An accurate estimate of k_{dis} from composition data allows one to predict the clearance of fibers and

TABLE 4.6

Dissolution Rate Coefficients for Use in Equation 4.4

Oxide	Coefficient P_i	Standard Error
SiO_2	−0.01198	0.00285
Al_2O_3	−0.21410	0.01102
CaO	0.10806	0.01119
MgO	0.13761	0.01262
Na_2O	0.09386	0.00867
B_2O_3	0.14669	0.00908
BaO	0.06921	0.03095
F	0.11867	0.06134
R^2	0.96	
Significance	0.97	
Degrees of Freedom	54	

Source: Eastes, W. and Hadley, J.G. 1996. A Mathematical Model of Fiber Carcinogenicity and Fibrosis in Inhalation and Intraperitoneal Experiments in Rats, *Inhalation Toxicology, 8*, 323–42.

the toxic effects associated with the fibers. Information of this type is of immense practical value to manufacturers and to the regulatory community. The following equation is used to determine the rate constant k_{dis}:

$$\log k_{dis} = \Sigma_{i=1}P_iW_i \qquad (4.4)$$

where W_i is the weight percent of oxide i and P_i is the coefficient associated with each oxide. These are summed over all oxides including SiO_2. The coefficients P_i (see Table 4.6) are determined by fitting measured k_{dis} to the weight percents for a range of compositions. For additional coefficients see Potter and Mattson 1991. The fact that various chemical species dissolve at different rates doesn't appear to affect the results. A separate set of coefficients is needed for each type of glass composition: borosilicate, aluminosilicate, phosphate, etc.

Therapeutic Materials

The manufacture of ceramic materials for the controlled release of drugs and radiation is a relatively new area. The materials that are being developed are both crystalline and glass and are in the form of both powders (i.e., microspheres, solid or hollow) and fibers. Nanosized materials are also being developed. The therapeutic material to be released is in some cases incorporated into the ceramic structure (e.g., glasses) and in other cases is located in the surface porosity. Radiation therapy materials are mostly glasses that allow the radiation to continue over a short period of time as the glass dissolves. Kawashita et al. 2005 reported the preparation of ferromagnetic (i.e., Fe_3O_4) microspheres for the thermal treatment of tumors. A magnetic field is required to produce the heat (about 43°C is needed to kill the cancer cells). The critical parameter for the ceramic is a controlled dissolution. In some cases the dissolution should be very low initially and then increase after drug or radiation therapy has completed, so that the ceramic is eventually completely resorbed. Therefore it is a must that the ceramic be nontoxic.

Rare earth aluminosilicate (REAS) glasses have been developed for applications as delivery agents for radiation* in the treatment of various cancerous tumors, White and Day 1994. These materials must be bioresorbable to avoid surgical removal and must also not exhibit cytotoxic effects. In the treatment of liver cancer, microspheres are injected into the liver via the heptic artery and follow the blood flow to the liver. Ten to 200 mg of glass are injected depending upon the amount of radiation desired. The amount of radiation that can be delivered is 5–7 times greater than can be delivered by external beam radiation. In these cases the glass must be sufficiently durable to allow the release of beta radiation over a specified period of time (about 2 weeks) while being lodged within the malignant tumor. High durability is very important so that radionuclides do not escape the targeted tissue. The radionuclides are significantly radioactive until 99.9% of the activity has decayed (after about ten half-lives). Once the radiation treatment has been completed (after about 3–4 weeks) the REAS can be resorbed into the body. It is important that these glasses not dissolve while being radioactive, which would release radioactive species into the other parts of the body damaging healthy tissue. Beta emitting rare earth isotopes of ^{90}Y, ^{166}Ho, ^{153}Sm, or ^{165}Dy are generally used.[†] These glasses are generally incorporated in the body as microspheres ranging from 15 to 35 μm in diameter. More recently Day 2002/03 developed borate glasses containing rare earths as biodegradable radiation delivery vehicles and as hollow microspheres for long-term drug delivery.

Additional Recommended Reading

Allsopp, D., Seal, K. and Gaylarde, C. 2004. *Introduction to Biodeterioration*, 2nd edition, Cambridge University Press, Cambridge, UK, 237 pp.

Banfield, J.F. and Navrotsky, A. (eds.). 2001. Nanoparticles and the Environment, *Reviews in Mineralogy and Geochemistry, Vol. 44*, Mineral. Soc. Am., Washington, DC.

Guthrie Jr., G.D. and Mossman, B. T. (eds.). 1993. *Reviews in Mineralogy, Vol. 28: Health Effects of Mineral Dusts*, Mineral. Soc. Am., Washington, DC.

---

* Beta radiation is desired since it is of high energy and has only a short range in living tissue (0.3–1.1 cm).

† Other beta emitters are ^{131}I, ^{64}Cu, ^{67}Cu, ^{89}Sr, ^{177}Lu, ^{186}Re, ^{188}Re, ^{117}Sn, ^{149}Pr, ^{32}P, and ^{199}Au.

Hochella Jr., M.F. and White, A.F. (eds.). 1990. *Reviews in Mineralogy, Vol. 23: Mineral-Water Interface Geochemistry*, Mineral. Soc. Am., Washington, DC.
Shackelford, J.F. (ed.). 1999. *Bioceramics, Applications of Ceramic and Glass Materials in Medicine*, Trans Tech Publications, Switzerland.

Exercises, Questions, and Problems

1. List as many forms of corrosion that you can that are either enhanced or started by biological effects. Discuss one in detail.

2. The accumulation of ions at the surface of a material through adsorption may cause the formation of a distinct surface phase or the incorporation of ions into a phase via co-precipitation. What factors affect adsorption? What other processes may occur simultaneously with adsorption?

3. Discuss the following terms and how they differ: biotransformation, bioaccessibility, bioavailability, and biopersistence. How are they related to corrosion?

4. List five body fluids that may be involved in corrosion of ceramics. What property of these fluids is the most important in corrosion?

5. Discuss one of the four types of interface reactions that may occur between a ceramic implant and human tissue.

6. Compare the rate constants of dissolution for mullite and wollastonite using equation 4.4 and the data of Table 4.5.

7. Why has chrysotile been confused as an amphibole?

8. Discuss the various aspects of dissolution of the glass beads used for drug delivery. Why is this dissolution important?

9. Discuss the importance of sulfate-reducing bacteria in the corrosion of stonework.

10. Discuss the four mechanisms that relate to the way microorganisms enhance or control corrosion reactions.

References

Allsopp, D., Seal, K. and Gaylarde, C. 2004. *Introduction to Biodeterioration*, 2nd edition, Cambridge University Press, Cambridge, UK, 237 pp.
Anon. 2009. Review of the Differences between Chrysotile and Amphibole Asbestos, Chrysotile Institute, http://www.chrysotile.com/data/encart_an_final.pdf, Aug.

Appleton, J.D., Fuge, R. and McCall, G.J.H. (eds.) 1996. Environmental Geochemistry and Health with Special Reference to Developing Countries, *Geological Society Special Publication* No 113, Geological Society, London.

Banfield, J.F. and Zhang, H. 2001. Nanoparticles in the Environment, pp. 1–58 in *Nanoparticles and the Environment. Reviews in Mineralogy & Geochemistry, 44,* Banfield, J.H. and Navrotsky, A. (eds.), Mineral Soc. Amer. Chantilly, VA.

Bauer, J., Mattson, S.M., and Eastes, W. In-Vitro Acellular Method for Determining Fiber Durability in Simulated Lung Fluid, accessed at http://fiberscience. owenscorning.com/prokdis/prokdis.html.

Berger, M. 2007. Nanoparticle Penetration of Human Skin—A Double Edged Sword, Nanowerk Spotlight, accessed at http://nanowerk.com/spotlight/spotid = 1820.php.

Bethke, C.M. 2002. The Geochemists Workbench—Release 4.0: A User's Guide to Rxn, Act2, Tact, React and Gtplot, Univ. Illinois.

Burello, E. and Worth, A. 2010. Accessed at http://www.rsc.org/chemistryworld/ News/2010/July/26071001.asp.

Davis, A., Ruby, M.V. and Bergstrom, P.D. 1992. Bioavailability of Arsenic and Lead in Soils from the Butte Montana Mining District, *Environ. Sci. Technol., 26,* 461–468.

Day, D.E. 2002/03. Reactions of Bioactive Borate Glasses with Physiological Liquids, *Glass Researcher, 12*(1&2), 21–2.

DOE. 1995. *Molecular Environmental Science: Speciation, Reactivity, and Mobility of Environmental Contaminants: An Assessment of Research Opportunities and the Need for Synchrotron Radiation Facilities,* Stanford Synchrotron Radiation Laboratory, SLAC-R-477.

Driskell, T.D. 1973. Calcium phosphate resorbable ceramic: A potential alternative for bone grafting, *J. Dent. Res.* 52, 123–131.

Eastes, W. and Hadley, J.G. 1996. A Mathematical Model of Fiber Carcinogenicity and Fibrosis in Inhalation and Intraperitoneal Experiments in Rats, *Inhalation Toxicology, 8,* 323–42.

Eastes, W., Potter, R.M. and Hadley, J.G. 2000. Estimating In-Vitro Glass Fiber Dissolution Rate from Composition, *Inhalation Toxicology, 12,* 269–80.

Finkelman, R.B., Skinner, H.C.W., Plumlee, G.S. and Bunnell, J.E. 2011. Medical Geology, *Geotimes,* Nov., htt:www.agiweb.org/geotimes/nov01/feature_med-geo.html

Fubini, B. and Areán, C.O. 1999. Chemical Aspects of the Toxicity of Inhaled Mineral Dusts, *Chem. Soc. Rev., 28,* 373–81.

Gunter, M.E., Bellos, E., and Mottana, A. 2007. Amphiboles: Environmental and Health Concerns, in *Reviews in Mineralogy & Geochemistry, Vol. 67: Amphiboles: Crystal Chemistry, Occurrence, and Health Issues,* Hawthorne, F., Oberti, R., Ventura, G.D., and Mottana, A. (eds.), pp. 453–516, Mineral. Soc. Amer., Chantilly, VA.

Guthrie Jr., G.D. 1992. Biological Effects of Inhaled Minerals, *Am. Miner., 77,* 225–43.

Guthrie Jr., G.D. and Mossman, B. T. (eds.). 1993. *Reviews in Mineralogy, Vol. 28: Health Effects of Mineral Dusts,* Mineral. Soc. Am., Washington, DC.

Hawthorne, F.C. and Oberti, R. 2007. Classification of the Amphiboles, *Reviews in Mineralogy and Geochemistry,* Vol. 67, pp. 55–88.

Hench, L.L. and Wilson, J. 1993. Introduction, pp. 1–24 in *An Introduction to Bioceramics,* Advanced Series in Ceramics, Vol. 1, World Scientific, River Edge, NJ.

Hueck, H.J. 1965. The Biodeterioration of Materials as Part of Hylobiology, *Mater. Org., 1*(1), 5–34.

Ironside, J.G. and Swain, M.V. 1998. Ceramics in Dental Restorations—A Review and Critical Issues, *J. Australasian Ceramic Society*, 34(2), 78–91.

Jolicoeur, C. and Poisson, D. 1987. Surface Physico-Chemical Studies of Chrysotile Asbestos and Related Minerals, *Drug and Chemical Toxicology*, 10(1&2), 1–47.

Jones, J.R., Gentlemen, E. and Polak, J. 2007. Bioactive Glass Scaffolds for Bone Regeneration, *Elements*, 3(6), 393–99.

Jurinski, J.B. and Rimstidt, J.D. 2001. Biodurability of Talc, *Amer. Mineral.*, 86, 392–399.

Kawashita, M. et al. 2005. Preparation of ferrimagnetic magnetite microspheres for in situ hyperthermic treatment of cancer, *Biomaterials*, 26, 2231–8.

Kwon, K.Y., Wang, E., Nofal, M. and Lee, S-W. 2011. Microscopic Study of Hydroxyapatite Dissolution as Affected by Fluoride Ions, *Langmuir*, 27(9), 5335–39.

Leineweber, J.P. 1982. Solubility of Fibres In Vitro and In Vivo. In *Biological Effects of Man-Made Mineral Fibres*, Proceedings of a WHO/IARC Conference, Vol. 2, pp. 87–101, Copenhagen, World Health Organization.

Leventouri, Th., Antonakos, A., Kyriacou, A., Venturelli, R., Liarokapis, E. and Perdikatsis, V. 2009. Crystal Structure of Human Dental Apatite as a Function of Age, *Int. J. Biomaterials*, Vol. 2009, Article ID 698547, 6 pp., doi: 10.1155/2009/698547.

Mathew, M. and Takagi, S. 2001. Structures of Biological Minerals in Dental Research, *J. Res. Natl. Inst. Stand. Technol.*, 106(6), 1035–44.

McNeil, M.B. and Odom, A.L. 1994. Thermodynamic Prediction of Microbiologically Influenced Corrosion (MIC) by Sulfate-Reducing Bacteria, pp. 173–9 in *Microbiologically Influenced Corrosion Testing, ASTM STP 1232*, Kearns, J.R. and Little, B.J. (eds.), ASTM, Philadelphia, PA.

Nel, A., Xia, T., Mädler, and Li, T. 2006. Toxic Potential of Materials at the Nanolevel, *Science*, 311(5761), 622–627.

Nolan, R.P. and Langer, A.M. 1993. Limitations of the Stanton Hypothesis, Chap. 9 in *Health Effects of Mineral Dusts*, G.D. Guthrie, Jr., and B.T. Mossman (eds.), *Reviews in Mineralogy*, Vol. 28, Min. Soc. Am., Washington, DC.

Oberdörster, G., Oberdörster, E. and Oberdörster, J. 2005. Nanotoxicology: An Emerging Discipline Evolving from Studies of Ultrafine Particles, *Environmental Health Perspectives*, 113(7), 823–39.

Parkhurst, D.L. and Appelo, C.A.J. 1999. User's Guide to PHREEQC (version 2)—A Computer Program for Speciation, Batch-Reaction, One-Dimensional Transport, and Inverse Geochemical Calculations, Report 99-4259, U.S. Geological Survey.

Plumlee, G.S., Morman, S.A. and Ziegler, T.L. 2006. The Toxicological Geochemistry of Earth Materials: An Overview of Processes and the Interdisciplinary Methods Used to Understand Them, pp. 5–57, in *Reviews in Mineralogy & Geochemistry, Vol. 64: Medical Mineralogy and Geochemistry*, Sahai, N. & Schoonen, M.A.A. (eds.), Mineral. Soc. Am., Chantilly, VA.

Potter, R.M. and Mattson, S.M. 1991. Glass Fiber Dissolution in a Physiological Saline Solution, *Glasstech. Ber.*, 64, 16–28.

Reeder, R.J., Schoonen, M.A.A. and Lanzirotti, A. 2006. Metal Speciation and Its Role in Bioaccessibility and Bioavailability, pp. 59–113 in *Reviews in Mineralogy & Geochemistry, Vol. 64: Medical Mineralogy and Geochemistry*, Sahai, N. and Schoonen, M.A.A. (eds.), Mineral. Soc. Am., Chantilly, VA.

Sahai, N. 2007. Medical Mineralogy and Geochemistry: An Interface Science, *Elements*, 3(6), 381–84.

Scholze, H. 1988. Durability Investigations on Siliceous Man-Made Mineral Fibers: A Critical Review, *Glastech. Ber., 61*, 161–71.

Smith, K.S. and Huyck, H.L. 1999. An Overview of the Abundance, Relative Mobility, Bioavailability, and Human Toxicity of Metals, pp. 29–70 in *The Environmental Geochemistry of Mineral Deposits* Part A: Vol 6A, Ch. 2, Plumlee, G.S. and Logsdon, M. (eds), Reviews in Economic Geology, Society of Economic Geologists, Inc., Chelsea, MI.

Tanaka, T.Y. and Chiba, M. 2006. A Numerical Study of the Contributions of Dust Source Regions to the Global Dust Budget, *Global and Planetary Change, 52*, 88–104.

Taylor, D.M. and Williams, D.R. 2005. Pharmaceutical Applications of Bioinorganic Chemistry, Ch. 17, pp. 617–42 in *Smith and Williams Introduction to the Principles of Drug Design and Action*, 4th edition, Smith, H.J. (ed.), CRC Press, Boca Raton, FL.

Thierry, D. 1988. The Application of Corrosion Monitoring Techniques to Study Microbial Corrosion in Cooling Water Systems, pp. 179–82 in *Microbial Corrosion—1*, Sequeira, C.A.C. and Tiller, A.K. (eds.), Elsevier Applied Science, London.

Wagner, P.A. and Ray, R.I. 1994. Surface Analytical Techniques for Microbiologically Influenced Corrosion—A Review, pp. 153–69 in *Microbiologically Influenced Corrosion Testing, ASTM STP 1232*, Kearns, J.R. and Little, B.J. (eds.), ASTM, Philadelphia, PA.

White, J.E. and Day, D.E. 1994. Rare Earth Aluminosilicate Glasses for In Vivo Radiation Delivery, pp. 181–208 in *Rare Elements in Glasses*, Shelby, J.E. (ed.), *Key Engineering Materials*, Vols. 94–95, Trans Tech Pub., Switzerland.

Yang, L., Sheldon, B.W. and Webster, T.J. 2010. Nanophase Ceramics for Improved Drug Delivery: Current Opportunities and Challenges, *Amer. Ceram. Soc. Bull., 89*(2), 24–31.

5

Acid/Base Effects

> The important thing in science is not so much to obtain new facts as to
> discover new ways of thinking about them.
>
> **Sir William Bragg**

The chemical species present in a liquid will determine whether it is of an acidic or basic character. Ceramics with a surface acid/base character similar to the liquid will tend to resist corrosion the best. In some cases, the secondary phases of a ceramic may be of a slightly different acid/base character than the major component and thus whether the major phase or the secondary bonding phase corrodes first will depend upon the acid/base character of the environment. In addition to corrosion, heterogeneous catalysis is also involved with the surface acidity of a ceramic material, Benesi and Winquist 1978.

Several acid-base reaction theories have been proposed. The Brönsted and Lowry Theory may be sufficient to explain those reactions in aqueous media where the acid/base character of a surface is determined by its zero point of charge *(zpc)* or the *p*H where the immersed surface has a zero net surface charge. This point of zero charge also relates to the situation where an equal number of positively and negatively charged dissolution species are present on the surface. This condition is called the *isoelectric point* or *iep*. The *iep* is the point of zero zeta potential. The *zpc* and the *iep* do not necessarily correspond to one another. The *zpc* and *iep* are often used interchangeably although they do refer to different surface conditions. The *zpc* refers to the absence of any type of charge, whereas the *iep* refers to the state of neutral surface charge, Jolivet 2000. For those materials that exhibit some solubility, it is more appropriate to use the *iep* value that represents the *p*H of minimum solubility. Cation substitutions or broken bonds at a material's surface cause Brönsted sites to form that are associated with underbonded oxygen atoms. These types of sites are quite prevalent, for example, in the amphiboles minerals.

In nonaqueous media, the Lewis Theory is probably more appropriate when acids are defined as those species that accept a pair of electrons thus forming a covalent bond with the donor, and bases are defined as those species that donate a pair of electrons thus forming a covalent bond with the acid. Lewis sites are associated with the presence of polyvalent cations. Ionization may follow formation of the covalent bonds. Those species that can both accept or donate electrons depending upon the character of its partner are called

TABLE 5.1

Point of Zero Charge for a Few
Selected Oxides Calculated from
Ionization Potentials

Oxide	zpc
SiO_2	3.4
TiO_2	4.4
ZrO_2	5.6
Al_2O_3	7.7
ZnO	9.9
MgO	10.3
CaO	10.7
Na_2O	11.8

Source: Carre, A. Roger, F. and Varinot, C.
1992. Study of Acid/Base
Properties of Oxide, Oxide Glass,
and Glass-Ceramic Surfaces, *J.
Colloid and Interface Sci.*, 154(1),
174–83.

amphoteric. Thus a particular species may act as an acid toward one partner but as a base toward another. Oxidizing agents are similar to acids since they tend to accept electrons; however, they keep the electrons to themselves rather than sharing.

Carre et al. 1992 have devised a simple approach to calculations of the *zpc* from ionization potentials of the metallic elements contained in pure oxides. Those values differ very little from those determined by Parks 1965. They used an additive method to calculate the *zpc* of multicomponent glasses. *The importance of the zpc in corrosion is that it is the pH of maximum durability.* The approach of Carre et al. is fundamentally very similar to that of Lewis, since oxide acidity depends upon the electron affinity of the metal, whereas O^{2-} anions act as the basic component. The *pH* of zero charge is given in Table 5.1 for a few selected oxides.

According to Carre et al. abrading or grinding the surface of various glasses increases the *zpc* (e.g., soda-lime glass *zpc* increased from about 8.0 to 12.0) supposedly by increasing the alkalinity at the surface. Acid washing produces just the opposite effect, decreasing *zpc* caused by leaching the alkali from the surface.

The biological activity of a mineral can be related to its properties through the characterization of the Lewis-acid/base sites per unit surface area and/or the Brönsted-acid sites per unit surface area. This is important when considering the health-related issues of inhaled mineral dusts.

Acid-base reactivity is closely related to the changing conditions created on a mineral surface by the surrounding environment. Atomic position relaxation on the surface relative to the bulk can create a completely new set of absorbed species on the surface. For example, alumina in a vacuum contains surface terminal Al ions that are pulled in tightly to the lower oxygens by coulombic forces. Consequently these Al ions all have three oxygen neighbors. In a non-vacuum environment containing some moisture hydrated alumina polyhedra ($Al(O,OH)_6$) are formed at the surface. The set of atomic position relaxations that form are completely different than in the vacuum case. In addition for the hydrous case an ordered overlayer of water molecules may form that create interesting consequences for sorption, Eng et al. 2000 and Waychunas et al. 1995.

Waychunas and Zhang 2008 have hypothesized that acid-base character and reactivity are both consequences of bonding deficits at edge and corner sites. Bond-deficient edge or corner sites would have fewer cations coordinated to an oxygen and thus would require protonation for stability. Additional basic sites are created by these oxygens since the protons would be essentially bonded more tightly. A comparison of the *zpc*s for an alumina single crystal (0001) face (6.3) and for an alumina powder (8.5–9.0) appears to confirm these ideas. Protons are therefore more difficult to remove from the larger number of edge and corner sites of the powder. Both the acid dissociation constant and the proton-transfer reactions were attributed by Waychunas and Zhang 2008 to be at least partially controlled by size and shape.

The acid-base character of two materials in contact is very important when determining whether a reaction will occur. The further apart the two materials are on the scale of acidity (or basicity), the greater the potential for reaction.

Exercises, Questions, and Problems

1. Explain what is meant by zero point of charge.
2. Explain what is meant by isoelectric point.
3. What is the zeta potential and how does it relate to the zero point of charge and the isoelectric point?
4. What is the relationship of zero point of charge to corrosion?

References

Benesi, H.A. and Winquist, B.H.C. 1978. Surface acidity of solid catalysts. *Adv. Catal.*, 27, 97–182.

Carre, A., Roger, F., and Varinot, C. 1992. Study of Acid/Base Properties of Oxide, Oxide Glass, and Glass-Ceramic Surfaces, *J. Colloid and Interface Sci.*, *154*(1), 174–83.

Eng, P.J., Trainor, T.P., Brown Jr., G.E., Waychunas, G.A., Newville, M., Sutton, S.R. and Rivers, M.L. 2000. Structure of the Hydrated Al_2O_3 (0001) Surface, *Science*, *288*, 1029–1033.

Jolivet, J.P. 2000. *Metal Oxide Chemistry and Synthesis. From Solution to Solid State.* Wiley, Chichester, UK (English translation of *De la Solution à l'Oxyde*, Intereditions et CNRS Edition, Paris, 1994).

Parks, G.A. 1965. The Isoelectric Points of Solid Oxides, Solid Hydroxides, and Aqueous Hydroxo Complex Systems, *Chem. Rev.*, *65*(2), 177–98.

Waychunas, G.A., Davis, J.A., Fuller, C.C. 1995. Geometry of Sorbed Arsenate on Ferrihydrite and Crystalline FeOOH: Re-evaluation of EXAFS Results and Topological Factors in Predicting Sorbate Geometry, and Evidence for Monodentate Complexes, *Geochimica et Cosmochimica Acta*, *59*, 3655–61.

Waychunas, G.A. and Zhang, H. 2008. Structure, Chemistry, and Properties of Mineral Nanoparticles, *Elements*, *4*, 381–7.

6

Diffusion

If you can't explain it simply, you don't understand it well enough.

Albert Einstein

When the transport of ions or molecules occurs in the absence of bulk flow it is called *diffusion*. Substances will spontaneously diffuse toward the region of lower chemical potential. This transport or flux of matter is represented by Fick's first law and is proportional to the concentration gradient. This is represented by:

$$J_{ix} = -D\left(\partial c_i / \partial x\right) \tag{6.1}$$

where:
  J_{ix} = flux of component i in the x direction,
  D = diffusion coefficient, and
  c_i = concentration of component i.

The flow of material is thus proportional to the concentration gradient and is directed from the region of high concentration to one of low concentration.

Fick's second law describes the nonstationary state of flow where the concentration of a fixed region varies with time:

$$\frac{\partial c}{\partial t} = \frac{\partial}{\partial x}\left(D\frac{\partial c}{\partial x}\right) \tag{6.2}$$

Since diffusion is directional, one must be aware of anisotropic effects. The rate of diffusion may be very different in different crystalline directions. In isometric crystals, the diffusion coefficient is isotropic, as it is in polycrystalline materials as long as no preferred orientation exists. The second-order tensor defined by the equations for the flux, J, in each of the x, y, and z directions, contains a set of nine diffusion coefficients designated D_{ij}. Due to the effects of the various symmetry operations in the tetragonal, hexagonal, orthorhombic, and cubic crystal classes, only a few of these D_{ij} have non-zero values. All the off-diagonal D_{ij} ($i = j$) are equal to zero. Thus only the three diagonal values are of any consequence; however, symmetry again causes some of these to be equivalent. In the remaining two crystal classes

TABLE 6.1

Effect of Symmetry upon the Second Rank Tensor Diffusion Coefficients

Crystal Class	Number of Independent Coefficients	Nonzero Coefficients
Cubic	1	$D_{11} = D_{22} = D_{33}$
Tetragonal and hexagonal	2	$D_{11} = D_{22} \neq D_{33}$
Orthorhombic	3	$D_{11} \neq D_{22} \neq D_{33}$
Monoclinic	4	$D_{11} \neq D_{22} \neq D_{33} \neq D_{31} = D_{13}$
Triclinic	6	$D_{11} \neq D_{22} \neq D_{33} \neq D_{21} = D_{12} \neq D_{31} = D_{13}$ $\neq D_{23} = D_{32}$

the number of independent coefficients increases; however, the total number is decreased somewhat, since $D_{ij} = D_{ji}$. The possible nonzero diffusion coefficients for each of the crystal classes is shown in Table 6.1.

A solution of equation 6.2 for nonsteady state diffusion in a semi-infinite medium (D is independent of concentration) is:

$$C(x,t) = \frac{C_o}{2} \left\{ 1 + erf\left(\frac{x}{2\sqrt{(Dt)}} \right) \right\} \quad (6.3)$$

where:
$C(x,t)$ = concentration after time t, and
C_o = initial concentration in the medium.

Solutions to equation 6.3 depend upon the boundary conditions that one selects in the evaluation. More than one set of boundary conditions have been selected by various investigators and thus several solutions to the equation exist in the literature that may provide some confusion to the uninitiated. In the above case (equation 6.3), which is appropriate for the diffusion between two solids, the boundary conditions were selected such that as time passes the diffusing species are depleted on one side of the boundary and increased on the other. This will yield a constant midpoint concentration at the boundary of $C_o/2$. In the case of corrosion of a solid by a liquid, one assumes that the concentration of diffusing species from the liquid into the solid remains constant at the boundary (C_s) at a value equal to that in the bulk. The solution to equation 6.2 is then:

$$C(x,t) = C_s \left\{ 1 - erf\left(\frac{x}{2\sqrt{(Dt)}} \right) \right\} \quad (6.4)$$

where C_s is the concentration at the surface. One should note that the sign within the brackets changes when the boundary conditions are changed.

TABLE 6.2

Diffusion Coefficients for Some Typical Ceramics

Diffusing Ion	System	D_o (cm²/sec)	Q (Kcal/mol)	Comment	Ref.
O=	Al_2O_3	1.9×10^3	152	Single Crystal, >1600°C	Oishi & Kingery 1960
O=	Al_2O_3	2.0×10^{-1}	110	Polycryst, >1450°C	Oishi & Kingery 1960
O=	Al_2O_3	6.3×10^{-8}	57.6	Polycryst, <1600°C	Oishi & Kingery 1960
O=	MgO	2.5×10^{-6}	62.4	1300–1750°C	Oishi & Kingery 1960
O=	SiO_2	1.5×10^{-2}	71.2	Vitreous, 925–1225°C	Sucov 1963
O=	$ZrO_2(Ca)$	1.0×10^{-2}	28.1	15 mol% CaO, 700–1100°C	Kingery et al. 1959
Al^{3+}	Al_2O_3	2.8×10	114	Polycryst, 1670–1905°C	Paladino & Kingery 1962
Ca^{2+}	$ZrO_2(Ca)$	3.65	109	16 mol% CaO	Rhodes & Carter 1962
Mg^{2+}	MgO	2.3×10^{-1}	78.7	1400–1600°C	Linder & Parfitt 1957
Mg^{2+}	$MgAl_2O_4$	2.0×10^2	78		Linder & Akerstrom 1956
Zr^{4+}	$ZrO_2(Ca)$	1.97	109	16 mol% CaO	Rhodes & Carter 1962
Pb?	$PbSiO_3$	1.0×10^{-4}	24.8	Glass, 300–600°C	Linder et al. 1960

$\sqrt{(Dt)}$ is a measure of the order of magnitude of the distance that an average atom will travel and thus approximates the distance over which the concentration will change during diffusion. The use of error functions (erf) in evaluating diffusion is relatively easy by use of published tables, Holman 1963, for various values of erf(z).

Most of the solutions to Fick's equations assume that D is constant; however, in most real cases, the diffusion coefficient can vary with time, temperature (see equation 3.1), composition, or position along the sample, or any combination of these. If these are included in the equation, the mathematics become very difficult if not impossible; thus the equations used to describe diffusion generally assume constant D. See Table 6.2 for some typical values of diffusion coefficients.

Several mechanisms for diffusion have been hypothesized and investigated. One of the more important in ceramic materials is diffusion by vacancy movement in nonstoichiometric materials. Another mechanism involves diffusion by movement from one interstitial site to another. The ease with which this mechanism can occur, however, is not as great as that by vacancy movement. Other mechanisms that provide high-diffusivity paths include diffusion aided by dislocations, free surfaces, or grain boundaries.

Permeability constants as a function of temperature give an indication of the ease of diffusion of a species through a material. Silica has the lowest permeability to oxygen. This has been attributed to the difference in mechanism of transport among silica and most other materials. Transport in silica is by molecular species, whereas in other materials it is by ionic species, Courtright 1993. *For this reason silica-forming reactions are the most desirable for protection against oxygen diffusion*, Tripp et al. 1973.

Since many applications of ceramics involve thermal gradients, some mention of thermal diffusion should be made. Based upon studies in liquids, this has been called the *Sorét effect*. To evaluate the effect using Fick's first law, an additional term must be added to equation 6.1 that involves the temperature gradient. The flux is then given by:

$$J_i = -D \frac{\partial c_i}{\partial x} - \beta_i \frac{dT}{dx} \qquad (6.5)$$

where β_i is a constant independent of the thermal gradient for component i and may be positive or negative depending upon whether diffusion is down or up the thermal gradient, respectively. This constant is proportional to D and is given by:

$$\beta_i = \frac{D_i Q_i^* c_i}{RT^2} \qquad (6.6)$$

where Q_i^* is an empirical parameter that describes the sign and magnitude of the thermal diffusion effect. It has also been called the *heat of transport*. One interesting phenomenon that comes from an analysis of thermal diffusion is that *a diffusion flux will set up a thermal gradient in an isothermal system*.

When an elastic stress gradient is present along with a concentration gradient, a potential term must be included in the equation for total flux, just as was necessary for the thermal gradient. Thus the total flux of atoms in a particular direction is increased (or decreased) over that due only to concentration differences. This effect is called *stress-assisted diffusion*.

Diffusion is probably the most important rate-controlling step when one is evaluating the kinetics of a reaction by thermal analysis. Diffusion in the gas phase is about 10^4 times greater than that in the liquid phase. For a more complete description of diffusion the reader is referred to any one of the texts on diffusion, Shewmon 1983 and Crank 1956.

Additional Recommended Reading

Shewmon, P.G. 1983. *Diffusion in Solids*, J. Williams, Jenks, OK.
Zhang, Y. and Cherniak, D.J. (eds.). 2010. Diffusion in Minerals and Melts, *Reviews in Mineralogy and Geochemistry*, Vol. 72, 1038 pp., Min. Soc. Am., Chantilly, VA.

Exercises, Questions, and Problems

1. Discuss the effects of crystal symmetry upon the diffusion coefficient.
2. Discuss the importance of boundary conditions upon the solution of equation 6.3.
3. What parameters affect the diffusion coefficient?
4. Discuss the various mechanisms of diffusion in ceramic materials.
5. What type of reaction is important for minimization of oxygen diffusion?
6. What is the Sorét effect?

References

Courtright, E.L. 1993. Engineering Limitations of Ceramic Composites for High Performance and High Temperature Applications, pp. 21–32 in *Proc. 1993 Conf. on Processing, Fabrication and Applications of Advanced Composites*, Long Beach, CA, 9–11 Aug, Upadhya, K. (ed.), ASM, OH.
Crank, J. 1956. *The Mathematics of Diffusion*, Oxford University Press, Fair Lawn, NJ.
Holman, J.P. 1963. *Heat Transfer*, McGraw-Hill, New York, NY, p. 271.
Kingery, W.D., Pappis, J., Doty, M.E. and Hill, D.C. 1959. Oxygen Ion Mobility in Cubic $Zr_{0.85}Ca_{0.15}O_{1.85}$, *J. Am. Ceram. Soc.*, 42(8), 393–8.
Paladino, A.E. and W.D. Kingery, 1962. Aluminum Ion Diffusion in Aluminum Oxide, *J. Chem. Phys.*, 37(5), 957–62.
Lindner, R. and Akerstrom, A. 1956. Self-Diffusion and Reaction in Oxide and Spinel Systems, *Z. Phys. Chem.*, 6, 162.
Lindner, R. and Parfitt, G.D. 1957. Diffusion of Radioactive Magnesium in Magnesium Oxide Crystals, *J. Chem. Phys.*, 26, 182.
Lindner, R., Hassenteufel, W. and Kotera, Y. 1960. Diffusion of Radioactive Lead in Lead Metasilicate Glass, *Z. Phys. Chem.*, 23, 408.
Oishi, Y. and Kingery, W.D. 1960. Self-Diffusion in Single Crystal and Polycrystalline Aluminum Oxide, *J. Chem. Phys.*, 33, 480.

Rhodes, W.H. and Carter, R.E. 1962. Ionic Self-Diffusion in Calcia Stabilized Zirconia, 64th Annual Mtg Abstracts, *Am. Ceram. Soc. Bull., 41*(4), 283.

Shewmon, P.G. 1983. *Diffusion in Solids,* J. Williams Book Co., Jenks, OK.

Sucov, E.W. 1963. Diffusion of Oxygen in Vitreous Silica, *J. Am. Ceram. Soc., 46*(1), 14–20.

Tripp, W.C., Davis H.H., and Graham, H.C. 1973. Effects of SiC Additions on the Oxidation of ZrB_2, *Ceram. Bull., 52*(8), 612–16.

7

Kinetics

Logic will take you from A to B. Imagination will take you everywhere.

Albert Einstein

It is normally expected that materials will corrode, and thus it is important to know the kinetics of the reaction so that predictions of service life can be made. *Thus the most important parameter of corrosion from the engineering viewpoint is the reaction rate.* Systems can often exist for extended periods of time in a state that is not the equilibrium state or the state of lowest free energy. These states are called *metastable states* and may occur for many reasons. One case is where a surface reaction forms a diffusion barrier that blocks or drastically diminishes further reaction. In another more important case, for the reaction to proceed to the lowest free energy state it must first pass through an intermediate state where the energy is higher than either the initial or final states. The energy required to overcome this barrier is called the *activation energy* (Q) and the net energy released is the *heat of reaction* (H). This is depicted in Figure 7.1 where the movement of an atom from an initial metastable state (a) to the final stable state (c) requires passage through the higher energy unstable state (b). The reaction is exothermic in going from (a) to (c) and endothermic in the reverse direction. The activation energy for the reverse direction obviously must be greater than for the forward direction. The speed of the reaction is dependent upon the total number of atoms in the metastable state, the vibration frequency of the atoms, and the probability that an atom during vibration will have the necessary energy to overcome the barrier. If sufficient energy is not acquired to overcome the activation energy barrier, the system will remain indefinitely in the metastable state. The number of atoms that pass over the barrier is then the rate of the reaction and is given by:

$$\text{Reaction rate} = Ae^{-Q/RT} \tag{7.1}$$

where A is a constant containing the frequency term and Q is the activation energy. Expressing this equation in logarithmic form one obtains:

$$\ln(\text{rate}) = \ln A - (Q/R)/T \tag{7.2}$$

A plot of ln (rate) versus reciprocal temperature yields Q/R as the slope and the intercept at $1/T = 0$ yields A.

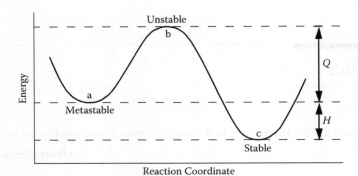

FIGURE 7.1
Energy barrier diagram. (Q = activation energy, and H = heat of reaction.)

The effect of temperature upon the reaction rate can be seen by the following example. Suppose that Q = 45 kcal/mol, a number not unreasonable for many ceramic reactions, and that R = 2.0 cal/mol K. Calculation of the exponential term yields a rate that is approximately 10^{23} times as fast at 1000 K as it is at 300 K. Thus if a reaction takes 1 second at 1000 K it takes on the order of 10^{12} years at room temperature. This is the basis of quenching and allows one to examine reactions at room temperature that have occurred at high temperature.

Quite often a plot of the logarithm of the corrosion rate versus the inverse temperature yields a straight line, indicating that corrosion is an activated process. Attempting to correlate various ceramic material properties to these activation energies, however, can be very misleading. *Generally, the range of activation energies experimentally observed for different materials is very large and any interpretation is difficult, since diffusion coefficients depend upon the composition and structure of the material through which diffusion occurs.* Since the interface composition generally changes with temperature, the driving force for diffusion also changes with temperature, neither of which has any relationship to an activated process.

When a substance increases the reaction rate but is not itself consumed in the reaction, it is called a *catalyst*. Catalysts operate by many different mechanisms, but all essentially go through a cycle where they are used and then regenerated. When a catalyst occurs in solution as a molecule or ion, it operates through a process called *homogeneous catalysis;* when the reaction occurs on a surface the process is called *heterogeneous catalysis*. The reaction path provided by the catalyst is one of lower activation energy and/or higher frequency factors. If the products of the reaction act as a catalyst, the reaction is said to be *autocatalytic*.

Reaction rates for condensed-phase processes normally involve the transport of products away from the boundary. Thus the rate of the overall process is determined by the rate of each individual step and by the reaction rate constant and concentration of reactants for that step. The reaction with the lowest rate determines the overall rate of the corrosion process. Some of the more important factors that may influence the rate of reaction are diffusion rates, viscosity, particle size, heat transfer, and the degree of contact or mixing.

The stoichiometric chemical equation of the overall process does not reveal the mechanism of the reaction. To determine the overall reaction rate one must determine all the intermediate steps of the process. Prediction or identification of the reaction mechanism is never certain, since other mechanisms could account for the experimental data.

The rate of the reaction expressed as the rate of change of concentration, dc/dt, depends upon the concentration of the reactants. Rates may also depend upon the concentrations of other substances not involved in the stoichiometric equation. The rate equation as a function of concentration of each substance that affects the rate is called the *rate law* for the reaction. When the rate equation contains powers of the concentration, the *order* of the reaction equals the exponent. Rate laws can be determined only experimentally and cannot be predicted from the chemical equation.

The first-order rate equation is given by:

$$dc/dt = -kc^n \tag{7.3}$$

where:

k = rate constant,
c = concentration of reacting species,
n = reaction order = 1 for first order, and
t = time.

If log c is plotted against time, a straight line is obtained for a first-order reaction. If the reaction is one of the first order, it will take twice as long for three-fourths to react as it will for one-half to react. A discussion of the order of reactions and the various equations can be found in any book on kinetics, Barret 1975.

Integration of equation 7.3 between concentration limits of c_1 and c_2 at time limits of t_1 and t_2 yields:

$$k = \frac{1}{(t_2 - t_1)} \ln(c_1 / c_2) \tag{7.4}$$

Thus it should be apparent from this equation that to determine k it is necessary to evaluate only the ratio of the concentrations at the two times. This can make analysis easier, since one can substitute any measurable property that is proportional to the concentration. Changes in properties such as volume, partial pressure of gases, light absorption, and electrical conductivity are often used.

Equation 7.3 is often written in a form relating the fraction of product formed to the reaction time:

$$d\alpha/dt = k(1 - \alpha)^n \qquad (7.5)$$

where α is now the total amount of product formed. According to Sharp et al. 1966 when $n = 1/2$ or $2/3$ the equations represent phase-boundary controlled reactions for circular disks (or cylinders) or spheres, respectively.

Diffusion controlled reactions have been represented by various functions of the amount of product formed given by the general equation:

$$F(\alpha) = kt \qquad (7.6)$$

The frequently cited article by Sharp et al. gives numerical data that allows one to evaluate $F(\alpha)$ from experimental data for the commonly used equations, which are given in Table 7.1. Since these equations have been derived for specific geometric mono-sized shapes, which are seldom present in actual cases, Sharp et al. concluded that considerable experimental accuracy was required to distinguish among the various possibilities.

TABLE 7.1

Kinetic Equations

Diffusion-Controlled

$\alpha^2 = (k/x^2)t$

    One-dimensional; $2x$ = reaction layer thickness.

$(1 - \alpha) \ln (1 - \alpha) + \alpha = (k/r^2)t$

    Two-dimensional; r = radius of cylinder.

$[1 - (1 - \alpha)^{1/3}]^2 = (k/r^2)t$

    Three-dimensional; r = radius of sphere

    Commonly called the Jander equation.

$(1 - 2\alpha/3) - (1 - \alpha)^{2/3} = (k/r^2)t$

    Three-dimensional; r = radius of sphere.

Phase-Boundary Controlled

$[1 - (1 - \alpha)^{1/2}] = (u/r)t$

    Circular disk or cylinder of radius = r and with u = velocity of interface; assuming an
    instantaneous nucleation.

$[-\ln(1 - \alpha)]^{1/2} = kt$

    Avrami-Erofe'ev random nucleation equation for the disk/cylinder case.

$[1 - (1 - \alpha)^{1/3}] = (u/r)t$

    Sphere of radius = r and u = velocity of interface; assuming an instantaneous nucleation.

$[-\ln(1 - \alpha)]^{1/3} = kt$

    Avrami-Erofe'ev random nucleation equation for the sphere case.

Source: Sharp, J.H., Brindley, G.W. and Narahari Achar, B.N. 1966. Numerical Data for Some Commonly Used Solid State Reaction Equations, *J. Am. Ceram. Soc.*, 49(7), 379–82.

It should be realized that the solutions to the kinetic equations discussed by Sharp et al. are only approximate. Frade and Cable 1992 pointed out that the deviation observed between experimental data and theoretical models are often due only to the approximations that were made in the original theoretical analysis. Frade and Cable 1992 reexamined the basic theoretical model for the kinetics of solid-state reactions by considering spherical particles, moving reaction boundaries, and changes in volume. *The discrepancies between the experimental data and the theoretical models are often due to nonspherical particles, a range in sizes, poor contact between reactants, formation of multiple products, and the dependency of the diffusion coefficient upon composition.* The commonly used Jander equation was originally derived for reactions between flat slabs and is therefore inappropriate for use with spherical particles, although the Jander model fits reasonably well for low values of conversion.

Many reactions are not simply zero-, first-, second-, or third-order reactions, since they proceed by a multistep mechanism. Multistep reactions may, however, behave as zero-order reactions, first-order reactions, etc. Some of the complexities that may be encountered are parallel steps, consecutive steps, and reversible steps or may even be other types of steps. Many times a complex reaction may appear to be zero, first, second, or third order only because the rate-limiting step is of that order and all other steps are very fast.

Nonisothermal thermogravimetry (TG) has been used by many investigators to study the kinetics of decomposition reactions. The amount of data that can be collected by dynamic methods is considerably more than by isothermal methods, which has led investigators to rely more heavily upon the dynamic method. The convenience today is so great with modern computerized thermal analyzers that one need only scan a sample at several different heating rates and then push the appropriate buttons to obtain the kinetic data! One must be extremely careful in collecting kinetic data in this fashion. A thorough understanding of the various effects that the sample characteristics, machine operation, etc., have upon the kinetics is important along with all the various assumptions that may have been made by the software programmer to use that data to calculate the kinetic parameters. Too often these *pseudokinetic* data are published in the technical literature and can be misleading to the unwary reader.

Although many advantages exist for the use of nonisothermal studies over isothermal studies, the main disadvantage is that the reaction mechanism usually cannot be determined, which leads to uncertainties in the activation energy, order of reaction, and frequency factor. There must be at least two dozen different methods and variations reported in the literature to calculate kinetic parameters from dynamic thermogravimetric studies. The most widely used is that of Freeman and Carroll 1958. Sestak 1966 performed a comparison of five methods and found a variation of approximately 10% in the calculated values of the activation energy. *Arnold et al. 1981 concluded that dynamic thermogravimetric studies provide insufficient data for calculation of reaction kinetics, that the data are influenced by the experimental procedures, and that the results are uncertain.*

Differential thermal analysis (DTA) and differential scanning calorimetry (DSC) have also been used to study reaction kinetics. The equation to evaluate the rate depends upon the mechanism and thus the mechanism must be known before these methods can be used. The rate of heat generation must also be proportional to the rate of reaction for these methods to be valid. An excellent review of the use of DTA to determine reaction kinetics has been given by Sestak 1984.

Probably the most important parameter that is uncertain in nonisothermal studies is the temperature of the sample. *The enthalpy of the reaction is often sufficient to raise or lower the sample temperature by as much as 1000°C.* This fact is overlooked or unavailable if one uses nonisothermal thermogravimetry, which is most often the case. DTA or DSC may be more appropriate than TG, since these techniques either determine the sample temperature or maintain the sample at a constant temperature relative to a reference material. Generally, the temperature range studied in thermal analysis to evaluate the kinetics of a reaction is on the order of 100–150°C. This range covers only about 10–20% of the total reaction and leads to excessive scatter in the calculated values of the activation energy and the pre-exponential term of the Arrhenius equation.

For corrosion rates to be useful to practicing engineers it is best that they be expressed in a useful manner. In most cases, the engineer is involved with the amount of material corroded away during a specified time period, or the depth of penetration per unit time. In the literature, corrosion rates are often given as the mass of material reacted per unit area for a unit time. These can easily be converted to the depth of penetration per unit time by dividing by the density of the material as shown below:

$$P = \frac{M}{\rho A t} \tag{7.7}$$

where:
   P = depth of penetration,
   M = mass loss,
   ρ = density,
   A = area of exposure, and
   t = time of exposure.

In using the above equation to calculate corrosion rates from laboratory experiments, one must be very conscious of the total surface area exposed to corrosion. This will include a determination of the open porosity of the specimen. Many investigators have attempted to compare corrosion resistance of various materials incorrectly by omitting the porosity of their samples. Omitting the porosity, although not giving a true representation of the material's corrosion, will give a reasonable idea of the corrosion of the as-manufactured material.

Additional Recommended Reading

Vetter, K.J. 1967. *Electrochemical Kinetics*, Academic Press, New York.

Exercises, Questions, and Problems

1. The speed of a reaction is dependent upon what parameters?
2. Explain what is meant by a metastable state.
3. Discuss the effect of temperature upon the reaction rate.
4. Discuss the discrepancies between the experimental data and the theoretical models for the kinetics of solid-state reactions.
5. Discuss the advantages and disadvantages of isothermal versus dynamic analysis for determining reaction kinetics.

References

Arnold, M., Veress, G.E., Paulik, J. and Paulik, F. 1981. The Applicability of the Arrhenius Model in Thermal Analysis, *Anal. Chim. Acta.*, *124*(2), 341–50.

Barret, P. (ed.). 1975. *Reaction Kinetics in Heterogeneous Chemical Systems*, Elsevier, Amsterdam.

Frade, J.R. and Cable, M. 1992. Reexamination of the Basic Theoretical Model for the Kinetics of Solid-State Reactions, *J. Am. Ceram. Soc.*, *75*(7), 1949–57.

Freeman, E.S. and Carroll, B. 1958. The Application of Thermoanalytical Techniques to Reaction Kinetics. The Thermogravimetric Evaluation of the Kinetics of the Decomposition of Calcium Oxalate Monohydrate, *J. Phys. Chem.*, *62*(4), 394–7.

Pourbaix, M. 1974. *Atlas of Electrochemical Equilibria in Aqueous Solution*, Eng trans by J.A. Franklin, NACE, Houston, TX.

Sestak, J. 1966. Errors of Kinetic Data Obtained from Thermogravimetric Curves at Increasing Temperature, *Talanta*, *13*(4), 567–79.

Sestak, J. 1984. *Thermophysical Properties of Solids*, Part D of *Thermal Analysis*, W.W. Wendlandt (ed.), Vol. XII of *Comprehensive Analytical Chemistry*, G. Svehla (ed.), Elsevier, Amsterdam.

Sharp, J.H., Brindley, G.W. and Narahari Achar, B.N. 1966. Numerical Data for Some
     Commonly Used Solid State Reaction Equations, *J. Am. Ceram. Soc.*, 49(7),
     379–82.

8

Surface Effects

The only thing that we know is that we know nothing and that is the highest flight of human wisdom.

Leo Tolstoy

Porosity and Surface Area

The corrosion of ceramics (i.e., weight gain/loss) is proportional to the porosity; the more porous the sample, the more corrosion that is exhibited. This is in reality related to the surface area exposed to corrosion. The fact that one material may yield a better corrosion resistance than another does not necessarily make it the better material, if the two materials have different porosities. This is very important, for example, when comparing different sintering aids for silicon nitride and their effects upon oxidation. The more oxidation-resistant material may not be due to the chemical species of the sintering aid used, but in actuality may be due to the fact that one particular sintering aid yields a denser sintered ceramic. One must remember that *it is not the total porosity that is important, but the surface area of the total porosity*, thus making the pore size distribution an important parameter to determine.

The porosity of a ceramic can affect the overall corrosion only if the attacking medium can penetrate the porosity. Washburn 1921 derived the following equation to determine the pore size distribution by mercury intrusion:

$$P = \frac{-2\gamma_{lv}\cos\varnothing}{r} \tag{8.1}$$

where P is the pressure required to force liquid into a cylindrical pore of radius r, γ is the surface tension of the liquid, and $\varnothing$ is the contact angle between the liquid and the ceramic. Although some have applied this equation to liquids other than mercury, the results are generally inaccurate due to the wetting of the solid by the liquid. *Several assumptions were made by Washburn: that the applied force required to force a nonwetting liquid into the pore is equal to the opposing capillary force, that the void space is one of nonintersecting*

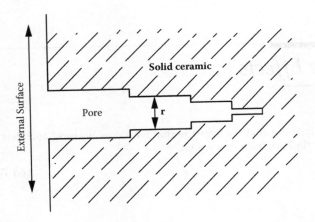

FIGURE 8.1
Nonintersecting cylindrical pores in a graded array becoming larger as the surface is approached, as assumed by Washburn.

cylindrical pores, and that the pores exist in a graded array with the largest ones toward the outside of the ceramic as shown in Figure 8.1.

One of the more controversial aspects of this technique is the discrepancy between intrusion and extrusion data, which has been explained by contact angle hysteresis by Smithwick and Fuller 1984. Conner et al. 1988 have shown the sensitivity of this technique to pore morphology. Moscou and Lub 1981 reported that the hysteresis stems from a combination of both contact angle differences for intrusion and extrusion and pore morphology.

Lapidus et al. 1985 and Conner and Lane 1984 have compared computer simulations of mercury flow through a pore space assumed to be a pore-throat network to actual porosimetry data and found that the throats determine the intrusion behavior and the pores determine extrusion behavior. The reader is referred to any of several review papers for more detailed information, Moscou and Lub 1981, Van Brakel et al. 1981, and Rootare and Nyce 1971.

One effect that is directly related to the pore size distribution is a phenomenon called *thermal transpiration*. This is the transport of gases through a ceramic caused by a thermal gradient. The relationship between pressure and temperature is given by

$$P_1 / P_2 = \sqrt{(T_1 / T_2)} \tag{8.2}$$

where the subscript 1 denotes the hot face. If the gas pressure is essentially the same on both sides, gases will migrate up the thermal gradient in an attempt to make the pressure on the hot face higher. The rate of migration is inversely proportional to the square root of the molecular weight of the gas. Pore size will affect the migration, since very fine pores create too great a resistance to flow and very large pores allow ordinary flow due to pressure differences. Thus at some intermediate pore size transpiration will occur. In ceramics with a large pore size distribution, ordinary flow tends to equalize the pressures, minimizing flow by transpiration. There are no known reports in the literature indicating that thermal transpiration influences corrosion of ceramics; however, it may suggest a means to minimize the effects from corrosive ordinary flow. *If sufficient flow of the transpiring gas is present, dilution of the corrosive gas at the hot face may lower the corrosion rate to an acceptable level.*

The manufacturers of flat glass by one of the float processes are well aware of the problems that thermal transpiration may cause. Although not a corrosion process, defective glass has been produced by gases transpiring up through the tin bath bottom blocks, rising through the tin, and then causing an indent in the bottom surface of the glass. In some cases the gas pressure has been sufficient to puncture completely through the glass ribbon. To eliminate this problem, bath bottom blocks are manufactured to a specific pore size distribution.

A purely surface area effect, which is very important in the corrosion of asbestos or chrysotile fibrils, is that related to ledge effects. As one can see from Figure 8.2b, ledges can greatly increase the exposed surface area. This is extremely important in the dissolution of spiral fibrils and their related health effects. Similar structural effects can be present due to dislocations and other defects (see Figure 8.2a). Chrysotile is a two-layer sheet silicate with a dimensional misfit between the octahedral and tetrahedral layers. This causes the sheet to curl, forming spiral fibrils. This property causes some confusion, since chrysotile is a sheet silicate, not a chain silicate, although both have properties related to fibrous materials. Surface areas determined from sample geometry are generally many times smaller than that determined from BET measurements. This difference can be attributed to the presence of microscopic surface features. Thus one must be careful how surface areas are determined and how these data are related to the subsequent dissolution data.

Lüttge 2005 described three different types of surface areas. The first, *geometric surface area,* is calculated from the dimensions of the crystals and ignores surface roughness. The second, *total surface area,* often used synonymously with specific surface area, is measured by BET. This area includes all those features that contribute to surface roughness such as etch pits, pores, steps, and kinks. Since the material surface is generally not homogeneous, some parts of a surface are significantly more reactive than others. This has led to the third type of surface area, *reactive surface area.* This type of area is very difficult or impossible to measure. For the determination of dissolution rates and rate constants, the quantification of surface area with respect to time is of extreme importance.

The surface area that can be exposed to a corroding medium is greatly increased if nanoparticles or nanomaterials are used. This is very important

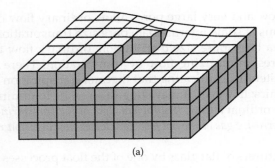

(a)

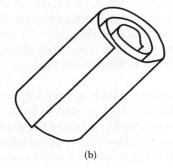

(b)

FIGURE 8.2
Surface area increase due to dislocations and defects (a) and ledge effects in fibrils (b).

in studies such as the effects of mineral dusts upon human health. In addition to the surface area of such small particles is the area exposed by nanopores. Nanopores are those generally with diameters less than 2 nm. Slightly larger mesopores have diameters between 2 and 50 nm. A large portion of the total porosity of consolidated materials is due to these small pores. The standard techniques of determining porosity described above do not account for these smaller pores. Whether or not these small pores are involved in corrosion is related to the ability of the corroding medium to enter the pore. This fact is obvious if one compares the results among Archimedes water adsorption, mercury porosimetry, and BET for the same material. This is why the corrosion by gases is generally more severe than that by liquids or solids.

Surface Charge

Gibson and LaFemina 1996 offered an excellent discussion of the various aspects of mineral surfaces and how these affect dissolution. Surfaces that have the same atomic structure (i.e., symmetry) as the bulk are termed *relaxed*. Those that are different are termed *reconstructed*. There is an excess electronic charge density associated with the broken or dangling bonds at the surface that is not present with bonds within the bulk. Different crystal faces exhibit different numbers of dangling bonds for the ions. If electrons can transfer between dangling bonds of anions and cations, then a situation arises where one ion has completely filled bonds and the other has completely empty bonds. When this occurs the surface is charge-neutral. This will occur for surfaces that have a stoichiometric ratio of anions to cations. If the contribution of electrons from the different ions causes an excess charge density, then the surface becomes charged. The atoms on the surface will move to minimize the excess charge density associated with the dangling bonds and thus cause a localized strain. Dissolution of mineral surfaces depends upon the surface structure or arrangement of the atoms on the surface, Lasaga 1990. Gibson and LaFemina 1996 reported that the exact chemical species forming the surface is of secondary importance and that it is the atomic connectivity that dominates surface relaxations. *This is important since one need not have data on a specific material, chemically, but only of identical structure to estimate its dissolution characteristics.*

Surface Energy

The surface energy of a material is the ratio of the potential energy difference obtained when moving an atom from the bulk to the surface to the area of the surface. A term that is closely related is the surface tension, which is the force required to move the atom to the surface divided by its diameter. Since liquids cannot maintain a shear stress, the surface energy and surface tension of liquids are equivalent. This is not the case for solids where the surface energy is generally greater than the surface tension. In general the symbol used to represent surface tension is σ, whereas the one used to represent surface energy is γ. The surface energy is greatest for the least densely packed crystallographic planes, which are also the least thermodynamically stable. For a face-centered-cubic material the number of nearest neighbors on the surface is 9 for the (111) plane, 8 for the (100) plane, and 6 for the (110) plane. In comparison, the number of nearest neighbors for the bulk is 12. Thus the surface energy decreases in the order (110) > (100) > (111). He et al. 2009 reported that corrosion rates are increased at nonstoichiometry sites, especially at grain boundaries.

One area where the surface energies play a very important role is in the movement of liquids into capillaries. For ceramics capillaries can be considered to be very small connected pores. The Laplace equation represents the equilibrium pressure difference at the interface (i.e., liquid-vapor) due to surface energy:

$$\Delta P = 2\gamma \, (\cos \phi)/R \qquad (8.3)$$

where:
ΔP = pressure drop across interface,
γ = surface energy of liquid,
ϕ = contact angle of meniscus at wall of capillary, and
R = radius of capillary.

In actual materials the porosity is not a cylindrical cavity and therefore one must use an effective radius that represents the weighted average of the contributing porosity.

The relationship of the surface energies among the solid-vapor interface, solid-liquid interface, and liquid-vapor interface is given by:

$$\cos \phi = \frac{\gamma_{sv} - \gamma_{sl}}{\gamma_{lv}} \qquad (8.4)$$

When the contact angle, ϕ, is less than 90°, capillary attraction will allow the liquid to fill the pores displacing the gas within without any applied force. When the contact angle is greater than 90° an applied force, P, is required to force the liquid into the pores. The pressure exerted upon a ceramic in service will depend upon the height and density of the liquid. When this pressure is greater than P, the liquid will enter the pores that have a radius greater than r.

Carrying this one step further, the penetration of liquids between like grains of a ceramic can be predicted from the interfacial surface energies of the liquid-solid and solid-solid interfaces according to Smith 1948, since if:

$$\gamma_{ss} \geq 2\gamma_{sl} \qquad (8.5)$$

complete wetting will occur. If:

$$\gamma_{ss} \leq 2\gamma_{sl} \qquad (8.6)$$

solid-solid contact is present and the liquid will occur in discrete pockets. A balance of forces exists when:

$$\gamma_{ss} = 2\gamma_{sl} \cos \phi/2 \qquad (8.7)$$

where ϕ is the dihedral angle between several grains and the liquid. Thus equation 8.5 is valid when $\phi > 60°$ and equation 8.6 is valid when $\phi < 60°$. For this reason 60° has been called the *critical dihedral angle* that separates the conditions

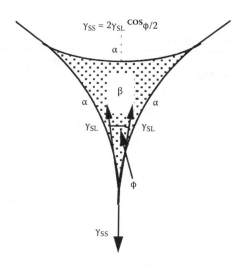

FIGURE 8.3
Balance of surface energy forces between a major and a secondary grain boundary phase. (α = solid grains; β = liquid grain boundary phase; γ_{ss} = surface energy between two solid grains; γ_{sl} = surface energy between solid and liquid; and ϕ = dihedral angle.)

of complete wetting and nonpenetration of the second phase between grains of the major phase. Although some data on dihedral angles exist as discussed later, very little actual data has been reported. The general factors that cause variation in the dihedral angle, however, are often mentioned.

The balance of forces (see Figure 8.3) holds well for grains that tend to be rounded. If marked crystallographic faces exist, equation 8.7 is no longer valid. Surface forces are then no longer tangential and isotropic, which was assumed in the derivation of equation 8.7. However, if:

$$\gamma_{ss} > \sqrt{3}\gamma_{sl} \tag{8.8}$$

the liquid occurs only at three grain intersections or triple points. Thus one would desire that γ_{ss} be $< 2\,\gamma_{sl}$ and at least $< \sqrt{3}\,\gamma_{sl}$ to minimize liquid penetration into the ceramic. This balance of forces is affected, however, by many things, one important factor being the temperature. Composition and grain size will also affect the overall balance of forces as discussed below.

Due to the random orientation of the three-grain junctions in polished sections, the determination of ϕ varies between 0 and 180°, even when it is constant throughout the structure. In this case the median of a large number of determinations is taken as the dihedral angle value.[*]

---

[*] The evaluation of a 3-D parameter by use of a 2-D array will always lead to a variation in values. Obtaining a representative value requires the determination of a large number of individual dihedral angles. The final average value is never the same as the actual value, but something less.

TABLE 8.1

Effects of Composition upon the Dihedral Angle

Substitution for MgO[a]	Amount (%)	Temperature (°C)	Dihedral Angle
Cr_2O_3	0 to 10	1550	25 to 45
Fe_2O_3	0 to 5	1550	25 to 20
Al_2O_3	0 to 5	1550	25 to 20
TiO_2	0 to 2	1550	25 to 15
Cr_2O_3	5	1550 to 1725	40 to 30

[a] Substitution for MgO in an 85% MgO–15% CMS composition.

Source: White, J. 1970. Magnesia-Based Refractories, Ch. 2 in *High Temperature Oxides, Part 1: Magnesia, Lime and Chrome Refractories,* Alper, A.M. (ed.), Vol. 5-1 of *Refractory Materials: A Series of Monographs,* Margrave, J.L. (series ed.), Academic Press, New York, pp. 77–141.

White 1970 in his studies of refractory systems has shown that as the temperature increases the dihedral angle decreases. He has also shown the effects of composition upon the dihedral angle in 85% MgO–15% Ca-Mg-silicate liquids at 1550°C in air. These effects are shown in Table 8.1. White reported that as the concentration of solid in the saturated liquid increased the dihedral angle decreased, which is the same as the effect of temperature. Since the curvature of the grains must decrease as the dihedral angle increases, larger grains will produce a smaller dihedral angle. In addition White showed that the dihedral angle between like grains was smaller than that between unlike grains, indicating that the penetration of liquid between unlike grains should be less than between like grains.

The nature of the bonding type of the solid being attacked compared to that of the attacking medium often can give an indication as to the extent of wetting that may take place. For example, transition metal borides, carbides, and nitrides, which contain some metallic bond character, are wet much better by molten metals than are oxides, which have ionic bond character, Inomata 1975. Various impurities, especially oxygen, dissolved in the molten metal can have a significant effect upon the interfacial surface energies. For example, Messier 1980 reported that silicon wet silicon nitride at 1500°C in vacuum but did not spread due to oxygen contamination. In most cases it is the nature of the grain boundary or secondary phases that is the controlling factor.

Puyane and Trojer 1980 examined the possibility of altering the wettability of alumina by using additives to their glass composition. They found that V_2O_5 and CeO_2 additions changed the surface tension of the glass in opposite directions, with V_2O_5 decreasing it and CeO_2 increasing it. They concluded that the *glass characteristics were more important than the solid parameters in corrosion.*

Banfield and Zhang 2001 have shown that for small enough particle sizes if the surface energies of two polymorphs are sufficiently different, their stability can be reversed. This has been demonstrated for TiO_2 when the particle size is below 14 nm where anatase becomes the stable phase with respect to rutile, Zhang and Banfield 1998. The excess pressure caused by the surface stress[*] is no longer negligible when particles reach the nanoscale. This surface stress is the reversible work per unit area needed to stretch the surface elastically. It must be incorporated into the calculations for the free energy of transformation between two polymorphs.

Additional Recommended Reading

Shaw, D.J. 1980. Charged Interfaces, Chp. 7 in *Introduction to Colloid and Surface Chemistry*, 3rd edition, Butterworths, London, pp. 148–82.

Exercises, Questions, and Problems

1. Explain why porosity is important to the interpretation of corrosion data.
2. Explain why the Washburn equation should not be used for liquids other than mercury.
3. What is the phenomenon called thermal transpiration?
4. How may thermal transpiration be used to control corrosion?
5. What parameters affect the dihedral angle in studies of liquid penetration between solid grains?
6. Look up the origins of the Laplace equation for fluid dynamics and compare it to the Washburn equation for fluid penetration into porous ceramics.

References

Banfield, J.F. and Zhang, H. 2001. Nanoparticles in the Environment, pp. 1–58 in *Nanoparticles and the Environment. Reviews in Mineralogy & Geochemistry, 44,* Banfield, J.H. and Navrotsky, A. (eds.), Min. Soc. Amer. Chantilly, VA.

---

[*] The surface stress and the surface free energy of liquids is equal, which is the reason that surface free energy is often called surface tension. These two are not generally equivalent for solids.

Conner, W.C. and Lane, A.M. 1984. Measurement of the Morphology of High Surface Area Solids: Effect of Network Structure on the Simulation of Porosimetry, *J. Catal.*, *89*, 217–25.

Conner Jr., W.C., Blanco, C., Coyne, K., Neil, J., Mendioroz, S. and Pajares, J. 1988. Measurement of Morphology of High Surface Area Solids: Inferring Pore Shape Characteristics, in *Characterization of Porous Solids*, Unger, K.K. et al. (eds.), Elsevier Science, Amsterdam, pp. 273–81.

Gibson, A.S. and LaFemina, J.P. 1996. Structure of Mineral Surfaces, pp. 1–62 in *Physics and Chemistry of Mineral Surfaces*, Brady, P.V. (ed.), CRC Press, New York.

He, H., Zhu, R.K., Qin, Z., Keech, P., Ding, Z. and Shoesmith, D.W. 2009. Determination of Local Corrosion Kinetics on Hyper-Stoichiometric UO_{2+x} by Scanning Electrochemical Microscopy, *J. Electrochem. Soc.*, *156*(3), C87–94.

Hochella Jr., M.F. and White, A.F. 1990. Mineral-Water Interface Geochemistry, *Reviews in Mineralogy*, vol. 23, Min. Soc. Am., Washington, DC.

Inomata, Y. 1975. Oxidation Resistant Si-Impregnated Surface Layer on Reaction Sintered Articles, *Yogyo Kyokaishi*, *83*(1), 1–3.

Lapidus, G.R., Lane, A.M., Ng, K.M. and Conner, W.C. 1985. Interpretation of Mercury Porosimetry Data Using a Pore-Throat Network Model, *Chem. Eng. Commun.*, *38*, 33–56.

Lasaga, A.C. 1990. Atomic Treatment of Mineral-Water Surface Reactions, pp. 17–85 in Mineral-Water Interface Geochemistry, *Reviews in Mineralogy*, Vol. 23, Hohella, M.F. and White, A.F. (eds.), Min. Soc. Am., Washington, DC.

Lüttge, A. 2005. Etch Pit Coalescence, Surface Area, and Overall Mineral Dissolution Rates, *Amer. Mineral.*, *90*, 1776–83.

Messier, D.R. 1980. Use of Ti to Enhance Wetting of Reaction-Bonded Si_3N_4 by Si, pp. 624–33 in *Ceramic Engineering and Science Proceedings*, Smothers, W.J. (ed.), Vol. 1, (7-8B).

Moscou, L. and Lub, S. 1981. Practical Use of Mercury Porosimetry in the Study of Porous Solids, *Powder Technol.*, *29*, 45–52.

Puyane, R. and Trojer, F. 1980. Refractory Wear and Wettability by Glass at High Temperatures, *Glass*, *57*(12), 5–8.

Rootare, H.M. and Nyce, A.C. 1971. The Use of Porosimetry in the Measurement of Pore Size Distribution in Porous Materials, *Int. J. Powder Metall.*, *7*(1), 3–11.

Smith, C.S. 1948. Grains, Phases, and Interpretation of Microstructure, *Trans. AIME*, *175*(1), 15–51.

Smithwick, R.W. and Fuller, E.L. 1984. A Generalized Analysis of Hysteresis in Mercury Porosimetry, *Powder Technol.*, *38*, 165–73.

Van Brakel, J., Modry, S. and Svata, M. 1981. Mercury Porosimetry: State of the Art, *Powder Technol.*, *29*, 1–12.

Washburn, E.W. 1921. Note on a Method of Determining the Distribution of Pore Sizes in a Porous Material, *Proc. Natl. Acad. Sci.*, *7*, 115–6.

White, J. 1970. Magnesia-Based Refractories, Ch. 2 in *High Temperature Oxides, Part 1: Magnesia, Lime and Chrome Refractories*, Alper, A.M. (ed.), Vol. 5-1 of *Refractory Materials: A Series of Monographs*, Margrave, J.L. (series ed.), Academic Press, New York, pp. 77–141.

Zhang, H. and Banfield, J.F. 1998. Thermodynamic Analysis of Phase Stability of Nanocrystalline Titania, *J. Materials Chem.*, *8*, 2073–6.

9

Thermodynamics

We cannot advance without new experiments in living, but no wise man tries every day what he has proved wrong the day before.

James Adams

Introduction

The driving force for corrosion is the reduction in free energy of the system. The reaction path is unimportant in thermodynamics; only the initial and final states are of concern. In practice, intermediate or metastable phases are often found when equilibrium does not exist and/or the reaction kinetics are very slow. In general, a reaction can occur if the free energy of the reaction is negative. Although the sign of the enthalpy (or heat) of reaction may be negative, it is not sufficient to determine if the reaction will proceed. The spontaneity of a reaction depends upon more than just the heat of reaction. There are many endothermic reactions that are spontaneous. To predict stability, therefore, one must consider the entropy. Spontaneous, irreversible processes are ones where the entropy of the universe increases. Reversible processes, on the other hand, are those where the entropy of the universe does not change. At low temperatures, exothermic reactions are likely to be spontaneous because any decrease in entropy of the mixture is more than balanced by a large increase in the entropy of the thermal surroundings. At high temperatures, dissociative reactions are likely to be spontaneous, despite generally being endothermic, because any decrease in the thermal entropy of the surroundings is more than balanced by an increase in the entropy of the reacting mixture.

In the selection of materials, an engineer wishes to select those materials that are thermodynamically stable in the environment of service. Since this is a very difficult task, knowledge of thermodynamics and kinetics is required so that materials can be selected that have slow reaction rates and/or harmless reactions. *Thermodynamics provides a means for the engineer to understand and predict the chemical reactions that take place.* The reader is referred to any of the numerous books on thermodynamics for a more detailed discussion of the topic, Gaskell 1981, Swalin 1962, and Bent 1965.

Mathematical Representation

The enthalpy and entropy are related through the free energy. The change in free energy of an isothermal reaction at constant pressure is given by:

$$\Delta G = \Delta H - T\,\Delta S \tag{9.1}$$

where:
  G = Gibbs free energy,
  H = enthalpy or heat of formation,
  T = absolute temperature, and
  S = entropy of reaction.

The change in free energy of an isothermal reaction at constant volume is given by:

$$\Delta F = \Delta E - T\,\Delta S \tag{9.2}$$

where:
  F = Helmholtz free energy, and
  E = internal energy.

From equations 9.1 and 9.2 it is obvious that the importance of the entropy term increases with temperature. The reactions of concern involving ceramic materials are predominately those at temperatures where the entropy term may have considerable effect on the reactions. In particular, species with high entropy values have a greater effect at higher temperatures.

Gibbs free energy is a more useful term in the case of solids, since the external pressure of a system is much easier to control than the volume. The change in free energy is easy to calculate at any temperature if the enthalpy and entropy are known. Evaluation of equation 9.1 will determine whether a reaction is spontaneous. *If the reaction is spontaneous, the change in free energy is negative, whereas if the reaction is in equilibrium, the free energy change is equal to zero.*

The free energy change for a particular reaction can be calculated easily from tabulated data, such as the JANAF Tables, Chase et al. 1985, by subtracting the free energy of formation of the reactants from the free energy of formation of the products. An example of the comparison of free energy of reaction and the enthalpy of reaction at several temperatures is given below for the reaction of alumina and silica to form mullite:

$$3\,Al_2O_3 + 2\,SiO_2 \rightarrow Al_6Si_2O_{13} \tag{9.3}$$

Using the following equations to calculate the enthalpy and free energy change from enthalpy and free energy of formation data given in the JANAF

TABLE 9.1

Enthalpy and Free Energy of
Formation at 1400 K

Material	ΔH_f (kJ/mol)	ΔG_f (kJ/mol)
Mullite	–6846.78	–5028.75
Alumina	–1688.91	–1229.39
Cristobalite	–899.81	–661.48

tables, assuming unit activity for all reactants and products, one can easily determine if the formation of mullite is a spontaneous reaction at the temperature in question:

$$\Delta G_r = \Sigma\ \Delta G_f\ (\text{products}) - \Sigma\ \Delta G_f\ (\text{reactants}) \tag{9.4}$$

$$\Delta H_r = \Sigma\ \Delta H_f\ (\text{products}) - \Sigma\ \Delta H_f\ (\text{reactants}) \tag{9.5}$$

Using the values from Table 9.1 one then calculates

$$\Delta H_r = (-6846.78) - \{3(-1688.91) + 2(-899.808)\}$$

$$\Delta H_r = +19.587\ \text{kJ/mol}$$

$$\Delta G_r = (-5028.75) - \{3(-1229.39) + 2(-661.482)\}$$

$$\Delta G_r = -17.609\ \text{kJ/mol}$$

It can be seen that although the enthalpy of reaction is positive the free energy of reaction is negative and the reaction is spontaneous at 1400 K and mullite is the stable phase, allowing one to predict that alumina will react with silica at that temperature.

Tabulations of the standard free energy, $\Delta G°$, at 1 bar and 298 K, as a function of temperature are available for the more common reactions, Chase et al. 1985 and Kubaschewski et al. 1967. For less common reactions, one must calculate the free energy of reaction by using values of $\Delta H°$, $\Delta S°$, and heat capacity data. Heat capacities can be experimentally determined by differential scanning calorimetry up to about 1000 K, Krupka et al. 1985, as can heats of reaction. The change in entropy cannot be obtained directly from thermal measurements. If one must do his own calculations, various computer programs are also available to aid the investigator, Gordon and McBride 1971, Eriksson 1975, and Allendorf 2003. The data of these tables are always in different stages of the confirmation process and can thus vary widely in accuracy. Therefore it is in the best interest of the user to check the source of the data. In addition, The Geochemist's Workbench®, which is a set of software tools for the researcher to calculate stability diagrams, manipulate reactions, and plot the results, is available through the website http://www.gwb.com. *The real problem with predicting whether a reaction may take place is in selecting the proper reaction to evaluate. Care must be taken not to overlook some possible reactions.*

Other forms of the free energy equation can be useful when evaluating corrosion by specific mechanisms. If the reaction is one of electrochemical nature, the free energy change for the reaction can be calculated using:

$$\Delta G = -nFE \tag{9.6}$$

where:

  n = number of electrons involved,
  F = Faraday constant (96,490 coulombs/mol), and
  E = standard cell potential (volts).

Tabulations of standard half-cell potentials (standard *emf* series) are available and are more commonly called *redox potentials*, Latimer 1952. The use of the *emf* series for studies in aqueous solutions has been established for a long time and has now been extended to nonaqueous electrolytes such as molten salt mixtures. According to Brenner 1975, who reported average errors of 32% between calorimetric and *emf* measurements, the use of equation 9.6 is not accurate and it should be modified as required for each galvanic cell evaluated.

Although industrial process gas streams are generally not in thermodynamic equilibrium, their compositions are shifting toward equilibrium at the high temperatures normally encountered. Using equilibrated gas mixtures for laboratory studies then is a basis for predicting corrosion but is not necessarily accurate. Which reaction products form at solid/gas interfaces can be predicted from free energy calculations using the following equation:

$$\Delta G^\circ = -RT \ln \left\{ \frac{(p_c)^u (p_d)^w}{(p_a)^x (p_b)^y} \right\} \tag{9.7}$$

where p = partial pressure of each component of the reaction:

$$xA + yB = uC + wD \tag{9.8}$$

The bracketed expression inside the logarithm in equation 9.7 is the equilibrium constant for the reaction; thus:

$$\Delta G^\circ = -RT \ln k_p \text{ (the well-known Nernst equation)} \tag{9.9}$$

When pure solids are involved in reactions with one or more nonideal gaseous species, it is more relevant to work with activities rather than compositions or pressures. Therefore the equilibrium constant can be expressed in terms of activities:

$$k = \frac{(a_c)^u (a_d)^w}{(a_a)^x (a_b)^y} \tag{9.10}$$

where the subscripts a and b denote reactants and c and d denote the products. The activity is the product of an activity coefficient and the concentration for a solute that does not dissociate. The solute activity coefficient is taken as approaching unity at infinite dilution. If the solute is an electrolyte that is completely dissociated in solution, the expression for the activity is more complicated. A few assumptions that are made in the use of equations 9.7 and 9.10 are that the gases behave as ideal gas mixtures, the activity of pure solids is equal to one, and the gas mixture is in equilibrium. In those cases where the ideal gas law is not obeyed, the fugacity is used in place of the activity to maintain generality. The assumption that the gases are ideal is not bad, since one is generally concerned with low pressures. The assumption of unity for the activity of solids is true as long as only simple compounds are involved with no crystalline solution. The assumption of equilibrium is reasonable near surfaces, since hot surfaces catalyze reactions.

If one is interested in the dissociation pressure of an oxide, equation 9.9 can be used where the equilibrium constant is replaced with the partial pressure of oxygen (pO_2), since for ideal gas behavior the activity is approximately equal to the partial pressure. If the oxide dissociates into its elements, the measured vapor pressure is equal to the calculated dissociation pressure. If the oxide dissociates into a lower oxide of the metal forming a stable gas molecule, the vapor pressure measured is greater than the calculated dissociation pressure. A compilation of dissociation pressures was given by Livey and Murray 1959. At moderate to high temperatures and atmospheric pressure, however, the fugacity and partial pressure are almost equal. Thus for most ceramic systems the partial pressure of the gas is used, assuming ideality.

An example where a pure solid reacts to form another pure solid and a gas is that of calcite forming lime and carbon dioxide. The equilibrium constant is then independent of the amount of solid as long as it is present at equilibrium.

$$CaCO_3 \rightarrow CaO + CO_2 \tag{9.11}$$

$$k = \frac{(a_{CaO})(a_{CO_2})}{a_{CaCO_3}} \tag{9.12}$$

Rearranging, we get:

$$a_{CO_2} = \frac{(a_{CaCO_3})k}{a_{CaO}} = k \qquad (9.13)$$

or

$pCO_2 = k_p$ (equilibrium reaction constant at constant pressure)   (9.14)

At constant temperature, if the partial pressure of CO_2 over $CaCO_3$ is maintained at a value less than k_p, all the $CaCO_3$ is converted to CaO. If the partial pressure of CO_2 is maintained greater than k_p, then all the CaO will react to form $CaCO_3$. This type of equilibrium, involving pure solids, is different from other chemical equilibria that would progress to a new equilibrium position and not progress to completion.

An example, similar to the above description for equation 9.9, for a reaction when both the reactants and products are all solid phases was given by Luthra 1988 for the reaction of an alumina matrix with SiC reinforcement fibers. The following equation depicts this reaction:

$$2Al_2O_3 + 3SiC \leftrightarrow 3SiO_2 + Al_4C_3 \qquad (9.15)$$

where the silica activity is dependent upon the alumina activity, assuming the activities of both SiC and Al_4C_3 are unity. This is given by:

$$a_{SiO_2} = [k(a_{Al_2O_3})^2]^{1/3} \qquad (9.16)$$

If the silica activity in the matrix is greater than the equilibrium silica activity, no reaction will occur between the matrix and the fiber. Since the activities of both silica and alumina are very small, minor additions of silica to the alumina matrix will prevent matrix/fiber reaction. Thus the use of small mullite additions prevents this reaction.

Since the corrosion of ceramics in service may never reach an equilibrium state, thermodynamic calculations cannot be strictly applied because these calculations are for systems in equilibrium. Many reactions, however, closely approach equilibrium, and thus the condition of equilibrium should be considered only as a limitation, not as a barrier to interpretation of the data.

Graphical Representation

The thermodynamics of reactions between ceramics and their environments can best be represented by one of several different types of stability diagrams. Graphs provide the same information as the mathematical equations; however, they can display unexpected relationships that provide new insight into solving a problem. Various types of graphical representations emphasize different aspects of the information and thus are well suited only to a specific problem. Figure 9.1 is a schematic representation for each of the various types of diagrams that one may find in the literature. Probably the most common type of graphical representation of thermodynamic data is the *equilibrium phase diagram*. These are based upon the Gibbs Phase Rule, which relates the physical state of a mixture with the number of substances or components that make up the mixture and with the environmental conditions of temperature and/or pressure. The region above the *solidus* is of greatest importance in most corrosion studies. The *liquidus* lines or the boundary curves between the region of 100% liquid and the region of liquid plus solid determines the amount of solid that can be dissolved into the liquid (i.e., saturation composition) at any temperature. For this reason these curves are also called *solubility curves*. Thus, these curves give the mole fraction (or weight fraction) at saturation as a function of temperature. To obtain concentrations one must also know the density of the compositions in question.

Another type of diagram is a graphical representation of the standard free energy of formation of the product between a metal and one mole of oxygen as a function of temperature at a constant total pressure. These are called *Ellingham diagrams*, Ellingham 1944. Richardson and Jeffes 1948 added an oxygen nomograph scale to the *Ellingham diagram* so that one could also determine the reaction for a certain partial pressure of oxygen in addition to the temperature. Since CO/CO_2 and H_2/H_2O ratios are often used in practice to obtain various partial pressures of oxygen (especially the very low values), Darken and Gurry added nomograph scales for these ratios, Darken and Gurry 1953. These diagrams now can be found in many places containing various numbers of oxidation/reduction reactions and have been referred to as *Ellingham, Ellingham-Richardson, Darken and Gurry*, or *modified-Ellingham diagrams*. On these plots (Figure 9.2), the intercept at $T = 0$ K is equal to $\Delta H°$ and the slope is equal to $-\Delta S°$.

To use the diagram shown in Figure 9.2 one needs only to connect the point representing zero free energy at the absolute zero of temperature (e.g., the point labeled O to the left of the diagram) and the point of intersection of the reaction and temperature in question. As an example, for alumina at 1400°C this line intersects the pO_2 scale at about 10^{-24} atmospheres, the equilibrium partial pressure of oxygen for the oxidation of aluminum metal to alumina. Any pressure lower than this will cause alumina to be reduced to the metal. This leads to the general tendency for oxides to be reduced at higher temperatures at constant oxygen partial pressures. One should also be aware that any metal will reduce any oxide above it in this diagram.

One should remember that all condensed phases of the reactions plotted on Figure 9.2 are assumed to be pure phases and therefore at unit activity. Deviations from unit activity are encountered in most practical reactions.

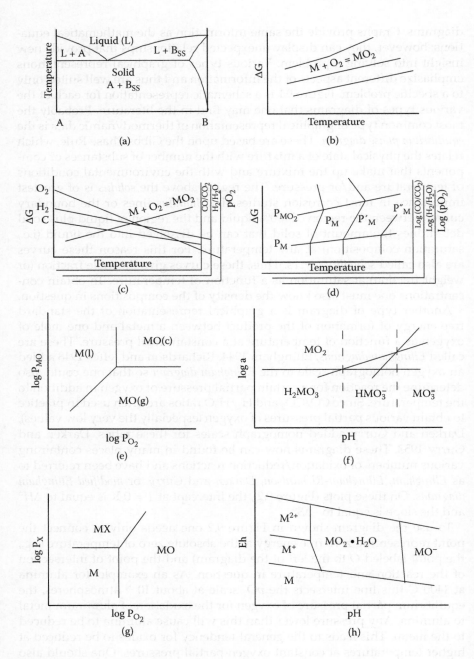

FIGURE 9.1
Representation of thermodynamic data: (a) phase equilibrium diagram, (b) Ellingham diagram, (c) Darken and Gurry modified Ellington diagram, (d) Lou et al. modified Ellingham diagram, (e) volatility diagram, (f) stability diagram, (g) phase stability or Kellogg diagram, (h) Pourbaix diagram.

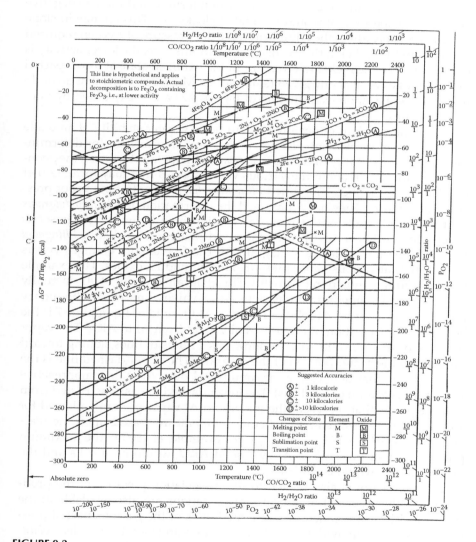

FIGURE 9.2

The standard free energy of formation of many metal oxides as a function of temperature. (Darken and Gurry 1953, reprinted with permission of McGraw-Hill.)

The correction that is applied is proportional to the activities of the products to that of the reactants by use of equations 9.7 and 9.10. As an example for the manufacture of glass containing nickel, the NiO activity is less than unity due to its solution in the glass. The correction term would then be negative and the free energy plot would be rotated clockwise. This change in slope can considerably affect the equilibrium partial pressure of oxygen required to maintain the nickel in the oxidized state. In this case, the lower activity is beneficial since the nickel will remain in the oxidized state at lower partial pressures of oxygen at any given temperature. Many reactions that do or do not occur based upon examination of Figure 9.2 can be explained by non-unit activities.

Since greater values of negative $\Delta G°$ indicate greater stability of an oxide with respect to its elements, *Ellingham diagrams* are excellent for determining the relative stability of oxides in contact with metals; however, they contain no information about the various vapor species that may form. Lou et al. 1985 have described a modified *Ellingham diagram* containing vapor pressure information. They have combined the information of *volatility diagrams* (isothermal plots of partial pressure relationships between two gaseous species in equilibrium with the condensed phases) with that of *Ellingham* type information to derive a diagram for the free energy changes versus temperature at various vapor pressures for individual oxides. The example for aluminum is shown in Figure 9.3. This diagram is a plot of pO_2 (actually $RT \ln pO_2$) and temperature for various $pAlO_x$ values. Line 6 is the boundary for the transition from Al solid or liquid to Al_2O_3 solid or liquid; line 7 is the boundary for transition of the principal vapors from Al to AlO_2. The vapor pressure of Al over solid Al_2O_3 is shown as a series of lines sloping toward the right in the center portion of the diagram. The upper dashed line is the isomolar line that defines the maximum pAl over Al_2O_3 in a nonreactive system (i.e., vacuum or inert gas). The lower dashed line is constructed from isobaric points that represent the maximum Al vapor pressure allowed for any hydrogen pressure at a particular temperature (based upon the reaction $Al_2O_3 + 3H_2 \rightarrow 2Al_{(g)} + 3H_2O_{(g)}$). For example, at 1800°C the maximum predicted vapor pressure of Al over solid Al_2O_3 would be 10^{-3} Pa and the maximum pO_2 would be $10^{-3.3}$ Pa.

The free energy is also related to the dissociation pressure of the product; thus other types of graphical representations are also available in the literature. These are generally isothermal plots of the gaseous partial pressures in equilibrium with the condensed phases and have been called *volatility diagrams, volatility maps,* or *phase stability diagrams,* Richardson and Jeffes 1948 and Darken and Gurry 1953. A similar type of diagram can be obtained when two oxidants are present (i.e., O_2 and N_2) as long as all possible condensed phases are known. Diagrams for systems such as metal-oxygen-carbon are available, Quets and Dresher 1969. An assumption that is usually made that is not always true is that the condensed phases are at unit activity. Unit activity should be applied only to species in the pure state. When more

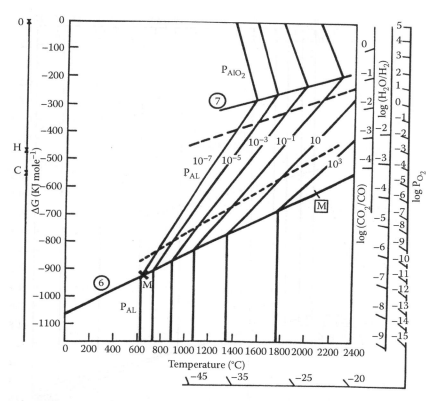

FIGURE 9.3
Ellingham type diagram for the Al-O system. (Lou et al. 1985, reprinted with permission of the American Ceramic Society/Wiley & Sons Copyright Clearance Center.)

than one gaseous species is involved in the reaction, *volatility diagrams* are more appropriate.

Many cases of corrosion of ceramic materials take place in an aqueous media (e.g., weathering of window glass). In these cases the pH of the system becomes important. Pourbaix 1974 first suggested the use of redox potential (*E*) versus pH plots to predict direction of reaction and the phases present. These plots, now called *Pourbaix diagrams,* are graphical representations of thermodynamic and electrochemical equilibria in aqueous systems. Figure 9.4 is a *Pourbaix diagram* of the system aluminum-water at 25°C. The two dashed lines labeled *a* and *b* in Figure 9.4 enclose the region where water is stable. At any potential and pH above the top line (b) water decomposes evolving oxygen. At any potential and pH below the lower line (a) water decomposes evolving hydrogen. These diagrams delineate three major regions of interest. The first is the region where no reaction occurs to the metal (i.e., the region of immunity), generally the lower portion of the diagram. The second is the region of corrosion where the metal reacts to form

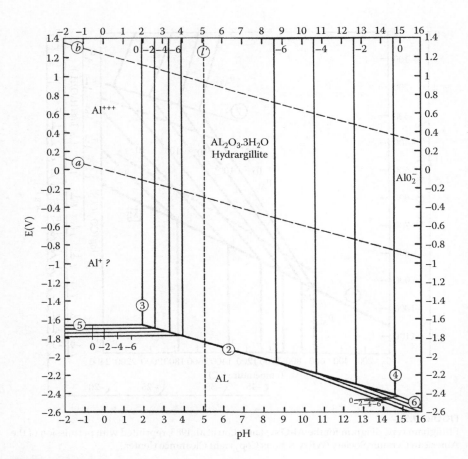

FIGURE 9.4
Potential-pH equilibrium diagram for the system alumina-water at 25°C. (Pourbaix 1974, reprinted with permission of CEBELCOR.)

an ion, generally the upper left region of the diagram. This second region is the one of most interest to the ceramist studying corrosion. The third is the region of passivity where the metal reacts to form an insoluble species that may be protective (generally an oxide), generally the upper right portion of the diagram. Garrels and Christ 1965 have extensively developed Pourbaix's concept for use in describing the action of water upon soils. These diagrams, related to soil-water systems, have been called *Garrels and Christ diagrams*. In aqueous dissolution studies it is also convenient to plot the pH of the solution versus the logarithm of the concentration of the species dissolved *(solubility diagrams)*.

Additional Recommended Reading

Bent, H.A. 1965. *The Second Law*, Oxford University Press, New York.
Gaskell, D.R. 1981. *Introduction to Metallurgical Thermodynamics*, 2nd edition, McGraw-Hill, New York.
Swalin, R.A. 1962. *Thermodynamics of Solids*, Wiley, New York.

Exercises, Questions, and Problems

1. Explain what is meant by a metastable phase. Why do these form?

2. Discuss the relationship of temperature and entropy and how they affect reactions.

3. Using equations 9.4 and 9.5 determine whether the following reactions will occur: $4Al + 3SiO_2 \rightarrow 2Al_2O_3 + 3Si$ and $2Al_2O_3 + 3SiC \rightarrow 3SiO_2 + Al_4C_3$.

4. Discuss the effects of reactions with nonideal gases and solids that are crystalline solutions.

5. Using the modified Ellingham diagram shown in Figure 9.2 determine whether a phosphate-bonded alumina brick will react with magnesium metal at 1000°C.

6. Discuss the assumptions that one must be aware of when using Figure 9.2 to determine whether various reactions will occur.

7. What is a volatility diagram?

8. Discuss the importance of Pourbaix diagrams.

References

Allendorf, M. 2003. New Hi-Temp Thermo Website Provides Scientists with Thermochemical Databases and Models for High-Temperature Materials Processing, Combustion, and Corrosion, *Refractory Appl. and News*, 8(4), 16–17.
Bent, H.A. 1965. *The Second Law*, Oxford University Press, New York.
Brenner, A. 1975. The Gibbs-Helmholtz Equation and the EMF of Galvanic Cells, II. Precision of Its Application to Concentration Cells, *J. Electrochem. Soc.*, 122(12), 1609–15.
Chase Jr., M.W., Davies, C.A., Downey Jr., J.R., Frurip, D.J.R., McDonald, R.A. and Syverud, A.N. 1985. *J. Phys & Chem Reference Data*, Vol. 14, Suppl. No. 1, *JANAF Thermochemical Tables*, 3rd edition, Parts I & II, Am. Chem. Soc. & Am. Inst. Phy.

Cussler, E.L. and Featherstone, J.D.B. 1981. Demineralization of Porous Solids, *Science*, 213, 28 Aug, 1018–19.

Darken, L.S. and Gurry, R.W. 1953. *Physical Chemistry of Metals*, McGraw-Hill, New York, pp. 348–49.

Ellingham, H.J.T. 1944. Reducibility of Oxides and Sulfides in Metallurgical Processes, *J. Soc. Chem. Ind.*, 63, 125.

Eriksson, G. 1975. Thermodynamic Studies of High Temperature Equilibria. XII. SOLGASMIX, A Computer Program for Calculation of Equilibrium Compositions in Multiphase Systems, *Chemica Scripta*, 8, 100–3.

Garrels, R.M. and Christ, C.L. 1965. *Solutions, Minerals, and Equilibria*, Harper and Row, New York.

Gaskell, D.R. 1981. *Introduction to Metallurgical Thermodynamics*, 2nd edition, McGraw-Hill, New York.

Gordon, S. and McBride, B.J. 1971. Computer Program for Calculation of Complex Chemical Equilibrium Compositions, Rocket Performance, Incident & Reflected Shocks, and Chapman-Jongnet Detonations, *NASA SP-273*, U.S. Printing Office, Washington, DC.

Guthrie Jr., G.D. and Mossman, B.T. (eds.). 1993. *Reviews in Mineralogy, Vol. 28: Health Effects of Mineral Dusts*, Mineral. Soc. Am., Washington, DC.

Hulbert, S.F., Bokros, J.C., Hench, L.L., Wilson, J. and Heimke, G. 1987. pp. 189–213 in *High Tech Ceramics*, Vincenzini, P. (ed.), Elsevier Science Pub. B.V., Amsterdam.

Janowski, K. and Rossi, R.C. 1968. Mechanical Degradation of MgO by Water Vapor, *J. Am. Ceram. Soc.*, 51(8), 453–55.

Kubaschewski, O., Evans, E.L. and Alcock, C.B. 1967. *Metallurgical Thermodynamics*, Pergamon Press, Oxford, UK.

Krupka, K.M., Hemingway, B.S., Robie, R.A. and Kerrick, D.M. 1985. High Temperature Heat Capacities and Derived Thermodynamic Properties of Anthophyllite, Diopside, Dolomite, Enstatite, Bronzite, Talc, Tremolite, and Wollastonite, *Am. Mineral.*, 70, 261–71.

Latimer, W.M. 1952. *The Oxidation States of the Elements and Their Potentials in Aqueous Solutions*, Prentice-Hall, Englewood Cliffs, NJ.

Livey, D.T. and Murray, P. 1959. The Stability of Refractory Materials, in *Physicochemical Measurements at High Temperatures*, Bockris, J. O'M. et al. (eds.), Butterworths Scientific Publications, London, pp. 87–116.

Lou, V.L.K., Mitchell, T.E. and Heuer, A.H. 1985. Review—Graphical Displays of the Thermodynamics of High-Temperature Gas-Solid Reactions and Their Application to Oxidation of Metals and Evaporation of Oxides, *J. Am. Ceram. Soc.*, 68(2), 49–58.

Luthra, K.L. 1988. Chemical Interactions in Ceramic and Carbon-Carbon Composites, in *Materials Research Society Symposium Proceedings, Vol. 125: Materials Stability and Environmental Degradation*, Barkatt, A., Verink Jr., E.D. and Smith, L.R. (eds.), Mat. Res. Soc., Pittsburgh, PA, pp. 53–60.

Nolan, R.P. and Langer, A.M. 1993. Limitations of the Stanton Hypothesis, Chap. 9 in *Health Effects of Mineral Dusts*, Guthrie Jr., G.D. and Mossman, B.T. (eds.), Reviews in Mineralogy, Vol. 28, Min. Soc. Am., Washington, DC.

Phase Diagrams for Ceramists, Vols. I–XII, Am. Ceram. Soc., Westerville, OH.

Pourbaix, M. 1974. *Atlas of Electrochemical Equilibria in Aqueous Solution*, Eng trans. by J.A. Franklin, NACE, Houston, TX.

Quets, J.M. and Dresher, W.H. 1969. Thermochemistry of the Hot Corrosion of Superalloys, *J. Materials*, 4(3), 583–99.

Reddy, M.M., Sherwood, S.I. and Doe, B.R. 1986. Limestone and Marble Dissolution by Acid Rain: An Onsite Weathering Experiment, pp. 226–38 in *Materials Degradation Caused by Acid Rain*, ACS Symposium Series 318, Baboian, R. (ed.), Am. Chem. Soc., Washington, DC.

Richardson, F.D. and Jeffes, J.H.E. 1948. The Thermodynamics of Substances of Interest in Iron and Steel Making from 0°C to 2400°C; I Oxides, *J. Iron and Steel Inst.*, 160, 261.

Sato, T., Ohtaki, S. and Shimada, M. 1985. Transformation of Yttria Partially Stabilized Zirconia by Low Temperature Annealing in Air, *J. Mater. Sci.*, 20(4), 1466–70.

Swalin, R.A. 1962. *Thermodynamics of Solids*, Wiley & Sons, New York.

Yoshimura, M., Hiuga, T. and Somiya, S. 1986. Dissolution & Reaction of Yttria-Stabilized Zirconia Single Crystals in Hydrothermal Solutions, *J. Am. Ceram. Soc.*, 69(7), 583–4.

Summary of Important Concepts Related to the Fundamentals of Corrosion

1. It is the total surface area exposed to attack that is important.

2. Grain boundary diffusion is more important at lower temperatures, and bulk diffusion is more important at high temperatures.

3. One need not have data on a specific material, chemically, but only on one of identical structure to estimate its dissolution characteristics.

4. One must remember that it is not the total porosity that is important, but the surface area of the total porosity.

5. If sufficient flow of a transpiring gas along a thermal gradient is present, dilution of the corrosive gas at the hot face may lower the corrosion rate to an acceptable level.

6. The characteristics of the corroding glass are more important than the solid parameters in corrosion.

7. Ceramics with an acid/base character similar to the liquid will tend to resist corrosion the best.

8. The importance of the zero point of charge (*zpc*) in corrosion is that it is the *p*H of maximum durability.

9. The spontaneity of a reaction depends upon more than just the heat of reaction. To predict stability, one must consider also the entropy.

10. If the reaction is spontaneous, the change in free energy is negative, whereas if the reaction is in equilibrium, the free energy change is equal to zero.

11. The real problem with predicting whether a reaction may take place is in selecting the proper reaction to evaluate. Care must be taken not to overlook some possible reactions.

12. Since the corrosion of ceramics in service may never reach an equilibrium state, thermodynamic calculations cannot be strictly applied because these calculations are for systems in equilibrium. Many reactions, however, closely approach equilibrium, and thus the condition of equilibrium should be considered only as a limitation, not as a barrier to interpretation of the data.

13. There is a general tendency for oxides to be reduced at higher temperatures at constant oxygen partial pressures. One should be aware that any metal will reduce any oxide above it in the Ellingham diagram.

14. Unit activity should be applied only to species in the pure state.

15. Thus the most important parameter of corrosion from the engineering viewpoint is the reaction rate.

16. Diffusion coefficients depend upon the composition and structure of the material through which diffusion occurs.

17. The rate of the reaction expressed as the rate of change of concentration, dc/dt, depends upon the concentration of the reactants.

18. The discrepancies between the experimental data and the theoretical models are often due to nonspherical particles, a range in sizes, poor contact between reactants, formation of multiple products, and the dependency of the diffusion coefficient upon composition.

19. Arnold et al. 1981 (Chapter 7) concluded that dynamic thermogravimetric studies provide insufficient data for calculation of reaction kinetics, that the data are influenced by the experimental procedures, and that the results are uncertain.

20. The enthalpy of the reaction is often sufficient to raise or lower the sample temperature by as much as 1000°C.

21. The flow of material by diffusion is proportional to the concentration gradient and is directed from the region of high concentration to one of low concentration.

22. In isometric crystals, the diffusion coefficient is isotropic, as it is in polycrystalline materials as long as no preferred orientation exists.

23. In most real cases, the diffusion coefficient can vary with time, temperature, composition, or position along the sample, or any combination of these.

24. Silica-forming reactions are the most desirable for protection against oxygen diffusion.

25. A diffusion flux will set up a thermal gradient in an isothermal system.

Section II

Corrosion Analysis

10

Assessment of Corrosion

> If we could first know where we are, and whither we are tending, we could then better judge what to do and how to do it.
>
> **Abraham Lincoln**

Introduction

Probably the most important thing that one can do in evaluating a corrosion problem is putting together a plan of attack. Quite often inaccurate or misleading results are obtained by not thinking through a plan thoroughly. With all the computer programs at one's disposal along with some knowledge of statistical design of experiments there is no reason for not performing the best possible experiment.

Many of us have fallen into the habit of performing a test only once and believing the results. This is probably one of the most important things not to do when evaluating a particular material for use under a certain set of conditions. The results of a test will generally vary to a certain degree and can vary considerably. It is up to the testing engineer to know or determine the test method variation. All ASTM standards now contain a statement of precision and bias to aid the test engineer in determining how his test fits into the overall imprecision of the procedure developed by the standards committee. In the development of an ASTM standard, a ruggedness test (ASTM Standard E-1169) is performed to determine the major sources of variation, Wernimont 1977. This test should be performed for any laboratory test that one might conduct to minimize the major sources of error. The idea of the ruggedness test is to determine the major sources of variation of a procedure and then minimize those variations to within acceptable limits.

Laboratory Test versus Field Trials

There are two general ways to approach a corrosion problem, either to conduct some laboratory tests to obtain information as to how a particular material will behave under certain conditions or to perform a postmortem examination of field trial samples. It is best to perform the laboratory test first to aid in making the proper selection of materials for a particular environment and then perform the field trial. Laboratory tests, however, do not always yield the most accurate information, since they rely on the investigator for proper setup; however, they are easier to control. The investigator must have a thorough understanding of the environment where the ceramic is to be used and must select the portions of the environment that may cause corrosion. For example, it is not sufficient to know that a furnace for firing ceramic ware is heated by fuel oil to a temperature of 1200°C. One most also know what grade fuel oil is used and the various impurities contained in the oil and at what levels. In addition, parameters such as partial pressure of oxygen, moisture content, etc., may be important. Once all these various parameters are known, the investigator can set up an appropriate laboratory test.

One must also understand all the various things that cannot be scaled down to a laboratory test, such as viscosity of liquids, time, temperature, etc. Care must be exercised when attempting to perform an accelerated laboratory test, which is usually accomplished by raising the temperature or increasing the concentration of the corrosive medium or both. *Since the mechanism of corrosion in the accelerated test may not be the same (generally it is not the same) as that under actual service conditions, erroneous conclusions and inaccurate predictions may be obtained.* The mechanisms must be the same for accurate application of laboratory test results to actual service conditions. Sample size is one parameter that is easily scaled; however, this can also cause problems. For example, when testing the corrosion of a ceramic by a liquid, the ratio of liquid volume present to the surface area of the exposed ceramic is extremely important. The investigator must remember that corrosion is controlled predominantly by thermodynamics and kinetics. Assuming that the proper laboratory tests have been conducted, the probability that any problems will arise are minimal.

The only way to analyze corrosion accurately is to conduct a field trial. This entails placing selected materials in actual service conditions, generally for an abbreviated time, and then collecting samples for analysis along with all the operational data of the particular environment. The size and amount of material or samples placed into actual service conditions for a field trial can be as little as one small laboratory test bar, or, for example, as large as a complete wall in a large industrial furnace. The larger the installation for the field trial, the more confidence one must have in the selection of materials. The larger installations are generally preceded by several laboratory tests and possibly a small-scale field trial. Abbreviated times may be as long as several years or as short as several days. Occasionally longer-term tests may be deemed necessary. Stutzman and Clifton 1997 reported that over 2000 samples of over 30 different types of stone were exposed to the environment for nearly 50 years in a test conducted at the Building and Fire Research Lab of NIST in an effort to evaluate the degradation of natural building stone.*

Data such as temperature and time are the obvious ones to collect, but there exists a large amount of other data that should be examined. Many times, however, some of the more important data do not exist for one reason or another. For example, maybe the oxygen partial pressure was not determined during the duration of the service life of the ceramic. In some cases, it may be impossible to collect certain pieces of data during the operation of the particular piece of equipment. At these times a knowledge of phase equilibria, thermodynamics, and kinetics can help fill in the gaps or at least give an indication as to what was present.

Sample Selection and Preparation

It should be obvious that powders will present a greater surface area to corrosion and thus will corrode more rapidly than a solid sample. This is especially true when one is evaluating nanomaterials. One may think this to be a good way to obtain a rapid test, but saturation of the corroding solution may cause corrosion to cease, or even cause a reverse-reaction (i.e., crystal growth), giving misleading results. This points to the extreme importance of the surface area-to-volume ratio (SA/V) of the ceramic to the corroding solution. It has been suggested that the volume-to-surface area[†] (V/SA) should be at least 40 mL/cm^2, NACE Standard TM0499. Another factor related to this is that during corrosion the surface may change, altering the SA/V ratio effect. Surface areas during dissolution have been reported to increase presumably due to opening of etch pits, microfissures, etc., Brady and House 1996.

Selecting samples for analysis provides another challenge to the investigator. Foremost in the selection process is selecting an area for analysis that is representative of the overall corrosion process. If this cannot be done, then many samples must be analyzed. Much of the modern analytical equipment necessitates the analysis of very small samples; thus one must be very sensitive to the selection of representative samples or at least evaluate multiple samples.

Much care must be given to preparing samples that contain an adherent reaction product surface layer. It is best to select a sample that is many times larger than required by the final technique and then mount this in some metallurgical mountant (e.g., epoxy). After the larger sample has been encased, then smaller samples can be safely cut from the larger piece.

---

* A complete description of the project can be found at http://stonewall.nist.gov.
† The V/SA is a more reasonable number.

Solid samples, when prepared for laboratory tests, should be cleaned in a noncorrosive solution to remove any loose particles adhering to the surface and any extraneous contamination. Brady and House 1996 have reported that initially an accelerated, nonlinear dissolution may occur from high-energy sites caused by grinding and incomplete removal of ultrafine particles. Best results are obtained if the cleaning is done in an ultrasonic cleaner. Cleaning solutions can be obtained from any of the metallographic supply companies. If the sample is mounted into one of the epoxy-type metallographic mountants, one must be aware that some cleaning solutions will react with the mountant. It is best to use supplies from one manufacturer to avoid these problems.

If as-manufactured samples are used for corrosion tests, one should remove a thin surface layer by grinding and cleaning before performing the corrosion tests. In this way, remnants from such things as powder-beds or encapsulation media used in the production of the material can be eliminated and therefore not interfere with the corrosion process.

Quite often the as-manufactured surface of a ceramic will have a different microstructure or even chemistry than the bulk. This often manifests itself as a thin surface layer (as much as several millimeters thick) that contains smaller grain sizes (more grain boundaries) and possibly a lower porosity. If the corrosion test corrodes only this thin surface layer, again misleading results will be obtained. One way to solve this problem is to remove the surface layer by grinding. Grinding, however, must be done with some thought to the final surface roughness, since again this will affect the SA/V ratio. Diamond-impregnated metal grinding discs should be used rather than silicon carbide paper discs or silicon carbide loose grit. Loose grit and the grinding media from paper discs have a tendency to become lodged within the pores and cracks of the sample being prepared. The final grinding media grit size should be no greater than 10 µm. It is best to clean samples after each grinding step in an ultrasonic cleaner with the appropriate cleaning solution.

Surface roughness of solid samples is an item that is often overlooked. Not only does a rough surface increase the area exposed to corrosion but it may also lead to problems with some analytical techniques. For example, when the surface roughness is on the order of the reaction layer thickness caused by corrosion, errors will be present in the depth profiles obtained by secondary ion mass spectrometry (SIMS). In those cases when surface analyses are planned, one should prepare solid samples to at least a 10 µm finish.

Grinding and polishing of samples that contain a reaction product surface layer should be done so that the reaction layer is not damaged, or the interface obscured. If part of the sample is metal, polishing should be done in the direction ceramic toward metal (i.e., hard toward soft) to eliminate smearing the metal over the ceramic. If very thin reaction layers are present, one can prepare taper sections to increase the area that is examined. Sample preparation of composites presents some additional problems, since materials of very different characteristics will be presented at the surface being polished. Chanat 1993 of Buehler Ltd. has offered some good advice for mounting, sectioning, and polishing. The most effective method involves the use of diamond abrasives with low nap cloths. High nap cloths can induce excessive relief at boundaries of different materials. This is especially important for composite materials of differing hardnesses.

Many tips on how to prepare samples can be obtained by reading the various technical journals published by the manufacturers of consumable grinding and polishing supplies. One particular article that offers some new ideas was that of Damgaard and Geels 2001. They emphasized the importance of polishing disk diameter and velocity, indicating that both are directly proportional to material removal rates. Although this may be true, one must be very aware of the amount of lubricant used, the pressures applied, and the area of the sample being polished. *If the lubricant supply rate is constant, which is generally the case, the material removal rate will peak at about 300–350 rpm.* Thus if one is thinking of purchasing automatic grinding and polishing equipment, he should look for something that has automatic lubrication flow rate control. Although many advances have been made in the grinding and polishing of ceramics, this area still is very much an art. Subtle changes in the procedure can make a major difference in the final finish of the sample.

Selection of Test Conditions

Although the selection of appropriate samples can be a major problem, the selection of the appropriate test conditions is an even more difficult task. The goal of the industrial corrosion engineer in selecting test conditions is to simulate actual service conditions. Selection of test conditions is much easier for the scientist, who is attempting to determine mechanisms. The major problem in attempting to simulate service conditions is the lack of detailed documentation. This is caused by not knowing the importance of such data in the corrosion of ceramics, the cost of collecting the data, or both. *Thus, if one wants to perform meaningful laboratory corrosion studies, it is imperative that the industrial environment of interest be accurately characterized.* One example that is probably the most difficult to duplicate is that of the human mouth for studies in dental restoration dissolution. Not only is dissolution in saliva and the various physicochemical reactions caused by food, chewing, and bacteria important but within a short period of implantation, restorations are covered with a biofilm called the pellicle.[*]

When conducting laboratory oxidation studies, a convenient way to obtain a range of oxygen partial pressures is desirable. Very low partial pressures

---

[*] The pellicle is a thin layer of salivary glycoproteins.

are never attained in practice by the use of a vacuum system. Instead, a mixture of gases in which oxygen is a component is used to establish the low partial pressure. The most important mixtures that are used are $CO_2 + CO$ and $H_2O + H_2$. Since the oxygen pressures are obtained through the equilibrium reactions:

$$CO_2 \leftrightarrow CO + 1/2\, O_2 \tag{10.1}$$

and

$$H_2O \leftrightarrow H_2 + 1/2\, O_2 \tag{10.2}$$

The partial pressure of oxygen is given by:

$$pO_2 = k_1 \left(\frac{pCO_2}{pCO} \right)^2 \quad \text{and} \quad pO_2 = k_2 \left(\frac{pH_2O}{pH_2} \right)^2 \tag{10.3}$$

where k_1 and k_2 are the equilibrium reaction constants. For constant ratios, the partial pressure of oxygen is independent of the total pressure. Thus these gas mixtures provide a means to obtain a range of oxygen pressures. Several techniques to mix these gases are discussed by Macchesney and Rosenberg 1970.

In the study of corrosion in coal gasification atmospheres, gas mixtures such as $CH_4 + H_2$ and $H_2S + H_2$ become important along with the ones listed above. As the gas mixture becomes more complex, the number of equations that must be solved to obtain the equilibrium gas composition at elevated temperatures and pressures also increases, making it convenient to use a program such as SOLGASMIX, Eriksson 1975, for the calculations. *One should not make the erroneous assumption that gas mixtures are the same at all temperatures, since the equilibrium mixture is dependent upon the equilibrium constant, which is temperature dependent.*

Data Reduction

The corrosion data that have been reported in the literature have been in many forms. This makes comparison between various studies difficult unless one takes the time to convert all the results to a common basis. Those working in the area of leaching of nuclear waste glasses have probably made the biggest effort in standardizing the reporting of data; however, a major effort is still needed to include the entire field of corrosion of ceramics. The reporting of corrosion data often omits important pieces of information about the material being tested, such as manufacturing process, density, porosity, etc. These are critical pieces of data when attempting to compare results among various investigators. The work and efforts of organizations like ASTM can aid in providing standard test procedures and standard data reporting methods. These are briefly described in Chapter 11.

The statistical analysis of data is of prime importance when attempting to determine what exactly affects the results. Two articles that are rather old but still appropriate are those of Westman 1927 and Ferguson 1930.

Exercises, Questions, and Problems

1. Discuss the importance of doing a ruggedness test before performing a series of experimental corrosion tests.

2. Discuss the problems that may arise when performing accelerated laboratory tests.

3. Field trials are an important way to evaluate materials. Discuss their advantages and disadvantages compared to laboratory tests.

4. Why is the ratio of the surface area of the material being corroded to the volume of the corroding medium important?

5. When grinding and polishing ceramics that have been in contact with a metal it is important to grind/polish in a particular direction. What is that direction and why is it important?

References

Brady, P.V. and House, W.A. 1996. Surface-Controlled Dissolution and Growth of Minerals, Ch. 4, pp. 225–305 in *Physics and Chemistry of Mineral Surfaces*, Brady, P.V. (ed.), CRC Press, New York.

Chanat, S. 1993. Preparation Techniques for Analysis of Fiber Reinforced Ceramic Matrix Composites, pp. 603–15 in *Ceramic Transaction, Vol. 38, Advances in Ceramic-Matrix Composites*, Bansal, N.P. (ed.), Am. Ceram. Soc., Westerville, OH.

Damgaard, M.J. and Geels, K. 2001. High Capacity Materialographic Specimen Preparation, Structure 38, *Struers Journal of Materialography*, 2001, 7–11.

Eriksson, G. 1975. Thermodynamic Studies of High Temperature Equilibria. XII. SOLGASMIX, A Computer Program for Calculation of Equilibrium Compositions in Multiphase Systems, *Chemica Scripta, 8*, 100–3.

Ferguson, R.F. 1930. Interpretation of Plant and Laboratory Test Data, *J. Amer. Ceram. Soc., 13*(5), 354–62.

Macchesney, J.B. and Rosenberg, P.E. 1970. The Methods of Phase Equilibria Determination and Their Associated Problems, Ch. 3 in *Phase Diagrams: Materials Science and Technology*, Alper, A.M. (ed.), Vol. 6-1 of *Refractory Materials*, Margrave, J.L. (ed.), Academic Press, New York, pp. 113–65.

Stutzman, P.E. and Clifton, J.R. 1997. Stone Exposure Test Wall at NIST, pp. 20–32 in *Degradation of Natural Building Stone*, Geotech. Special Pub. No. 72, Labuz, J.F. (ed.), ASCE, Reston, VA.

Wernimont, G. 1977. Ruggedness and Evaluation of Test Procedures, *ASTM Standardization News*, March, 13–16.

Westman, A.E.R. 1927. Statistical Methods in Ceramic Research, *J. Amer. Ceram. Soc.*, 10(3), 133–47.

11

Corrosion Test Procedures

When you can measure what you are speaking about and express it in numbers you know something about it; but when you cannot measure it, when you cannot express it in numbers, your knowledge is of a meager and unsatisfactory kind.

Lord Kelvin

Introduction

The American Society for Testing and Materials (ASTM) was formed in 1898 through the efforts of Andrew Carnegie and the chief chemist of the Pennsylvania Railroad, Charles Dudley, who were both convinced that a solution was necessary to the unexplainable differences of testing results that arose between their laboratories. These early efforts were focused upon improving the understanding between seller and buyer of the quality of their products. Although ASTM and other organizations have made considerable progress in eliminating the unexplainable differences in testing results between laboratories, new materials and new applications continue to present new and exciting challenges to the corrosion engineer. These challenges, however, are ones that must be overcome if there is to be honest competition in the world market of materials.

Many standard tests have been developed through ASTM to evaluate the corrosion resistance of various ceramic materials. These various tests have been listed in Tables 11.1 and 11.2 and can be found in the *Annual Book of ASTM Standards*, Volumes 2.05, 4.01, 4.02, 4.05, 14.04, 15.01, and 15.02. A brief summary of each of these is given below. Standards that are in the process of being developed have not been listed in Tables 11.1 and 11.2. These draft standards can be found on the ASTM website. ASTM designates some procedures as standard test methods and others as standard practices. The distinction between these two is best given by their definitions. ASTM defines test method as *a definitive procedure for the identification, measurement, and evaluation of one or more qualities, characteristics, or properties of a material, product, system, or service that produces a test result,* and practice as *a definitive procedure for performing one or more specific operations or functions that does not produce a test result,* ASTM 1986. Standard practices provide the user with accepted

TABLE 11.1

ASTM Test Methods Related to Corrosion of Ceramics

ASTM Designation	Vol. No.	Title
C-151	4.01	Autoclave Expansion of Portland Cement
C-157	4.02	Length Change of Hardened Hydraulic-Cement Mortars and Concrete (reapproved 2008)
C-225	15.02	Resistance of Glass Containers to Chemical Attack (reapproved 2009)
C-267	4.05	Chemical Resistance of Mortars, Grouts, and Monolithic Surfacings (reapproved 2006)
C-282	2.05	Acid Resistance of Porcelain Enamels (reapproved 2010)
C-283	2.05	Resistance of Porcelain Enameled Utensils to Boiling Acid (reapproved 2008)
C-288	15.01	Disintegration of Refractories in an Atmosphere of Carbon Monoxide (reapproved 2009)
C-370	15.02	Moisture Expansion of Fired Whiteware Products (reapproved 2012)
C-413	4.05	Absorption of Chemical-Resistant Mortars, Grouts, and Monolithic Surfacings (reapproved 2006)
C-452	4.01	Potential Expansion of Portland Cement Mortars Exposed to Sulfate (reapproved 2010)
C-456	15.01	Hydration Resistance of Basic Brick and Shapes (reapproved 2008)
C-492	15.01	Hydration of Granular Dead-Burned Refractory Dolomite (reapproved 2008)
C-544	15.01	Hydration of Magnesite or Periclase Grain (reapproved 2008)
C-556	15.02	Resistance of Overglaze Decorations to Attack by Detergents (withdrawn 1994)
C-577	15.01	Permeability of Refractories (reapproved 2007)
C-614	2.05	Alkali Resistance of Porcelain Enamels (reapproved 2010)
C-620	15.01	Hydration Resistance of Pitch-Bearing Basic Refractory Brick (withdrawn 1996)
C-621	15.01	Isothermal Corrosion Resistance of Refractories to Molten Glass (reapproved 2009)
C-622	15.01	Corrosion Resistance of Refractories to Molten Glass Using the Basin Furnace (withdrawn in 2000)
C-650	15.02	Resistance of Ceramic Tile to Chemical Substances (reapproved 2004)
C-675	15.02	Alkali Resistance of Ceramic Decorations on Returnable Beverage Glass Containers (reapproved 2011)
C-676	15.02	Detergent Resistance of Ceramic Decorations on Glass Tableware (reapproved 2004)
C-724	15.02	Acid Resistance of Ceramic Decorations on Architectural Type Glass (reapproved 2010)
C-735	15.02	Acid Resistance of Ceramic Decorations on Returnable Beer and Beverage Glass Containers (reapproved 2004)
C-738	15.02	Lead and Cadmium Extracted from Glazed Ceramic Surfaces (reapproved 2011)

TABLE 11.1 (*Continued*)

ASTM Test Methods Related to Corrosion of Ceramics

ASTM Designation	Vol. No.	Title
C-777	15.02	Sulfide Resistance of Ceramic Decorations on Glass (reapproved 2004)
C-863	15.01	Evaluating Oxidation Resistance of Silicon Carbide Refractories at Elevated Temperatures (reapproved 2010)
C-872	2.05	Lead and Cadmium Release from Porcelain Enamel Surfaces (reapproved 2010)
C-895	15.02	Lead and Cadmium Extracted from Glazed Ceramic Tile (reapproved 2007)
C-927	15.02	Lead and Cadmium Extracted from Lip and Rim Area of Glass Tumblers Externally Decorated with Ceramic Glass Enamels (reapproved 2009)
C-1012	4.01	Length Change of Hydraulic-Cement Mortars Exposed to a Sulfate Solution (reapproved 2004)
C-1034	15.02	Lead and Cadmium Extracted from Glazed Ceramic Cookware (withdrawn in 2001)
C-1106	4.05	Chemical Resistance and Physical Properties of Carbon Brick (reapproved 2005)
C-1203	15.02	Quantitative Determination of Alkali Resistance of a Ceramic-Glass Enamel (reapproved 2004)
C-1220	12.01	Static Leaching of Monolithic Waste Forms for Disposal of Radioactive Waste (reapproved 2010)
C-1285	12.01	Chemical Durability of Nuclear, Hazardous, and Mixed Waste Glasses and Multiphase Glass-Ceramics: The Product Consistency Test (PCT) (reapproved 2008)
C-1308	12.01	Accelerated Leach Test for Diffusive Releases from Solidified Waste and a Computer Program to Model Diffusive, Fractional Leaching from Cylindrical Waster Forms
C-1370	4.05	Determining the Chemical Resistance of Aggregates for Use in Chemical-Resistant Sulfur Polymer Cement Concrete and Other Chemical-Resistant Polymer Concretes

procedures for the performance of a particular task. Test methods provide the user with an accepted procedure for determination of fundamental properties (e.g., density, viscosity). These standards must be updated or reapproved by the end of the eighth year after the last approval. If not reapproved, the standard is then withdrawn.

The Materials Characterization Center (MCC)[*] of the Pacific Northwest Laboratories is another organization that has developed standard test procedures, Mendel 1981. Several of these tests have been used extensively by those investigating the leaching of nuclear waste glasses. Test MCC-1 (now

---

[*] The MCC was created in 1980 by the U.S. Dept. of Energy to assist in projects responsible for nuclear waste form production, transportation, and disposal.

TABLE 11.2

ASTM Practices Related to Corrosion of Ceramics

ASTM Designation	Vol. No.	Title
C-454	15.01	Disintegration of Carbon Refractories by Alkali (reapproved 1998)
C-768	15.01	Drip Slag Testing Refractory Brick at High Temperature
C-874	15.01	Rotary Slag Testing of Refractory Materials
C-987	15.01	Vapor Attack on Refractories for Furnace Superstructures
C-1109	12.01	Analysis of Aqueous Leachates from Nuclear Waste Materials Using Inductively Coupled Plasma-Atomic Emission Spectroscopy
C-1174	12.01	Prediction of the Long-Term Behavior of Materials, Including Waste Forms, Used in Engineering Barrier Systems for Geological Disposal of High-Level Radioactive Waste
C-1463	12.01	Dissolving Glass Containing Radioactive and Mixed Waste for Chemical and Radiochemical Analysis
C-1662	12.01	Measurement of the Glass Dissolution Rate Using the Single-Pass Flow-Through Test Method
G-7	14.04	Atmospheric Environmental Exposure Testing of Nonmetallic Materials
G-90	14.04	Accelerated Outdoor Weathering of Nonmetallic Materials Using Concentrated Natural Sunlight

an ASTM standard test C-1220) involves a procedure for testing the durability of monolithic glass samples in deionized or simulated ground water at 40, 70, and 90°C for 28 days. One disadvantage of this test is that no standard glass is used, thus eliminating corrections for bias. It does, however, require the reporting of mass loss normalized to the fraction of the element leached in the glass sample allowing one to make comparisons between glasses. Test MCC-3 in contrast evaluates an agitated crushed glass sample to maximize leaching rates. Test temperatures are extended to 110, 150, and 190°C. Again a standard glass is not used.

With the global economy of today the engineer must be familiar with standards from countries other than the United States. A list of the various acronyms used for these various organizations is given in Table 11.3. In addition to the individual countries that maintain standards there is also the International Organization for Standardization (ISO) and the International Electro Technical Commission (IEC). A list of the standards from these organizations is given in Table 11.4. The development of international standards is indispensable for the future progress of international trade. In addition to the ISO and IEC the National Association of Corrosion Engineers (NACE) also has many standards relating to corrosion of materials; however, very few relate to ceramics. One that does is Standard TM0499: Immersion Corrosion Testing of Ceramic Materials.

TABLE 11.3

Standards Organizations and Their Acronyms

ACNOR	Association Canadienne de Normalisation
ABNT	Associaçâo Brasileira de Normas Técnicas
AENOR	Asociacion Espanola de Normalizacion y Certificacion (Spain)
AFNOR	Association Française de Normalisation
ANSI	American National Standards Institute
ASA	Australian Standards Association
ASI	Austrian Standards Institute
ASRO	Asociatia de Standardizare din România
ASTM	American Society for Testing and Materials
BDS	Bulgarian Institute for Standardization
BIS	Bureau of Indian Standards
BNQ	Bureau de Normalisation du Quebec
BNSI	Barbados National Standards Institution
BOBS	Botswana Bureau of Standards
BPS	Bureau of Product Standards (Philippines)
BSI	British Standards Institution
BSJ	Bureau of Standards Jamaica
BSMI	Bureau of Standards, Metrology and Inspection (Taiwan)
BSN	Badan Stardardisasi Nasional (Indonesia)
BSTI	Bangladesh Standards and Testing Institute
CAS	China Association of Standardization
CEN	Comite Europeen de Normalisation
CCN	Conseil Canadien des Normes
CNIS	China National Institute of Standardization
COGUANOR	Comisión Guatemalteca de Normas
COHCIT	Consejo Hondureño de Ciencia y Tecnología
CONACYT	Consejo Nacional de Ciencia y Tecnologia (El Salvador)
CSA	Canadian Standards Association
CSNI(CSI)	Czech Standards Institute
CYS	Cyprus Organization for Standardization
DBOS	Dominican Bureau of Standards
DGN	Direccion General de Normes (Mexico)
DIN	Deutsches Institut für Normung (Germany)
DS	Danish Standards Bureau
DSM	Department of Standards of Malaysia
DTNM	Dirección de Tecnología Normalización y Metrología (Nicaragua)
ELOT	Hellenic Organization for Standardization (Greece)
ENS	European Standards
EOS	Egyptian Organization for Standardization
ESMA	Emirates Standardization & Metrology Authority
EUS	Estonia Standards

Continued

TABLE 11.3 (*Continued*)

Standards Organizations and Their Acronyms

FSA	Finland Standards Asociation
GOST	Gosudarstvenii Stardart (Russian Federation)
GSB	Ghana Standards Board
HSI	Hungarian Standards Institution
HZN	Croatian Standards Institute
IAN	Institute Algerien de Normalisation
ICONTEC	Instituto Colombiano de Normas Tecnicas y Certificacion
IBN	Institute Belge de Normalisation (Belgium)
IBNORCA	Instituto Boliviano de Normalización y Calidad
INEN	Instituto Ecuatoriano de Normalizacion (Ecuador)
INN	Chile Instituto Nacional de Normalizacion
INTECO	Instituto de Normas Tecnicas de Costa Rica
IIRS	Irish Institute for Industrial Research and Standards
IRAM	Inst. Argentino de Racionalizaion de Materiale
ISBH	Institute for Standardization of Bosnia and Herzegovina
ISI	Indian Standards Institute
ISO	International Organization for Standardization
IST	Icelandic Standards
INDECOPI	Instituto Nacional de Defensa de la Competencia y de la Protección de la Propiedad Intelectual (Peru)
JBS	Jamaica Bureau of Standards
JSA	Japanese Standards Association
JISC	Japanese Industrial Standards Committee
JISM	Jordan Institution for Standards and Metrology
KATS	Korean Agency for Technology and Standards
KAZINST	Kazakhstan Institute for Standardization
KEBS	Kenya Bureau of Standards
KSA	Korean Standards Association
MASM	Mongolian Agency for Standardization and Metrology
MBS	Malawi Bureau of Standards
MCC	Materials Characterization Center (USA)
MSB	Mauritius Standards Bureau
MSZ	Magyar Szabvány (Hungarian Standards)
NBSM	Nepal Bureau of Standards and Metrology
NBN	Belgian Standards
NBR	Brazilian Standards
NC	Oficina Nacional de Normalizacion de Cuba
NEN	Netherlands Standards
NIST	National Institute for Standards and Technology (USA)
NSF	Norges Standardiseringsforbund (Norway)
OCC	Office Congolais de Controle (Congo)

TABLE 11.3 (*Continued*)

Standards Organizations and Their Acronyms

ON	Osterreichisches Normungsinstitut (Austria)
PKN	Polski Komitet Normalizacyjny (Poland)
PSI	Palestine Standards Institute
PSQCA	Pakistan Standards and Quality Control Authority
QSAE	Quality and Standards Authority of Ethiopia
SABS	South African Bureau of Standards
SAC	Standards Association of China
SANZ	Standards Association of New Zealand
SASO	Saudi Arabian Standards Organization
SAZ	Standards Association of Zimbabwe
SCC	Standards Council of Canada
SFS	Finnish Standards Association
SII	Standards Institution of Israel
SIRIM	Malaysia
SIS	Swedish Institute for Standardization
SIST	Slovenian Institute for Standardization
SLBS	Saint Lucia Bureau of Standards
SLSI	Sri Lanka Standards Institution
SNIMA	Service de Normalisation Industrielle Marocaine
SNV	Schweizerische Normenvereinigung (Switzerland)
SON	Standards Organization of Nigeria
STAMEQ	Directorate for Standards and Quality (Vietnam)
STN	Slovak Standards Institute
SZS	Savezni Zavod za Standardizaciju (Yugoslavia)
TISI	Thai Industrial Standards Institute
TSE	Turk Standardslari Enstitusu
TTBS	Trinidad & Tobago Bureau of Standards
UNI	Ente Nationale Italiano di Unificazione
UNIT	Uruguayo Normas Instituto de Técnicas
VSC	Vietnam Standards Centre
ZABS	Zambia Bureau of Standards

ASTM Standards

Autoclave Expansion of Portland Cement, C-151

Samples of portland cement are exposed to water vapor at 2 MPa and 23°C for 3 hours in an autoclave. The test evaluates the potential for delayed expansion caused by the hydration of CaO or MgO or both. The percent linear expansion change is reported.

TABLE 11.4

ISO and IEC Test Methods Related to Corrosion of Ceramics

Designation	Title
ISO 695	Glass—Resistance to attack by a boiling aqueous solution of mixed alkali.
ISO 7086	Glass hollowware in contact with food—Release of lead and cadmium.
ISO 8391	Ceramic cookware in contact with food—Release of lead and cadmium.
ISO 9689	Raw optical glass—Resistance to attack by aqueous alkaline phosphate-containing detergent solutions at 50°C.
ISO 10629	Raw optical glass—Resistance to attack by aqueous alkaline solutions at 50°C.
ISO 17092	Determination of corrosion resistance of monolithic ceramics in acid and alkaline solutions.
ISO 20509	Fine ceramics (advanced ceramics, advanced technical ceramics)—Determination of oxidation resistance of nonoxide monolithic ceramics.
IEC 60793-1-33	Optical fibers—Part 1-33: Measurement methods and test procedures—Stress corrosion susceptibility.

Length Change of Hardened Hydraulic-Cement Mortars and Concrete, C-157

Samples of hardened cement or concrete are tested in lime-saturated water at 23°C for 15 or 30 minutes depending upon sample size. The samples are then dried at 23°C and a relative humidity of 50%. The length change is recorded after 4, 7, 14, and 28 days and 8, 16, 32, and 64 weeks.

Resistance of Glass Containers to Chemical Attack, C-225

Attack by dilute sulfuric acid (representative of products with *p*H less than 5.0) or distilled water (representative of products with *p*H greater than 5.0) on glass bottles and the attack by pure water upon powdered glass (for containers too small to test solubility by normal methods) all at 121°C are covered in this standard test method.

Chemical Resistance of Mortars, Grouts, and Monolithic Surfacings, C-267

This method tests the resistance of resin, silica, silicate, sulfur, and hydraulic materials, grouts, and monolithic surfacings to a simulated service environment. Any changes in weight, appearance of the samples or test medium, and the compressive strength are recorded.

Acid Resistance of Porcelain Enamels, C-282

This test method was developed to test the resistance of porcelain enamel coatings on stoves, refrigerators, tabletops, sinks, laundry appliances, and so on, to 10% citric acid at 26°C. Several drops of acid solution are placed onto a flat area about 50 mm in diameter. After 15 minutes the samples are cleaned and evaluated for changes in appearance and cleanability.

Resistance of Porcelain Enameled Utensils to Boiling Acid, C-283

Test samples 82 mm in diameter make up the bottom of glass tube that is filled with 150 ml of a solution prepared from 6 grams of citric acid in 94 grams of distilled water. The test cell is placed onto a hot plate and the solution is allowed to boil for 2 1/2 hours. The results are reported as the change in weight.

Disintegration of Refractories in an Atmosphere of Carbon Monoxide, C-288

Providing a higher than expected amount of carbon monoxide normally found in service conditions, this method can be used to obtain the relative resistance of several refractory products to disintegration caused by exposure to CO. Samples are heated in nitrogen to the test temperature of 500°C and then held in an atmosphere of 95% CO for times sufficient to produce complete disintegration of half the test samples.

Moisture Expansion of Fired Whiteware Products, C-370

Unglazed, rod-shaped samples are tested for their resistance to dimensional changes caused by water vapor at elevated temperatures and pressures. Five samples are placed into an autoclave for 5 hours in an atmosphere of 1 MPa of steam. The amount of linear expansion caused by moisture attack is then recorded.

Absorption of Chemical-Resistant Mortars, Grouts, and Monolithic Surfacings, C-413

Silica and silicate samples, in addition to other materials, are tested for absorption in boiling xylene after 2 hours. The percent absorption is recorded.

Potential Expansion of Portland-Cement Mortars Exposed to Sulfate, C-452

Samples of portland cement are mixed with gypsum and then immersed into water at 23°C for 24 hours and 14 days or more. The change in linear expansion is recorded.

Disintegration of Carbon Refractories by Alkali, C-454

Carbon cubes with a hole drilled into them to form a crucible are used as the samples to test their resistance to attack from molten potassium carbonate at approximately 1000°C for 5 hours. The results of this standard practice are reported as visual observations of the degree of cracking. Variations of this procedure have been used by many to investigate the resistance of refractories to attack by molten metals and molten glasses.

Hydration Resistance of Basic Brick and Shapes, C-456

One-inch cubes cut from the interior of basic brick are tested in an autoclave containing sufficient water to maintain a pressure of 552 kPa at 162°C for 5 hours. This test is repeated for successive 5-hour periods to a maximum of 30 hours or until the samples disintegrate. The results are reported as visual observations of hydration and cracking.

Hydration of Granular Dead-Burned Refractory Dolomite, C-492

A 100-gram dried powder sample of dolomite that is coarser than 425 μm is tested by placing it into a steam-humidity cabinet that is maintained at 71°C and 85% humidity for 24 hours. The sample is then dried at 110°C for 30 minutes and the amount of material passing a 425 μm sieve is determined.

Hydration of Magnesite or Periclase Grain, C-544

A carefully sized material that is between 425 μm and 3.35 mm is tested by placing a dried 100-gram sample into an autoclave maintained at 162°C and 552 kPa for 5 hours. The sample is then weighed after removal from the autoclave and dried at 110°C. The hydration percentage is calculated from the weight difference between the final dried weight and the weight of any material coarser than 300 μm.

Resistance of Overglaze Decorations to Attack by Detergents, C-556; Withdrawn 1994

Overglaze decorations on pieces of dinnerware are tested by submerging the samples into a solution of sodium carbonate and water at a temperature of 95°C. Samples are removed after 2, 4, and 6 hours and rubbed with a muslin cloth. The results are reported as visual observations of the degree of material removed by rubbing.

Permeability of Refractories, C-577

Although not a corrosion test, C-577 is important in determining the ease of flow of various gases through a material. This test method is designed to determine the unidirectional rate of flow of air or nitrogen through a 2-inch cube of material at room temperature.

Alkali Resistance of Porcelain Enamels, C-614

The coatings on washing machines, dishwashers, driers, etc., are tested for their resistance to solution containing 260 grams of tetrasodium pyrophosphate dissolved in 4.94 liters of distilled water. The loss in weight is determined after exposure for 6 hours at 96°C.

Hydration Resistance of Pitch-Bearing Refractory Brick, C-620

Full-sized pitch-containing bricks are placed into a steam humidity cabinet and tested for 3 hours at 50°C and 98% humidity. The test is repeated for successive 3-hour periods until visually affected. The results are reported as visual observations of hydration and disintegration.

Isothermal Corrosion Resistance of Refractories to Molten Glass, C-621

This method compares the corrosion resistance of various refractories to molten glass under static, isothermal conditions. Samples approximately 1/2 inch square by 2 inches long are immersed into molten glass and then heated to a temperature that simulates actual service conditions. The duration of the test should be sufficient to produce a glass line cut of 20–60% of the original sample thickness. After the test, samples are cut in half lengthwise and the width or diameter is measured at the glass line and halfway between the glass line and the bottom of the sample before testing.

Corrosion Resistance of Refractories to Molten Glass Using the Basin Furnace, C-622; Withdrawn in 2000

This standard practice determines the corrosion of refractories by molten glass in a furnace constructed of the test blocks with a thermal gradient maintained through the refractory. Because of the cooling effects of the thermal gradient, the duration of this test is 96 hours. Since the glass is not replaced during the test, solution products may modify the results of the test. The depth of the glass line cut is determined across the sample and the volume corroded is determined by filling the corroded surface with zircon sand and determining the volume of sand required.

Resistance of Ceramic Tile to Chemical Substances, C-650

This method is designed to test plain colored, glazed, or unglazed impervious ceramic tile of at least 4 1/4 by 4 1/4 inches to the resistance against attack by any chemical substance that may be of interest. The test conditions may be any combination of time and temperature deemed appropriate for the expected service conditions. Hydrochloric acid or potassium hydroxide at 24°C for 24 hours is the recommended exposure. The results are reported as visually affected or not affected but also the calculated color difference may be reported.

Alkali Resistance of Ceramic Decorations on Returnable Beverage Glass Containers, C-675

Two ring sections cut from each container and representative of the label to be evaluated are placed into the test solution at 88°C of sodium hydroxide, trisodium phosphate, and tap water for successive 2-hour intervals. The results are reported as the time required for 90% destruction of the label. A variation of this method conducted at 60°C for 24 hours in a mixture of sodium hydroxide, trisodium phosphate, and distilled water determines the reduction in thickness of the label.

Detergent Resistance of Ceramic Decorations on Glass Tableware, C-676

In this standard method glass tableware with ceramic decorations is immersed in a solution of sodium pyrophosphate and distilled water at 60°C for successive 2-hour periods. The samples are then rubbed with a cloth under flowing water, dried, and evaluated as to the degree of loss of gloss up to complete removal of the decoration.

Acid Resistance of Ceramic Decorations on Architectural Type Glass, C-724

A citric acid solution is placed onto the ceramic decoration of the architectural glass for 15 minutes at 20°C and the degree of attack after washing is determined visually.

Acid Resistance of Ceramic Decorations on Returnable Beer and Beverage Glass Containers, C-735

Representative containers are immersed into hydrochloric acid solution such that half the decoration is covered for 20 minutes at 25°C. The results are reported as the visually observed degree of attack.

Lead and Cadmium Extracted from Glazed Ceramic Surfaces, C-738

This standard method determines quantitatively by atomic absorption the amount of lead and cadmium extracted from glazed ceramic surfaces when immersed into 4% acetic acid solution at 20–24°C for 24 hours.

Drip Slag Testing Refractory Brick at High Temperature, C-768

Test samples of this standard practice are mounted into the wall of a furnace such that their top surface slopes down at a 30° angle. Rods of slag are placed through a hole in the furnace wall such that when the slag melts it will drip and fall 2 inches to the surface of the refractory test piece. Slag is fed continuously to maintain consistent melting and dripping onto the sample. Test temperatures are about 1600°C and the duration of the test is from 2 to 7 hours. The volume of the corroded surface is determined by measuring the amount of sand required to fill the cavity. In addition, the depth of penetration of slag into the refractory is determined by cutting the sample in half.

Sulfide Resistance of Ceramic Decorations on Glass, C-777

Decorated ware is immersed into a solution of acetic acid, sodium sulfide, and distilled water at room temperature for 15 minutes such that only half the decoration is covered by the test solution. The results are reported as visually observed deterioration of the decoration.

Evaluating Oxidation Resistance of Silicon Carbide Refractories at Elevated Temperatures, C-863

The volume change of one-fourth of a 9-inch straight is evaluated in an atmosphere of steam and at any three temperatures of 800, 900, 1000, 1100, and 1200°C. The duration of the test is 500 hours. In addition to the average volume change of three samples, any weight, density, or linear changes are also noted in this standard method.

Lead and Cadmium Release from Porcelain Enamel Surfaces, C-872

Samples cut from production parts or prepared on metal blanks under production conditions are exposed to 4% acetic acid at 20–24°C for 24 hours. Samples 26 cm^2 are placed into a test cell similar to the one used in C-283 and covered with 40 mL of solution for each 6.45 cm^2 of exposed surface area. The Pb and Cd released into solution are determined by atomic absorption spectrophotometry.

Rotary Slag Testing of Refractory Materials, C-874

This standard practice evaluates the resistance of refractories to flowing slag by lining a rotary furnace, tilted at 3° axially toward the burner, with the test samples. The amount of slag used and the temperature and duration of the test will depend upon the type of refractory tested. The results are reported as the percent area eroded.

Lead and Cadmium Extracted from Glazed Ceramic Tile, C-895

This standard method determines quantitatively by atomic absorption the amount of lead and cadmium extracted from glazed ceramic tile when immersed into 4% acetic acid solution at 20–24°C for 24 hours.

Lead and Cadmium Extracted from Lip and Rim Area of Glass Tumblers Externally Decorated with Ceramic Glass Enamels, C-927

This standard method determines quantitatively by atomic absorption the amount of lead and cadmium extracted from the lip and rim area of glass tumblers when immersed into 4% acetic acid solution at 20–24°C for 24 hours.

Alkali Vapor Attack on Refractories for Glass-Furnace Superstructures, C-987

This standard practice evaluates the resistance to alkali attack of refractories by placing a 55 mm square by 20 mm thick sample over a crucible containing molten reactant such as sodium carbonate at 1370°C. A duration at test temperature of 24 hours is recommended, although other times can be used to simulate service conditions. The results are reported as visual observations of the degree of attack.

Length Change of Hydraulic-Cement Mortars Exposed to a Sulfate Solution, C-1012

Samples are tested in a solution of Na_2SO_4 or $MgSO_4$ in water (50 g/L) at 23°C for times initially ranging from 1 to 15 weeks. Extended times may be used if required. The percent linear expansion is recorded.

Lead and Cadmium Extracted from Glazed Ceramic Cookware, C-1034; Withdrawn in 2001

This standard test method determines quantitatively by atomic absorption the amount of lead and cadmium extracted from glazed ceramic cookware when immersed into boiling 4% acetic acid solution for 2 hours.

Chemical Resistance and Physical Properties of Carbon Brick, C-1106

At least three 2-inch cubes per test medium and per test temperature are immersed into approximately 150 mL of the desired test liquid. The closed containers are placed into a constant temperature oven or bath and then examined after 1, 7, 14, 28, 56, and 84 days. The samples are evaluated for weight change and compressive strength change.

Quantitative Determination of Alkali Resistance of a Ceramic-Glass Enamel, C-1203

The chemical dissolution of a ceramic-glass enamel decorated glass sample is determined by immersing it into a 10% alkali solution near its boiling point (95°C) for 2 hours. The dissolution is determined by calculating the difference in weight losses between the decorated sample and an undecorated sample, normalized for the differences in areas covered and uncovered by the decoration.

Static Leaching of Monolithic Waste Forms for Disposal of Radioactive Waste, C-1220

The chemical durability of a simulated or radioactive waste form is evaluated at temperatures of 40, 70, and 90°C in high-purity water and silicate/bicarbonate and brine solutions. Test duration ranges from 7 days to 1 year or longer.

Chemical Durability of Nuclear, Hazardous, and Mixed Waste Glasses and Multiphase Glass-Ceramics: The Product Consistency Test (PCT), C-1285

Evaluation of chemical species released to a test solution from homogeneous, phase-separated, and devitrified glasses, and glass-ceramic waste forms. Samples are crushed glasses tested under varying test conditions.

Accelerated Leach Test for Diffusive Releases from Solidified Waste and a Computer Program to Model Diffusive, Fractional Leaching from Cylindrical Waste Forms, C-1308

This test evaluates the leaching rates from a solidified matrix, determines if the leach rates are mass diffusion controlled, determines diffusion constants, and verifies projected long-term release.

Determining the Chemical Resistance of Aggregates for Use in Chemical-Resistant Sulfur Polymer Cement Concrete and Other Chemical-Resistant Polymer Concretes, C-1370

This standard test method determines the chemical resistance of at least three 200 gm samples of aggregate immersed into 400 mL of the desired solution, covered, and held at 60°C for 24 hours. The resistance to attack is determined by the change in weight during the test.

Atmospheric Environmental Exposure Testing of Nonmetallic Materials, G-7

This standard practice evaluates the effects of climatic conditions upon any nonmetallic material. Samples are exposed at various angles to the horizon and generally are faced toward the equator. It is recommended that temperature, humidity, solar radiation, hours of wetness, and presence of contaminants be recorded.

Performing Accelerated Outdoor Weathering of Nonmetallic Materials Using Concentrated Natural Sunlight, G-90

This standard practice describes the use of a Fresnel-reflector to concentrate sunlight onto samples in the absence of moisture. A variation in the procedure allows the spraying of purified water at regular intervals on the samples.

Nonstandard Tests

Many individual laboratories use test procedures that are similar to ASTM standard procedures; however, they have been modified to suit their own particular needs or capabilities. Even though a particular ASTM test was developed for a certain material under specific conditions, it does not imply that other materials cannot be tested in the same manner. For example, C-621 for corrosion of refractories by molten glass could be used to test non-refractories by various other liquids. A variation of this test has been used by some glass technologists where the refractory samples are rotated to simulate a forced convection situation. The real problem with this test is that one generally does not know the glass velocity distribution along the sample with sufficient accuracy to extrapolate laboratory results to commercial furnaces. A more appropriate test to evaluate forced convection upon dissolution is the rotating disk test, shown in Figure 11.1. In this setup the diffusion boundary layer across the lower disk face has a constant value for any experimental temperature and rotational velocity. The dissolution of the solid disk is therefore constant, a situation that does not occur in the finger test (see also Chapter 1 section on Attack by Molten Glasses). Any test that is used should be subjected to ruggedness testing first to determine the important variables.

It is almost impossible to test the corrosion of ceramics and maintain all samples equivalent, since variations in density and porosity are generally present. Thus it is important to test more than one sample under a particular set of conditions and average the results or normalize the test results to constant porosity.

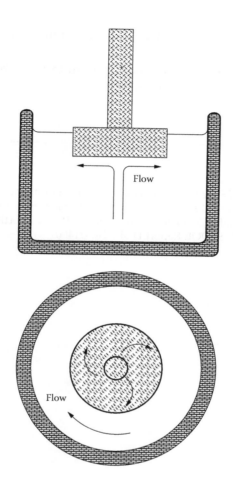

FIGURE 11.1
Rotating disk setup.

Additional Recommended Reading

Baboian, R. (ed.). 2004. *Manual 20, Corrosion Tests and Standards*, 2nd edition, ASTM, Philadelphia, 882 pp. (mostly metals).

Exercises, Questions, and Problems

1. Why is it important to determine the factors that cause variation during testing? What ASTM Standard addresses these factors?

2. What is the difference between a standard method and a standard practice?

3. Are there situations when a nonstandard test may be used? If yes, what precautions should be taken?

References

ASTM, 1986. *Form and Style for ASTM Standards*, 7th edition, Philadelphia, PA, March.

Mendel, J.E. (compiler). 1981. *Nuclear Waste Materials Handbook-Waste Form Test Methods*, Materials Characterization Center, Pacific Northwest Laboratories, Richland, WA, U.S. DOE Report DOE/TIC-11400.

12

Characterization Methods

> Every experiment proves something. If it doesn't prove what you wanted it to prove, it proves something else.
>
> **Anon**

Introduction

According to Weisser and Bange 2000 it was Lavoisier who in 1770 first recorded the aqueous corrosion of a silicate glass predominantly by use of an analytical balance. The analysis of corrosion has been changing over the years with the greatest changes probably taking place within the last 30 years. These changes have been due mostly to the availability of sophisticated computerized analytical tools. It has taken many years for investigators to become familiar with the results obtained and how to interpret them. In some cases, special sample preparation techniques had to be perfected. Although one could conceivably employ all the various characterization methods described below, in most cases only a few are needed to obtain sufficient information to solve a particular problem. The determination of the overall mechanism of corrosion requires a thorough detailed investigation using several characterization methods. Many times, though, the investigator has a limited amount of time and/or funds to obtain his data and thus must rely on a few well-chosen tools. It should be obvious that considerable thought should be given to the selection of samples, test conditions, characterization methods, and interpretation of the results, especially if the data are to be used for prediction of lifetimes in actual service conditions.

Multiple technologies today are being integrated together to give the investigator a wealth of data from one instrument. For example, AFM, confocal Raman and fluorescence microscopy, and scanning near-field optical microscopy (SNOM) are integrated into one platform yielding physical and chemical information on the nanometer scale.[*]

Rather than discussing characterization methods instrument-by-instrument the following discussion is grouped by category of the type

---

[*] http://www.ntmdt.com.

of information desired. In some cases this may not be optimal since some individual instruments offer multiple techniques, as mentioned above.

Chemical Analysis

Bulk Analysis

The bulk chemical analysis of a corroded material is a widely used tool in the evaluation of corrosion. In most cases it is the minor constituents that will be most important. It may even be necessary to examine the trace element chemistry. When corrosion has taken place through reaction with a liquid, it is important to analyze the chemistry of the liquid. In this way, it is possible to establish whether it is the bulk or the bonding phases that are being corroded. Probably one of the most sensitive methods in analytical chemistry is that of spectrophotometry. Gajić-Krstajić et al. 2004 determined very low concentrations of ruthenium in 0.5 mol/dm^3 (1 dm = 100 mm) H_2SO_4 solution at high anodic potentials after the corrosion of Ti/RuO_2 electrodes. Concentrations as low as 8×10^{-10} mol/dm^3 were determined. In another method Schmidt and Rickers 2003 determined the concentration of chemical species in corroding fluids and melts by synchrotron radiation x-ray fluorescence (SR-XRF). The studies of Schmidt and Rickers are quite interesting, since they were performed *in situ* at pressures up to 1.1 GPa and temperatures up to 800°C in a hydrothermal diamond-anvil cell.

A chemical analysis that is normally not done is that of the gaseous phases produced during corrosion. This is not an easy task for large-scale experiments but can be accomplished on the micro scale, such as that done with the aid of a thermobalance connected to a gas chromatograph, mass spectrometer, or infrared absorption spectrometer.

Fourier transform infrared (FTIR)* spectroscopy is a technique that uses the absorption of infrared energy to create a molecular fingerprint of the sample. The absorption peaks correspond to the vibrational frequencies of the mineral-specific chemical bonds between the atoms of the material. Various modes of vibration include stretching, bending, and rotation, all of which are unique to a particular material, thus making positive identification possible. The latest instruments are self-calibrating and are therefore very accurate and reproducible. An interesting study involving the use of FTIR in the analysis of mortars can be found in Silva et al. 2005.

---

* Another meaning for FTIR is Frustrated Total Internal Reflection. This technique is used in multitouch sensing devices, such as fingerprint image acquisition.

Surface Analysis

Since corrosion takes place through reaction with the surface of a material, it is easier to determine mechanisms when the chemistry of the surfaces involved are analyzed. In this way one may no longer be confronted with evaluation of minor constituents and trace elements, since the corrosive reactants and products are more concentrated at the surface. The only drawbacks to surface analysis are that of the cost of the equipment and the necessity of a skilled technician. Secondary ion mass spectroscopy (SIMS) is a technique that currently receives wide use, since it provides element detection limits in the sub-ppm range and very good spatial resolution. Profiling of the various elements, another form of surface analysis, in question can be a very enlightening experiment. In this way, the depth of penetration can be determined and the elements that are the more serious actors can be evaluated. Although somewhat dated, the article by Lodding 1992 has provided an excellent review of the use of SIMS to the characterization of corroded glasses and superconductors.

Determination of surface structures of ceramics for corrosion studies is most likely best accomplished by techniques such as Auger photoelectron diffraction (APD), x-ray photoelectron diffraction (XPD), scanning electrochemical microscopy (SECM), or atomic force microscopy (AFM). These techniques are all useful for studies of surfaces or interfaces on the submicron level. The investigator must be aware of the sample requirements for each of these techniques, which is true for all techniques, especially if there may be any interaction between the sample and the environment of the equipment. For example, the SECM requires the sample to be immersed into an electrolyte solution. Other techniques are available (e.g., LEED), but they are better suited to other materials or suffer from various limitations. Gibson and LaFemina 1996 offer an excellent discussion of the various aspects of mineral surfaces and how these affect dissolution.

Microstructure and Phase Analysis

Visual Observation

The most obvious method of analysis is that of visual observation. The human eye is excellent at determining differences between a used and an unused ceramic. Such things as variations in color, porosity, and texture should be noted. If no obvious changes have taken place, one should not assume that no alteration has occurred. Additional examination on a much finer scale is then required. Many times visual observation can be misleading. For example, a sample may exhibit a banded variation in color, indicating a possible chemical variation. On closer examination, however, the color differences may be due only to porosity variation. An aid to visual observation is the dye penetration test. In this method a sample is immersed into a solution such as methylene blue and then examined under a stereo microscope.

Optical Microscopy

A complement to visual observation is that of optical microscopy. Many people have devoted their entire lives to the study of ceramic microstructures through the examination of various sample sections and the use of some very sophisticated equipment. A preliminary examination should be conducted with a stereo microscope and photographs taken. It is sometimes difficult to remember what a particular sample looked like after it has been cut into smaller pieces and/or ground to a fine powder for further analysis. A photographic record solves that problem.

The ceramics community has fallen into the habit of making only polished sections for observation by reflected light, when a tremendous amount of information can be obtained by observing thin sections with transmitted light. This trend has been brought about by the presence of many other pieces of equipment. Polished sections must be supplemented by x-ray diffractometry (XRD) and also energy dispersive spectroscopy (EDS) and/or scanning electron microscopy (SEM) to obtain a full identification. A full identification can be made, however, with the use of thin sections. The only drawback is that an expert microscopist is required who understands the interaction of polarized and unpolarized light with the various features of the sample. It is true that the preparation of a thin section is more tedious than that of a polished section but with today's automatic equipment there is not much difference. In addition, a thin section does not require the fine polishing (generally down to submicron grit sizes) that a polished section does. The problem of pullouts does not interfere with the interpretation of the microstructure in transmitted light like it does in reflected light. The major drawback of a thin section is that it must be not greater than one crystal thick. Historically this has been 30 μm[*], but newer advanced ceramics contain much finer crystal sizes. With today's advanced ceramics being produced from submicron-sized powders, many products do not lend themselves to thin section examination. In those cases, polished sections must suffice. Since the smallest grain size observable by optical microscopy is on the order of 0.5 μm, samples composed of nano-sized grains yield little information when examined by polished or thin section. A technique that allows one to observe features down to the nanometer level is that of atomic force microscopy (AFM). With the use of AFM, for example, surface morphologies can be observed during etching processes of the exact same area. AFM also allows observation of samples in contact with liquids, something that is not

---

[*] The 30 μm thickness was first used by petrologists in the study of rocks by thin section.

possible with conventional techniques. Watari 2005 used AFM to observe the continuous etching of human teeth in various acids.

Composite and radioactive samples present special problems in preparation for optical microscopy. The differences in hardness are the most obvious problem, especially when metals are present, since softer materials tend to smear over harder ones. Gammon and Hayes 2001 have reported methods for preparation of a carbon-fiber-reinforced polymer. The preparation of radioactive samples must be done in remotely operated hot cells. Oberländer et al. 2001 have reported a technique for preparation of these radioactive samples.

The preservation of images is very important not only for communication and the preparation of printed reports but also for historical records and for comparison studies. With modern-day digital microscopy the investigator has the option to save images in several different formats. Knowledge of these various formats and what they contain is especially important for archiving. An excellent review of this topic was given by Ying and Cheung 1999. The most common formats used are BMP, GIF, JPEG, and TIFF,[*] with TIFF probably being the most popular. Ying and Cheung concluded that for archiving and quantitative analysis one should use formats that do not remove any information (e.g., RAW file format). Some of these formats do contain algorithms for uncompressing the image files.

One major advantage of the light microscope over most electron microscopes is the ability to observe dynamic processes. Time-lapse video microscopy can be used to follow real-time corrosion processes. Obviously, room-temperature processes and those in aqueous media are the easiest to observe. Much of the latest work in the area of video microscopy has taken place in cell biology. Anyone interested in additional reading in this area should read the book by Cherry 1991. The newer[†] environmental scanning electron microscope (ESEM) can now do everything that was previously done by optical microscopy and even on a smaller scale.

Electrochemical Impedance Spectroscopy

Electrochemical impedance spectroscopy (EIS), Lasia 1999, a form of dielectric spectroscopy, measures the dielectric properties of a medium as a function of frequency. EIS can reveal the energy storage and dissipation properties of a material. Xu et al. 2010 utilized EIS in an attempt to characterize the bacterial demineralization of human dentine. This is difficult to detect by other methods since the dentine is covered by an envelope of enamel, which demineralizes much slower than dentine. Since the EIS method could detect

---

[*] BMP—bitmap (Windows based); GIF—graphics interchange format; JPEG—joint photographic experts group; TIFF—tagged image file format.
[†] The first commercial ESEM was available in 1988.

only porosity changes in the dentine, other methods such as SEM/EDX, XRD, etc., were necessary to fully characterize the demineralization process.

Nuclear Magnetic Resonance

Nuclear magnetic resonance (NMR) is not commonly used for corrosion analysis. It can provide information, however, on lattice defects, such as stacking faults. Large numbers of lattice defects can enhance the crystal surface reactivity, leading to increased dissolution. It is not necessary to examine samples of long-range order, since NMR spectra are influenced by the local environment of only a few spheres of coordination. Thus it is a good technique for those samples of low crystallinity.

Scanning Electron Microscopy/Energy Dispersive Spectroscopy

If an evaluation of the corroded surface is required and one does not want to destroy the sample totally, then an examination by scanning electron microscopy/energy dispersive spectroscopy (SEM/EDS) can yield valuable information. With most ceramics, however, the sample requires a conductive coating of carbon or gold before examination. In the use of field emission scanning electron microscopy (FESEM) most samples can be examined without the conductive coating, since the instrument operates at very low accelerating voltages (generally <3 kV). Materials that exhibit some conductivity (e.g., SiC) work quite well in the FESEM. Even insulating materials offer excellent results at the low accelerating voltages used. Within the last 10 years the FESEM has become the workhorse for materials characterization. In addition, the FESEM offers a higher spatial resolution on the nanometer level.

If the same sample is to be used for both optical reflected-light microscopy and SEM/FESEM, the optical work should be done first. Quite often the polished section prepared for optical examination is too large for the SEM/FESEM and the coating required for SEM may interfere with optical examination.

Chemical analysis by EDS can be quite useful in identifying phases observed in reflected-light optical microscopy. Even though the resolution of topographic features can be as good as several hundred angstroms in the SEM, the resolution of the EDS data is generally on the order of one micron. The EDS data also come from a small volume of sample and not just the surface. This may lead to the EDS signal originating from several overlapping features and not just what one observes from the topographic features. Although SEM/FESEM can be performed on as-received or rough surfaces, EDS is best performed on polished or flat surfaces. The analysis by SEM/EDS in combination with XRD and optical microscopy is a powerful tool in the evaluation of corrosion. See Figure 15.3, which shows optical, SEM/EDS, and XRD data for the corrosion of a mullite refractory, and the corresponding text for an example of the use of EDS in phase identification.

Transmission Electron Microscopy

Transmission electron microscopy (TEM) can be used to evaluate the corrosion effects upon grain boundary phases. It is also useful for the identification of lattice defects. TEM is a very time-consuming method and quite often the samples are not representative due to their small size (several mm or less) and the thinning process. TEM does not lend itself to the observation of porous samples and thus is confined to observation of dense regions of corroded samples.

X-Ray Diffractometry

Phase analysis is normally accomplished through the use of x-ray diffractometry (XRD), although optical microscopy can also be used. XRD is generally best done on powdered samples; however, solid flat surfaces can also be evaluated. Generally a sample of about one and one-half grams is necessary but sample holder designs vary considerably and various sample sizes can be accommodated. Solid flat samples should be on the order of about one-half inch square. Powder camera techniques are available that can be used to identify very small quantities of powders. In multiphase materials, the minor components must be present in amounts greater than about 1–2 wt% for identification. Once the mineralogy of the corroded ceramic is known, a comparison with the original uncorroded material can aid in the determination of the mechanism of corrosion.

Although quantitative XRD can be performed, the accuracy is dependent upon sample preparation (crystal orientation plays a major role), the quality of the standards used, and the care taken in reducing various systematic and random errors. Several articles have been published in the literature that the interested reader may want to consult before taking on the task of quantitative XRD, Alexander and Klug 1948, Chung 1974, Dickson 1969, and Brime 1985. The one by Brime 1985 is especially good since it compares several techniques.

Although the author is unaware of the use of high-temperature XRD in the evaluation of corrosion, there is no technical reason why it could not be useful. The major problem with high-temperature XRD is the identification of multiple phases at temperatures where the peaks become sufficiently broadened to obscure one another.[*]

---

[*] High-temperature XRD also has many problems related to sample holders, sample temperature determination, and thermal expansion effects. Anyone considering this technique should have ample time to obtain results.

Physical Property Measurement

Gravimetry and Density

The evaluation of weight change, either positive or negative, during a reaction in many cases is sufficient to determine that corrosion has taken place. Weight change in itself, however, is not always detrimental. In the case of passive corrosion, a protective layer forms on the exposed surface, generally as a weight gain. This would indicate that corrosion had taken place, but it is not necessarily detrimental, since the material is now protected from further corrosion.

If at all possible, one should perform weight change experiments in a continuous manner on an automated thermal analyzer (i.e., TGA) rather than performing an interrupted test where the sample is removed from the furnace after each heat treatment and weighed. In the interrupted test one runs the risk of inaccurate weight measurements due to handling of the sample. Mitra et al. 2004 stated, "Although the weight measurements were taken before and after oxidation, yet the recordings of weight gain were inaccurate due to spallation." In addition, the change in temperature may cause mechanical disruption due to expansion and contraction that may lead to crumbling of the sample. Keuleers et al. 2002 reported on the comparison of methods for determination of activation energies of thermal decomposition by gravimetric methods, TGA and modulated TGA. They also discussed the problems associated with side reactions. Thus they concluded that one must fully understand the decomposition process before evaluating activation energies.

Density measurements are another form of gravimetry, but in this case the volume change is also measured. Many times volumetric changes will take place when a material has been held at an elevated temperature for an extended time. This implies that additional densification or expansion has taken place. Additional densification, although not necessarily a form of corrosion, can cause serious problems in structural stability. Expansion of a material generally implies that corrosion has taken place and that the reactions present involve expansion. Again these may not be degrading to the material but may cause structural instability.

One must exercise care in comparing density data obtained by different methods. Generally the apparent density obtained from helium pycnometry is slightly higher than that obtained from water absorption. For example, the data for a sample of fusion cast α/β alumina gave 3.47 g/cc by water absorption compared to 3.54 g/cc by helium pycnometry.* Helium pycnometry lends itself to the determination of densities of corroded samples. Helium,

---

* Data from author's laboratory.

being a gas, has better access to very small pores, crevices, cracks, etc., than does water.

Mechanical Property Tests

Probably the most widely used mechanical property test is that of modulus of rupture (MOR). One generally thinks of corrosion as lowering the strength of a material; however, this is not always the case. Some corrosive reactions may in fact raise the strength of a material. This is especially true if the MOR test is done at room temperature. For example, a high-temperature reaction may form a liquid that more tightly bonds the material when cooled to room temperature. A method that is often used is first soaking the samples in a molten salt and then performing an MOR test. This evaluates both the high-temperature strength and the effects of corrosion upon strength. Long-term creep tests or deformation under load tests can yield information about the effects of alteration upon the ability to resist mechanical deformation. There is a lot of literature data for hot MOR tests of materials such as Si_3N_4. In this technique the temperature is raised to some set value and held for a short time for equilibration, and then the MOR test is done. The temperature is then raised to the next level and the MOR test repeated. For a more detailed discussion of the effects of corrosion upon mechanical properties see Chapters 18 and 19.

Porosity—Surface Area

The evaluation of the porosity of a corroded sample generally presents the investigator with a rather difficult task. Most often the best method is a visual one, since porosity differences generally reveal themselves as banding of either various shades or of different colors. Determination of the variations in pore size distribution in different zones of the sample may be a significant aid to the analysis. With modern computerized image analysis systems, one has the capability of evaluating porosity and pore size distributions rather easily, Exner and Hougardy 1988. One must be aware of the fact that sample preparation techniques can greatly affect the results obtained by image analysis. Grinding and polishing can produce pullouts, rounding of grain edges, and smearing of softer phases over harder ones, all of which affect the final image.

The determination of the porosity of an uncorroded specimen, however, is extremely important in determining the surface area exposed to corrosion. Two samples identical in every way except porosity will exhibit very different corrosion characteristics. *The one with the higher porosity or exposed surface area will exhibit the greater corrosion.* This is therefore not a true test of corrosion but is valuable in the evaluation of a particular as-manufactured material. Not only is the value of the total volume of porosity important but the size distribution is also important. Different environments can access a different set of pore sizes and thus produce different results.

The porosity test by water absorption is not sufficient, since the total porosity available for water penetration is not equivalent to the total porosity available for gaseous penetration. Although water absorption is a convenient method to determine porosity, it yields no information about pore size, pore size distribution, or pore shape. Mercury intrusion, however, does yield information about pore size distribution in the diameter range between 500 and 0.003 μm. *One must remember that the size distribution obtained from mercury intrusion is not a true size distribution but one calculated from an equivalent volume.* By assuming the pores to be cylindrical, one can calculate an approximate surface area from the total volume intruded by the mercury. *A sample that has been used for mercury intrusion should not be subsequently used for corrosion testing, since some mercury remains within the sample after testing.* For applications involving gaseous attack, a method that measures gas adsorption may be more appropriate, such as the permeability test that better evaluates the passage of gas through a material. Permeability tests, however, are not as easy to perform as porosity tests. A major problem with the permeability test is sealing the edges of the sample against gas leakage.

Determination of the surface area directly by gas adsorption or indirectly by mercury intrusion may not correlate well with the surface area available to a corrosive liquid, since the wetting characteristics of the corrosive liquid is quite different from that of an adsorbed gas or mercury. Thus one should exercise caution when using data obtained by these techniques. See Chapter 8 for a more detailed discussion of pore size determination.

Reaction Rates

The rate of corrosion is probably the parameter of greatest concern to production engineers. How long will a particular item last? This parameter can be quite elusive and is one that takes great care in determining. The goal is to determine the amount or concentration of a particular chemical species or compound at some particular time.

The reader is referred to Chapter 7 on kinetics for a review of the fundamentals of thermogravimetric analysis (TGA). Probably the most important parameter that is uncertain in nonisothermal studies is the temperature of the sample. The enthalpy of the reaction is often sufficient to raise or lower the sample temperature by as much as 1000°C. This fact is overlooked or unavailable if one uses nonisothermal thermogravimetry, which is most often the case. Differential thermal analysis (DTA) and differential scanning calorimetry (DSC) may be more appropriate than TG, since these techniques either determine the sample temperature or maintain the sample at a constant temperature relative to a reference material. A thorough understanding of the various effects that the sample characteristics, machine operation, etc., have upon the kinetics is important along with all the various assumptions that may have been made by the software programmer to use that data to calculate the kinetic parameters.

The one area where reaction kinetics is of great concern is that of radio-nuclide release from spent nuclear fuel. The repository redox conditions can greatly affect the release of radionuclides. In oxidizing conditions the solubility of UO_2 is about five orders of magnitude greater than in reducing conditions, He et al. 2009. He et al. studied the reaction kinetics by scanning electrochemical microscopy (SECM). The scanning probe, in addition to mapping topography, evaluates the electrochemical reactivity of surface species on the substrate. The current variation of the small electrode tip near the sample surface represents changes in topography or conductivity (i.e., reactivity). He et al. concluded that corrosion was enhanced at nonstoichiometric sites (probably grain boundaries). SECM is useful for evaluating chemistry in high resolution near interfaces.

When the corrosion process can be represented by a redox reaction[*] (i.e., galvanic corrosion), polarization methods can be used to determine reaction rates, gain information about mechanisms, and determine the susceptibility of certain materials to corrosion in a particular environment. Anodic or cathodic polarization can be used by changing the potential at either of the electrodes or cyclic polarization can be used by cycling between the two. The most common method, called potentiodynamic polarization, is where the potential of an electrode is varied at a predetermined rate by applying current through the electrode. Cyclic polarization is often used to evaluate pitting (see ASTM Standard G61). The presence of hysteresis is indicative of pitting and the size of the loop is related to the amount of pitting. Although generating scans with the various equipment available is relatively easy, interpreting the scans can be quite difficult. Three electrodes are used: the working electrode made of material of interest, a counter electrode made of an inert material (e.g., platinum), and a stable reference electrode. These techniques were first used in the early 1960s by the metallurgical field and are used extensively today. Although not common in the general ceramics field, potentiometric methods are used to evaluate protective ceramic coatings on metals, Ibrahim et al. 2002 and Yao et al. 2006.

Electrochemical impedance spectroscopy (EIS) often reveals information about the reaction mechanism of an electrochemical process. At certain frequencies different reaction steps will dominate, and the frequency response shown by EIS can help identify the rate-limiting step. A mathematical development of the fundamental equations of EIS can be found in Lasia 1999.

Thermal Property Measurement

An obvious thermal property related to corrosion that may be of interest is the melting point. Melting points of single cation oxides and nonoxides can

---

[*] This requires that the ceramic be electrically conductive and/or contain multivalent cations.

be found in various handbooks or databases, but mixed cation compounds may be difficult to find. In this case one must use a technique such as differential thermal analysis (DTA) or differential scanning calorimetry (DSC) to determine the melting point. A hot-stage microscope could also be used to determine melting points. Hot-stage microscopy is more dependent upon the operator than DTA or DSC. These techniques can also be used to determine polymorphic transition temperatures.

Calorimeters of various types have been used to determine heats (enthalpy) of dissolution. Differential scanning calorimeters, accelerated rate calorimeters, isothermal microcalorimeters, and titration calorimeters are among the most common types used. In many cases only very small quantities of material (i.e., milligrams) are available for examination. In those cases the microcalorimetric techniques are preferable, Bastos et al. 2003. The accuracy of a micro-sized sample, however, is lower than that of macro-sized ones.

Additional Recommended Reading

Cherry, R.J. (ed.). 1991. *New Techniques of Optical Microscopy and Microspectroscopy*, 1 vol. in *Topics in Molecular and Structural Biology*, Neidle, S. and Fuller, W. (series eds.), CRC Press, Boca Raton, FL.

Gibson, A.S. and LaFemina, J.P. 1996. Structure of Mineral Surfaces, pp. 1–62 in *Physics and Chemistry of Mineral Surfaces*, P.V. Brady (ed.), CRC Press, New York, NY.

Mason, C.W. 1983. *Handbook of Chemical Microscopy*, Vol. 1, 4th edition, Wiley, New York.

Riga, A.T. and Patterson, G.H. (eds.). 1997. *Oxidative Behavior of Materials by Thermal Analytical Techniques*, ASTM STP 1326, ASTM, West Conshohocken, PA, 247 pp.

Scully, J.R., Silverman, D.C. and Kendig, M.W. (eds.). 1993. *Electrochemical Impedance: Analysis and Interpretation*, ASTM STP 1188, ASTM, Philadelphia, PA, 480 pp.

Wachtman, J.B. 1993. *Characterization of Materials*, Butterworth-Heinemann, Boston, MA.

Zipperian, D.C. 1993. Microstructural Analysis Using Image Analysis, pp. 631–51 in *Ceramic Transaction, Vol. 38, Advances in Ceramic-Matrix Composites*, Bansal, N.P. (ed.), Am. Ceram. Soc., Westerville, OH.

Exercises, Questions, and Problems

1. List all the possible techniques that one may use to analyze a corroded sample and the type of information obtained.

2. Describe the differences between laboratory tests and field trials.

3. List the various parameters of a laboratory test that can be scaled from the actual environment and list those that cannot. How will this affect the overall interpretation of the results of a lab test?

4. Discuss the errors that may arise when performing an accelerated laboratory test. In addition, what characteristics of a small lab sample lead to errors compared to the full-size installation?

5. Calculate the increased interface surface exposed by polishing a sample at a 45° taper, if the original perpendicular cross section had a 1 μm thick interface.

6. What parameters are important in the grinding and polishing of a sample and how do they affect the final result?

7. Discuss the information that one may obtain by examining a corroded sample with the unaided human eye.

8. Discuss the importance of the surface area of the corroded sample to the volume of the corroding liquid.

9. How does an interrupted weight change test vs. temperature interfere with the results? How can this problem be overcome?

References

Alexander, L. and Klug, H.P. 1948. Basic Aspects of X-Ray Absorption, *Anal. Chem.*, 20, 886–89.

Bastos, M., Bai, G., Qvarnstrom, E. and Wadso, I. 2003. A New Dissolution Microcalorimeter: Calibration and Test, *Thermochimica Acta*, 405, 21–30.

Brime, C. 1985. The Accuracy of X-Ray Diffraction Methods for Determining Mineral Mixtures, *Mineral. Mag.*, 49(9), 531–38.

Cherry, R.J. (ed.). 1991. *New Techniques of Optical Microscopy and Microspectroscopy*, 1 vol. in *Topics in Molecular and Structural Biology*, Neidle, S. and Fuller, W. (series eds.), CRC Press, Boca Raton, FL.

Chung, F.H. 1974. Quantitative Interpretation of X-Ray Diffraction Patterns of Mixtures: I. Matrix-Flushing Method for Quantitative Multicomponent Analysis, *J. Appl. Cryst.*, 7, 519–25.

Dickson, M.J. 1969. The Significance of Texture Parameters in Phase Analysis by X-Ray Diffraction, *J. Appl. Cryst.*, 2, 176–80.

Exner, H.E. and Hougardy H.P. (eds.). 1988. *Quantitative Image Analysis of Microstructures*, DGM Informationsgesellschaft mbH., Germany, p. 235.

Gajić-Krstajić, Lj.M., Trišović, T.Lj. and Krstajić, N.V. 2004. Spectrophotometric Study of the Anodic Corrosion of Ti/RuO$_2$ Electrode in Acid Sulfuric Solution, *Corrosion Sci.*, 46, 65–74.

Gammon, L.M. and Hayes, B.S. 2001. Microscopy of Composite Materials, Structure 38, *Struers J. Materialography*, 16–18.

Gibson, A.S. and LaFemina, J.P. 1996. Structure of Mineral Surfaces, pp. 1–62 in *Physics and Chemistry of Mineral Surfaces*, Brady, P.V. (ed.), CRC Press, New York, NY.

He, H., Zhu, R.K., Qin, Z., Keech, P., Ding, Z. and Shoesmithl, D.W. 2009. Determination of Local Corrosion Kinetics on Hyper-Stoichiometric UO_{2+x} Scanning Electron Microscopy, *J. Electrochem. Soc., 156*(3), C87–94

Ibrahim, M.A.M., Korablov, S.F. and Yoshimura, M. 2002. Corrosion of Stainless Steel Coated with TiN, (TiAl)N and CrN in Aqueous Environments, *Corrosion Sci., 44*, 815–28.

Keuleers, R.R., Janssens, J.F. and Desseyn, H.O. 2002. Comparison of some methods for activation energy determination of thermal decomposition reactions by thermogravimetry, *Thermochimica Acta*, 385 (1–2) 1–194.

Lasia, A. 1999. Electrochemical Impedance Spectroscopy and Its Applications, in *Modern Aspects of Electrochemistry, Vol. 32*, Conway, B.E., Bockris, J. and White, R.E. (eds.), Kluwer Academic/Plenum, New York, NY, pp. 143–248.

Lodding, A. 1992. Characterization of Corroded Ceramics by SIMS, Ch. 4 in *Corrosion of Glass, Ceramics and Ceramic Superconductors*, Clark, D.E. and Zoitos, B.K. (eds.), Noyes Publications, Park Ridge, NJ, pp. 103–21.

Mitra, R., Khanna, R. and Rao, V.V.R. 2004. Microstructure, Mechanical Properties and Oxidation Behavior of a Multiphase (Mo,Cr)(Si,Al)$_2$ Intermetallic Alloy–SiC Composite Processed by Reaction Hot Pressing, *Materials Sci. and Engr., A382*, 150–61.

Oberländer, B.C., Espeland, M., Solum, N.O. and Bøe, A. 2001. A New Preparation Concept for Radioactive Materialography Samples, Structure 38, *Struers J. Materialography*, 12–15.

Schmidt, C. and Rickers, K. 2003. In-Situ Determination of Mineral Solubilities in Fluids Using a Hydrothermal Diamond-Anvil Cell and SR-XRF: Solubility of AgCl in Water, *Amer. Mineralogist, 88*(2–3), 288–92.

Silva, D.A., Wenk, H.R. and Monteiro, P.J.M. 2005. Comparative Investigation of Mortars from Roman Colosseum and Cistern, *Thermochemica Acta, 438*, 35–40.

Wachtman, J.B. 1993. *Characterization of Materials*, Butterworth-Heinemann, Boston, MA.

Watari, F. 2005. In-Situ Quantitative Analysis of Etching Process of Human Teeth by Atomic Force Microscopy, *J. Electron Micros., 54*(3), 299–308.

Weisser, M. and Bange, K. 2000. Sophisticated Methods Available to Analyze Glass Corrosion, *The Glass Researcher, 9*(2), 16–17, 21.

Xu, Z., Neoh, K.G., Amaechi, B. and Kishen, A. 2010. Monitoring Bacterial-Demineralization of Human Dentine by Electrochemical Impedance Spectroscopy, *J. of Dentistry, 38*, 138–48.

Yao, Z., Jiang, Z. and Zhang, X. 2006. Effect of Na_2SO_4 on Structure and Corrosion Resistance of Ceramics Coatings Containing Zirconium Oxide on Ti-6Al-4V Alloy, *J. Am. Ceram. Soc., 89*(9), 2929–32.

Ying, X. and Cheung, L. 1999. Image File Formats for Digital Microscopy, *The Americas Microscopy and Analysis*, Jan., 15–17.

Section III

Corrosion of Specific Materials

13

Architectural Materials

The most unforeseen circumstances will swamp you and baffle the wisest calculations. Only vitality and plenty of it helps you.

Washington A. Roebling

Cement, Concrete, Limestone, and Marble

All the building materials contain some lime, several being mostly calcium carbonate (limestone and marble*). Thus these calcareous materials are attacked by acids. Generally any aqueous liquid with a pH less than 6 will exhibit some attack. The corrosion of concrete takes place by the leaching of water-soluble salts that are formed by the reaction of the acid and the calcium-containing compounds of the concrete, Webster and Kukacka 1986. For example, the reaction of calcium carbonate with SO_2-containing acids (sulfuric acid has been reported to make up 60–70% of the acidity in acid rain for the northeastern United States) forms calcium sulfate or gypsum ($CaSO_4 \cdot 2H_2O$). The gypsum, being much more soluble than the carbonate, is then washed away. Runoff water has been shown to be more important than the pH of the solution for $pH > 3$. This causes etching of the surface in addition to pitting and scaling. Webster and Kukacka 1986 gave the following mechanisms for the dissolution of concrete:

1. dissolution of hydrated cement compounds,
2. dissolution of anhydrous cement compounds,
3. dissolution of calcareous aggregates in the mix,
4. deposition of soluble sulfate and nitrate salts,
5. formation of new solid phases within pores, and
6. production of stresses from numbers 4 and 5.

---

* Limestone is a sedimentary rock composed mostly of calcite and aragonite from the skeletal remains of marine organisms. Secondary minerals include silica and clay. The Great Pyramid of Giza is constructed entirely of limestone. Marble, in contrast, is a more dense compact variety of limestone containing some dolomite remains of marine organisms.

Generally a combination of the above parameters is operative. In addition there is the deterioration of any reinforcing steel.

The most common soluble salts that are involved in efflorescence are the alkali and calcium and magnesium sulfates, chlorides, nitrates, and carbonates. Sodium sulfate generally leaves a fluffy deposit behind whereas gypsum deposits are more adhesive and scaly. The chlorides are much more mobile in solution than the sulfates and therefore tend to penetrate deeper into the corroding material. The disruptive action of these soluble salts is mostly through hydration-dehydration-rehydration. Thus the damage caused by these salts is due to their having several hydration states. Therefore sodium chloride is not applicable to this type of disruption since it expands only during dehydration, Amoroso and Fassina 1983. Sodium exists in two hydrate forms, the mono- and the decahydrate. Amoroso and Fassina 1983 reported that the monohydrate exists only above 32°C with the decahydrate being the stable form below that temperature. A heptahydrate may also exist at intermediate temperatures.

One area where there is nothing in the technical literature is that of residential home foundations. Many of these homes were built with concrete block foundations (basements) and many were built where the groundwater surrounding the foundation does not drain off rapidly. Rainwater, as it seeps through the soil, dissolves various soluble salts. In soils containing clay or slate (vs. sand) the groundwater can and will find its way to the home foundation. The soil around a foundation is backfill and very porous, thus allowing the water to easily fill that area around the house. Depending upon the amount of rain, the water against the foundation (below ground level) can be several feet high. This water containing dissolved species from the soil is very acidic (from dissolved carbonates and sulfates)[*] and it will attack the concrete blocks, generally near the very bottom of the foundation. Very small pinholes form first with larger pores forming later in the exterior coating. Concrete building blocks contain vertical holes (several inches in size) that can fill with this acidic water (see Figure 13.1). As it seeps through the lower foundation blocks it leaves deposited scales (efflorescence) on the inside of the foundation wall. This efflorescence is direct evidence that the above process has taken place. It takes about 10–15 years for this process to exhibit efflorescence (depending on the amount of rain). Once the efflorescence is present (if the interior walls are painted, scaling and blistering of the paint will occur) moisture will continue to enter the basement, causing additional problems of mildew and mold. Even water on the floor can occur. Sump pumps and dehumidifiers have been the accepted answer to this problem; however, sumps do not remove the water as well as a planned drainage system, either inside or outside of the foundation, would. Over an extended period of time this process will compromise the concrete block foundation, leading eventually to complete failure.

---

[*] Biogenic mechanisms may also be present (see Fundamentals, Chapter 4).

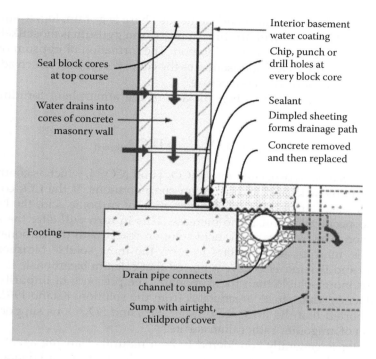

FIGURE 13.1
Water seepage through foundation blocks and method to remove. (H_2O Basement Waterproofing, Inc. 2008; accessed through http://leakfreehomes.com.)

Amoroso and Fassina 1983 discussed the mechanism of limestone deterioration by sulfate pollutant attack. This deterioration starts by the deposition of either dry or wet SO_2. They discuss the effects of atmospheric conditions (e.g., wind) and the limestone surface characteristics upon the degree and rate of deterioration. At low relative humidity the reaction favors the formation of calcium sulfite according to the following equations:

$$CaCO_3 + SO_2 \rightarrow CaSO_3 \cdot 1/2\ H_2O + CO_2 \tag{13.1}$$

or

$$CaCO_3 + SO_2 \rightarrow CaSO_3 \cdot 2\ H_2O + CO_2 \tag{13.2}$$

If water is present, the sulfite can be oxidized to sulfate according to:

$$CaSO_3 \cdot 2\ H_2O + 1/2\ O_2 \rightarrow CaSO_4 \cdot 2\ H_2O \tag{13.3}$$

This later reaction is one thought to be not of major consequence since a large amount of water would not allow a sufficient deposition of dry SO_2 to occur.

Oxidation of SO_2 in an atmospheric aqueous phase (e.g., rain, fog) to sulfuric acid that will then react with the limestone to form gypsum is most likely the predominant mechanism of deterioration. The formation of gypsum on the surface is generally rough and facilitates the collection of dirt, which adds to the visual degradation of the structure.

Rainwater can dissolve carbon dioxide of the atmosphere forming the weak carbonic acid:

$$CO_{2(gas)} + H_2O_{(liquid)} \rightarrow H_2CO_{3\,(liquid)} \qquad (13.4)$$

This then becomes dissociated into [HCO_3^-] and [CO_3^{2-}], which is controlled by the carbonic acid and hydrogen ion concentrations. If the CO_2 concentration increases, increasing the concentration of carbonic acid, the hydrogen ion concentration must also increase. This in turn will alter the pH of the solution. The weak carbonic acid solution reacts with the carbonates of limestones, marbles, and mortars to form the more soluble bicarbonates. Amoroso and Fassina 1983 have reported that calcium bicarbonate is about 100 times more soluble than the carbonate. The presence of impurity ions will influence what phase precipitates from the solution. Kitano 1962a and 1962b reported that the presence of Mg^{2+}, Sr^{2+}, and SO_4^{2-} ions supports the formation of aragonite[*] rather than calcite.

Rainwater is generally considered to be a natural cleansing agent; however, on taller buildings it can wash dirt from higher levels and deposit that dirt at lower levels, especially if the flow of rainwater is slow. Many factors affect the flow of rainwater over a building surface. Some of these are controllable; others are not. The amount of dirt or pollutants in the air along with wind direction and velocity are a few that are not controllable. The texture, porosity, and architectural features (e.g., ledges, corrugations, sculptures) of the building surface are but a few of the controllable factors. All these factors must be considered by the building designer to minimize corrosion problems.

Nitrogen oxides when in contact with water form nitrous and nitric acid. Although nitric acid is a weaker acid than sulfuric acid it can cause more extensive damage to concrete since it reacts with $Ca(OH)_2$ to form the more soluble calcium nitrate. A contributor to various nitrogen oxides not commonly considered is that of bat guano.[†] Bird excrement is also a source of nitrogen oxides. Limestone and calciferous sandstone are the most susceptible to the effects of acids released from bird excrement.[‡] In addition to the nitric acid formed by the reaction of water with the excrement, organisms that live on the excrement can also contribute to stone decay, Bassi and Chiantante 1976.

---

[*] Aragonite is a metastable polymorph of calcite.
[†] Guano consists of ammonia, nitrates, and carbonic, oxalic, phosphoric, and uric acids.
[‡] http://www.buildingconservation.com/articles/birddamage/birddamage.htm.

The dissolution of limestones and marbles by atmospheric corrosion takes place by one or a combination of the following reactions, Skoulikidis 1982:

$$H_2O_{(l)} + CO_{2(g)} \rightarrow H_2^{2+}CO_{3(aq)}^{2-} \tag{13.5}$$

$$CaCO_{3(s)} + H_2^{2+}CO_{3(aq)}^{2-} \rightarrow Ca^{2+}(HCO_3)_{2(aq)}^{2-} \tag{13.6}$$

Since all atmospheres contain carbon dioxide, these reactions will take place in the presence of rainwater. The temperature and the CO_2 partial pressure in the solution were reported by Reddy et al. 1986 also to affect the weathering rate. If any sulfur is present in the atmosphere (generally from industrialization that also increases the carbon dioxide concentration), the following reactions take place:

$$SO_{2(g)} + 1/2\ O_{2(g)} \rightarrow SO_{3(g)} \tag{13.7}$$

$$SO_{3(g)} + 2H_2O_{(l)} \rightarrow H_2^{2+}SO_{4(aq)}^{2-} \tag{13.8}$$

$$CaCO_{3(s)} + H_2^{2+}SO_{4(aq)}^{2-} \rightarrow Ca^{2+}SO_{4(aq)}^{2-} + H_2^{2+}CO_{3(aq)}^{2-} \tag{13.9}$$

These reactions are more severe than those shown in 13.5 and 13.6 above and the subsequent dissolution is more rapid.

$$CaCO_{3(s)} + SO_{3(g)} + 2H_2O_{(l)} \rightarrow CaSO_4 \cdot 2H_2O_{(s)} + CO_2 \tag{13.10}$$

This final reaction is one that leaves a white scum on the surface of the limestone or marble through the evaporation of the water and formation of gypsum. Vanadium oxides (a pollutant contributed by internal combustion engines) can act as catalysts for reaction 13.7, along with colloidal particulates suspended in the atmosphere and photochemical reactions, Skoulikidis 1982. Due to the variations of the catalysts upon the oxidation of SO_2 the solid that forms is generally not completely gypsum ($CaSO_4 \cdot 2H_2O$) but contains as much as 15% $CaSO_3 \cdot 2H_2O$.

Skoulikidis has reported that the sulfation of marble monuments is through galvanic corrosion, where the anode is the environment (SO_2, air, and water vapor), the cathode is marble ($CaCO_3$), and the electrolyte is the gypsum ($CaSO_4 \cdot 2H_2O$) formed during reaction. He also reported that the formation of gypsum was linear up to a thickness of about 300 Å and was thereafter parabolic. This was attributed to the pores of the original film being filled with new material blocking the diffusion of SO_2, O_2, and H_2O to the marble/gypsum interface. Subsequent formation of gypsum (greater than 300 Å) was due to diffusion of Ca^{2+} and CO_3^{2-} to the surface, forming more gypsum at the gypsum/environment interface.

Winkler 1973a has described salt weathering as a primary cause of damage to historical buildings (an example of salt weathering can be seen in Figure 1.2). The damage to porous stone has been attributed to the properties of the salt solution including surface tension, vapor pressure, and viscosity, which control the flow, evaporation, precipitation, and growth within the stone, Puehringer and Engstrom 1985. Important parameters in the damage caused by salt weathering were found by Rodriguez-Navarro and Doehne 1999 to be environmental factors, such as relative humidity, supersaturation ratio, and the kinetics of nucleation and growth. How supersaturation was reached and where it was located were also found to be important. Although several mechanisms have been proposed for salt damage, crystallization due to evaporation appears to be the most important. Rodriguez-Navarro and Doehne 1999 experimentally found that mirabilite and thenardite* were susceptible to subflorescence, and therefore more damaging than halite, which was susceptible to efflorescence.

In the early 1500s the great sculptor and artist Michelangelo was having problems with a form of salt weathering. Although several techniques were used at that time for the creation of frescos,[†] they generally were paintings placed upon wet plaster. Actually the word *fresco* means fresh. A top coat about one-half inch thick of a smooth paste made of lime (actually calcium hydroxide) and sand was placed over a dried plaster wall. This top coat was permeable and provided an excellent surface for the acceptance of pigments and also sealed in the colors as the plaster dried. The mixture of calcium hydroxide and sand once applied to the wall and during the evaporation of the water reacted with atmospheric carbon dioxide, forming calcium carbonate. Thus these frescos when finished were actually paintings locked into essentially limestone or marble. The artists were forced to work rapidly to avoid dried-out plaster. Michelangelo's problem was the result of salts in rainwater that leached through the plaster (from the back), dissolving the calcium carbonate and causing the pigments to blister and flake. Michelangelo solved this problem by mixing a volcanic ash known as pozzolana instead of sand with the lime (there was no need for reaction with atmospheric CO_2), forming a strong bond and an essentially impervious plaster. Michelangelo apparently got this idea from the Romans, who used pozzolana in their extensive construction work. Even with this

---

[*] mirabilite = $Na_2SO_4 \cdot 10\,H_2O$; thenardite = Na_2SO_4, named after Louis Thénard, occurs in five polymorphs.
[†] The art of painting on wet plaster was known as far back as the second millennium BCE. Centuries later the Etruscans and then the Romans decorated walls and tombs with this technique. The art of fresco flourished from the last half of the 13th century (about 1270) in central Italy, King 2003.

new mixture Michelangelo's frescos were prone to efflorescence, which was corrected by the use of less water in the mix, King 2003.

Temperature, pO_2, humidity, the concentration of acidic components, and the condensation of moisture all play a role in the degradation of concrete by flue gases in exhaust stacks of coal-fired power plants, Pavlik et al. 2007. The corrosive reactions are primarily those of gaseous CO_2, SO_2, SO_3, NO_x, and condensates of sulfuric acid with the products of cement hydration. CO_2 converts calcium hydroxide and hydrated calcium silicates and aluminates into calcium carbonate. SO_2, SO_3, and sulfuric acid react with all hydrated calcium cement compounds to form sulfur-bearing compounds ($CaSO_3 \cdot 1/2$ H_2O, $CaSO_4$, $CaSO_4 \cdot 1/2 \, H_2O$, $CaSO_4 \cdot 2H_2O$ (gypsum), $3CaO \cdot Al_2O_3 \cdot CaSO_4 \cdot 12H_2O$, and $3CaO \cdot Al_2O_3 \cdot 3CaSO_4 \cdot 31\text{-}32H_2O$ (ettringite)). The particular compound that forms depends upon the specific conditions present, generally pH and temperature. These reactions are also relevant to historical monuments.

Of great concern to historians is the environmental effect upon national monuments. Researchers have shown some interest in the formation of crusts and patinas[*] for a long time. The formation of these crusts and patinas are from biogenic origin, interaction with atmospheric pollutants, attempts at cleaning and protection, and any combination of these processes. Many marble monuments develop a golden patina on the surface due to the selective dissolution of the calcium contained in marbles with a small content of iron, Skoulikidis 1982. This causes an enrichment of iron compounds at the surface. Garcia-Vallès et al. 1998 defined the crust as the whole surface deposit. This crust is often composed of different layers or patinas of various colors, microtextures, and composition. Each separate patina may also exhibit a microlaminated texture. The monuments studied by Garcia-Vallès et al., all from around the Mediterranean basin, exhibited crusts of predominately calcite and gypsum with additions of calcium oxalates, calcium phosphates, quartz, and small amounts of clay minerals. The local water availability (humidity, runoff, etc.) determines whether a continuous biofilm forms. The local environmental conditions also determine the thickness of each layer. Garcia-Vallès et al. divided the overall crust into two sections: The first, being the outermost, was composed of a gypsum-rich black layer between 40 and 400 µm thick. The second or innermost 10 to 600 µm thick layer was up to 95% calcite with smaller amounts of calcium oxalates and calcium phosphates. The color differences were due to exact location in the monument, the composition of the minerals and organic matter, the organic matter degree of evolution, and the distribution of the organic matter among the various mineral phases. Gypsum dissolution in either dew or raindrops created deliquescent forms and running water determined the color and thickness variations. Where monuments were close to marine environments NaCl was also found in the pores and fissures. In the process of developing

---

[*] Patinas are the thin surface layers of corrosion, generally colored and most often associated with metals (e.g., the green color of the Statue of Liberty).

these crusts some substrate alteration is inevitable. The conversion of any dolomite crystals near the inner patina to calcite occurred in some cases. This process was activated by the dissolution of gypsum from the surface that diffused through the crust to the substrate rock, Evamy 1967. The presence of calcium sulfate in solution enhanced this conversion. In this process of dedolomitization, magnesium sulfate formed as efflorescence and/ or subflorescence.

Clay-Based Materials

The deterioration of brick masonry is a combination of corrosive attack upon the brick and the mortar that is used to bond the brick together. Mortars are generally of a calcareous type and as such their corrosion is very similar to that described above for concrete and limestone. The actual corrosion of the mortar depends upon the type used: clay-based, lime-sand, or Portland cement. The most susceptible to acid rain are those containing calcium of one form or another. The formation of gypsum from the reaction of acid rain upon calcium carbonate produces local stresses sufficient mechanically to disrupt the mortar. Any calcium hydroxide present from the hydration of portland cement is especially susceptible to acid attack. Additional reactions such as the reaction of calcium sulfate with tricalcium aluminate forming *ettringite* ($3CaO \cdot Al_2O_3 \cdot 3CaSO_4 \cdot 32H_2O$) also produces large local stresses due to the associated expansive crystallization.

It has been noted that 12th-century Venetian bricks are more durable than modern ones, even though they were fired at lower temperatures. This has been attributed to a very uniform microstructure, the lack of a highly vitrified surface, and a larger pore size (avoiding the < 1 µm critical pore size) when compared to modern bricks, Charola and Lazzarini 1986. Bricks may also contain Na_2SO_4 if fired in a kiln using sulfur-rich fuel and at a sufficiently low temperature. The sodium sulfate, dissolved by water, can recrystallize after evaporation, forming the anhydrous salt *thenardite* or the decahydrate *mirabilite*. The specific environmental conditions of temperature and relative humidity will change the degree of hydration and the amount of mechanical disruption. The crystallization of sodium sulfate within the pores of various materials causes more damage than any other salt, Flatt and Scherer 2002. At temperatures less than 32°C in rising humidity large volume changes occur going from thenardite (density = 2.68 g/cc) to mirabilite (density = 1.47 g/cc). That is equivalent to almost a doubling of the volume. According to Flatt and Scherer a tensile hoop stress is formed that is sufficient to cause failure. This tensile hoop stress is caused by a radial compressive stress that is exerted by the spherical pore walls on the crystal within the pore. The temperature dependence of the hoop stress is about 1.5 MPa/°C below 32°C. Most stone, brick, and masonry have strengths in the 3–10 MPa range; thus it is easy to see why materials containing thenardite that may get wet can generate sufficient stresses to cause failure.

Ranogajec et al. 2008 showed that the genesis and growth of a fungus on clay roofing tiles was a function of the pore characteristics, surface porosity and roughness, and microstructure. The firing process during manufacture determined these parameters. Roofing tiles fired to 1100°C, forming some mullite from the clay decomposition, exhibited the best resistance to fungus attack. Pores greater than 3 μm that accommodate water expelled from smaller pores provide a medium adequate for microorganism growth. The resistance of the various crystalline phases to microorganism attack was in the order quartz (most resistant), K-feldspar, Na-feldspar, amphibole, dolomite, pyroxene, apatite, calcite, and feldspathoids.

Clay minerals such as chlorite, illite, illite/smectite mixed layers, nontronite, palygorskite, and vermiculite are all reducible by microorganisms with nontronite being the most reducible and illite being the least, Dong et al. 2009. Nontronite is the most reducible due to its structure being the most expandable, which allows easier access of electron donor compounds to the Fe^{3+} sites. In nonexpandable clay minerals Fe^{3+} reduction can be enhanced by electron shuttling compounds. The rate of reduction depends upon the type of microorganism, the specific clay mineral, solution chemistry and pH, and temperature. The reductants are sulfate-reducing bacteria or dissimilatory iron-reducing prokaryotes,[*] also a bacteria.

Sandstone

Mansch and Bock 1994 examined the effects of nitrifying bacteria upon the corrosion of sandstone. Gaseous pollutants, once dissolved in water, react with a sandstone surface forming sulfates, nitrites, or nitrates. Gaseous NO is less soluble than NO_2, both of which are less soluble than SO_2. It was found that nitrifying bacteria preferred sandstones containing calcareous binders (as opposed to a siliceous binder), a pH range of 8.0 to 8.5, and a pore size between 1 and 10 μm. Mansch and Bock concluded that corrosion caused by nitrifying bacteria was more severe than that caused by the gaseous pollutants alone.

Portland brownstone[†] is widely used as a building material, especially in New York City. This material is composed mostly of sandstone but also

---

[*] Dissimilatory Fe^{3+} reduction is the process in which microorganisms transfer electrons to external ferric iron, reducing it to ferrous iron without assimilating the iron. Prokaryotes are a group of organisms that lack a cell nucleus.
[†] Portland brownstone is a variety of sandstone that was quarried in Portland, Connecticut.

contains some albite all in a matrix of kaolinite, mica, iron oxides, illites, and smectites. The deterioration that is exhibited by this material is related to the dilation of the clay with changes in humidity. Gonzalez et al. 2002 studied the stress/strain relationships developed from wetting by use of a novel method that evaluated the warping caused by differential strain due to the expansion of the clay. This differential strain was attributed to nonuniform wetting and was as much as twice the strength of the brownstone.

Glass

For a thorough discussion of the corrosion and especially weathering of glass see Chapter 16.

Torge 2001 reported that the chemical composition of flat glass used for windows in the Middle Ages was highly susceptible to weathering. Medieval stained glass church windows were made of a colored blown flat glass painted with a dark brown enamel. The glass was then cut into various sized pieces that were held together with leaden cames.* Corroded glass of this type exhibited a 1 to 4 mm thick core of unreacted glass, then a thin gel layer 0.01 to 0.1 mm thick, followed on the exterior by a corroded heterogeneous layer several millimeters thick. The exterior corroded layer consisted of gypsum, syngenite (a potassium calcium sulfate), and some amorphous silica. To restore these windows to their original beauty, the corroded exterior layer must be removed. The inner gel layer should not be removed since this provides a barrier to further corrosion.

Additional Recommended Reading

Schaffer, R.J., 2004. *The Weathering of Natural Building Materials*, 208 pp., Donhead Publishing, Shaftesbury, UK.

Winkler, E.M. 1973. *Stone: Properties, Durability in Man's Environment*, 230 pp., Springer-Verlag, New York, NY.

---

* Cames are H-shaped flexible lead strips.

Exercises, Questions, and Problems

1. Discuss the six mechanisms for the dissolution of concrete.
2. What is efflorescence? How does this compare to subflorescence?
3. Why is the content of SO_2 in the atmosphere important for limestone dissolution?
4. Describe the electrochemical setup that represents the galvanic corrosion of marble by sulfation.
5. What is a patina? How is it formed?
6. Describe the characteristics of 12th-century Venetian bricks that make them more durable than modern-day bricks.
7. How does acid rain form and what is its effect upon limestone?
8. Describe salt weathering.

References

Amoroso, G.G. and Fassina, V. 1983. Stone Decay and Conservation, *Materials Science Monographs 11*, Elsevier, Amsterdam, p. 453.

Bassi, M. and Chiantante, D. 1976. The Role of Pigeon Excrement in Stone Biodeterioration. *International Biodeterioration Bull.*, *12*(3), 73–79.

Charola, A.E. and Lazzarini, L. 1986. Deterioration of Brick Masonry Caused by Acid Rain, pp. 250–58 in *Materials Degradation Caused by Acid Rain*, ACS Symposium Series 318, Baboian, R. (ed.), Am. Chem. Soc., Washington, DC.

Dong, H., Jaisi, D.P., Kim, J. and Zhang, G. 2009. Microbe-Clay Mineral Interactions, *Amer. Mineral.*, *94*, 1505–19.

Evamy, B.D. 1967. Dedolomitization and the Development of Rhombohedral Pores in Limestones, *J. Sediment Petrol.*, *37*, 1204–15.

Flatt, R.J. and Scherer, G. 2002. Hydration and Crystallization Pressure of Sodium Sulphate: A Critical Review, pp. 29–34 in *Mat. Res. Soc. Symp. Proc. Vol. 712, Materials Issues in Art & Archaeology VI*, Symp. Nov. 2001, Boston, Vandiver, P.B., Goodway, M. and Mass, J.L. (eds.), MRS, Warnerdale, PA.

Garcia-Vallès, M., Vendrell-Saz, M., Molera, J. and Blazquez, F. 1998. Interaction of Rock and Atmosphere: Patinas on Mediterranean Monuments, *Environmental Geology*, *36*(1–2), 137–49.

Gonzalez, I.J., Higgins, M. and Scherer, G. 2002. Hygric Swelling of Portland Brownstone, pp. 21–27 in *Mat. Res. Soc. Symp. Proc. Vol. 712, Materials Issues in Art & Archaeology VI*, Symp. Nov. 2001, Boston, Vandiver, P.B., Goodway, M. and Mass, J.L. (eds.), MRS, Warnerdale, PA.

King, R. 2003. *Michelangelo and the Pope's Ceiling*, Penguin Books, New York, NY, 373 pp.

Kitano, Y. 1962a. Behavior of Various Inorganic Ions in the Separation of Calcium Carbonate from a Bicarbonate Solution, *Bull. Chem. Soc. Japan*, *35*, 1973.

Kitano, Y. 1962b. Polymorphic Formation of Calcium Carbonite in Thermal Springs with an Emphasis on the Effect of Temperature, *Bull. Chem. Soc. Japan*, 35, 1980.

Mansch, R. and Bock, E. 1994. Simulation of Microbiologically and Chemically Influenced Corrosion of Natural Sandstone, pp. 203–16 in *Microbiologically Influenced Corrosion Testing*, ASTM STP 1232, Kearns, J.R. and Little, B.J. (eds.), ASTM, Philadelphia, PA.

Pavlik, V., Bajza, A., Rouseková, I., Unčík, S. and Dubík, M. 2007. Degradation of Concrete by Flue Gases from Coal Combustion, *Cement and Concrete Research*, 37, 1085–95.

Puehringer, J. and Engstrom, L. 1985. Unconventional Methods for the Prevention of Salt Damage, pp. 241–50 in *Proceedings 5th Inter. Cong. on Deterioration and Conservation of Stone*, Furlan, V. (ed.), LCP, Laussane.

Ranogajec, J., Markov, S., Kiurski, J. and Radeka, M. 2008. Microbial Deterioration of Clay Roofing Tiles as a Function of the Firing Temperature, *J. Am. Ceram. Soc.*, 91(11), 3762–67.

Rodriguez-Navarro, C. and Doehne, E. 1999. Salt Weathering: Influence of Evaporation Rate, Supersaturation and Crystallization Pattern, *Earth Surf. Process. Landforms*, 24, 191–209.

Skoulikidis, T.N. 1982. Atmospheric Corrosion of Concrete Reinforcements, Limestones, and Marbles, pp. 807–25 in *Atmospheric Corrosion*, Ailor, W.H. (ed.), Wiley, New York, NY.

Torge, M., Müller, W. and Adam, K. 2001. Scientific Examinations of Historical Stained Glass, Structure 38, *Struers J. of Materialography*, 25–29.

Webster, R.P. and Kukacka, L.E. 1986. Effects of Acid Deposition on Portland Cement Concrete, pp. 239–49 in *Materials Degradation Caused by Acid Rain*, ACS Symposium Series 318, Baboian, R. (ed.), Am. Chem. Soc., Washington, DC.

Winkler, E.M. 1973. *Stone: Properties, Durability in Man's Environment*, Springer-Verlag, New York, NY, 230 pp.

14

Biologically Corroded Materials

> The scientist is not a person who gives the right answers; he's one who asks the right questions.
>
> **Claude Lévi-Strauss**

Introduction

When one thinks of ceramic materials that are used in the human body the first that comes to mind are dental applications—false teeth, caps, and crowns. The use of ceramics in the body, however, is becoming much more widespread than dental applications. Some of these now include bone and joint replacements, heart valves, eye lenses, and therapeutic glasses for treatment of tumors. The materials that are most often used include hydroxyapatite,[*] alumina, zirconia, bioactive phosphate glasses, and rare earth alumino-silicate glasses. Some applications require the bioceramic to react with and bond to the bone or be resorbed and replaced by the bone. Resorbable materials obviate the need for removal. An interesting review of the history, although dated, of bioceramics can be found in the article by Hulbert et al. 1987.

The first prosthesis was most likely that used for human eye replacement. The earliest artificial eye, made of bitumen paste covered with a thin layer of gold, was discovered in Iran and dated to 2900–2800 BC.[†]

The Venetians in 1579 invented thin shells to cover the eye socket opening. These were obviously very uncomfortable. The first glass sphere used as an eye prosthesis was in 1884 and implanted in the scleral cavity shortly after evisceration. Most are made of acrylic today. The first plastic eye was made by U.S. Army dental technicians in 1943.[‡] Some are made of cryolite[§] glass containing arsenic invented by the Germans in 1835. No data could be found on problems related to the corrosion of artificial ceramic eyes. More recent studies have indicated a desire to implant subretinal micro-photodiode

---

[*] *Hydroxy* is the most commonly used prefix for this mineral, although *hydroxyl* is sometimes used.

[†] http://en.wikipedia.org/wiki/glass_eyes.

[‡] http://www.eyeprosthetics.com (homepage of Denver Optic Co.).

[§] Cryolite—Na_3AlF_6.

arrays[*] to replace degenerated photoreceptor cells in the retina. Although *in vitro* experiments have been promising, *in vivo* tests have shown dissolution of the silica passivation layer on the semiconductor chips. To solve this problem Hämmerle et al. 2002 have investigated coating of the micro-electrodes with TiN with promising results.

Inhaled, ingested, or adsorbed mineral dusts have been studied for over 50 years, but again it is only recently that the interactions with body fluids are becoming understandable. Those minerals that are soluble are generally not a problem unless they contain a toxic chemical species (Pb, Cr, etc.). Those that are inert or nearly so can cause various forms of cancer.

Inert Materials

Alumina and Zirconia

Although alumina and zirconia are considered essentially inert bioceramics, they are truly not 100% inert. The human body considers the implant as a foreign material and isolates it by formation of a fibrous capsule. This encapsulation is apparently not a problem for applications as joint replacements; however, it can be for wear particles formed due to joint movement.

Shikha et al. 2008 studied the effects of nitrogen ion implantation on the dissolution of polycrystalline alumina in Ringer's[†] solution. They found that at low ion doses (10^{15} ions/cm^2) the dissolution resistance increased due to compound formation. At higher doses (10^{17} ions/cm^2) dissolution resistance decreased due to damage accumulation at the surface.

Zirconia for femoral head prostheses[‡] has received some bad press in the late 1990s due to apparently some changes in the manufacturing process. These materials are a toughened partially stabilized zirconia containing yttrium (about 3 mol%). The steam sterilization procedure led to the transformation of the tetragonal polymorph to the monoclinic polymorph causing a roughening of the surface. This roughening caused increased wear on the polyethylene acetabular component and eventual failure.[§]

A new femoral head material called Biolox® delta has been developed by Ceram Tec AG that is a composite of alumina (75%), yttria-stabilized zirconia

---

[*] These are essentially arrays of tiny solar cells first developed by Chow 1993.

[†] Ringer's solution is a solution of generally Na, K, and Ca chlorides and Na bicarbonate in water.

[‡] Ivory, a form of dentine which is about 70% hydroxyapatite, was used as the first femoral head replacement in 1891 in Germany, Gomez and Morcuende 2005.

[§] http://www.fda.gov/MedicalDevices/Safety/AlertsandNotices/PublicHealthNotifications/default.htm.

(24%), and 1% trace elements, mostly Cr_2O_3. This material is an attempt to take advantage of the properties of all the combined materials.

Bioactive Implants

Bioactive Glasses

Bioactive glasses were first discovered by Hench in 1969. The special chemistry of these glasses allowed them to bond to living bone. These Na_2O–CaO–P_2O_5–SiO_2 glasses have been trademarked as Bioglass® and marketed under several other names depending upon the application.[*] *The beneficial effect of these glasses is their controlled release of soluble silicon and calcium ions.* In this way the glass acts as a substrate for the growth of new cells. Newer forms of these glasses have been prepared via sol-gel routes that contain numerous very fine interconnected pores. Dissolution kinetics are a function of the following variables, Hench 2003/04:

1. composition,
2. particle size,
3. pore size distribution, average size, and volume%,
4. surface area,
5. thermal stabilization temperature, and
6. chemical stabilization temperature.

The alumina content of bioactive glasses is very important in controlling the durability of the glass surface. The bioactivity, although dependent upon the bulk composition of the glass, decreased beyond acceptable levels once the alumina content rose above 1.0 to 1.5 wt%, Hench 1993. This same phenomenon was present for glass compositions containing cations such as Ta_2O_5 except higher levels were tolerable (1.5–3.0 wt%).

Bioactive glass and glass-ceramics bond to living bone through an apatite (more specifically hydroxyl carbonate apatite, HCA) layer that forms at the interface. See Table 14.1 for a list of acronyms used in the literature to describe hydroxyapatite. It has been postulated by Kokubo et al. 1991 that silica from the glass provided the sites for nucleation of the apatite and that calcium increased the supersaturation of the fluid with respect to apatite.

According to Pantano et al. 1974, Andersson et al. 1991, and Hench 1998, bioactive glasses react by the following:

---

[*] Bioglass® particulate used in periodontal applications is called Perioglas® and used in orthopedic applications is called NovaBone®.

TABLE 14.1

Acronyms Used to Describe Hydroxyapatite

Acronym	Meaning
HA, HAp, or OHAp	Hydroxyapatite
FHA	Fluorohydroxyapatite
CHA or HCA	Carbonated hydroxyapatite
HCFA	Hydroxl-carbonate-fluorapatite
CDHA	Calcium-deficient hydroxyapatite
GR-HA	Glass-reinforced hydroxyapatite

1. ion exchange (leaching),
2. dissolution, and
3. precipitation.

The three reactions listed above do not necessarily occur in succession but may overlap. Andersson et al. 1991 reported that the leaching of Na, Ca, and P was not selective and that the ratios of these cations in the bulk glass were proportional to that of the solution (a *tris*–citric acid buffer solution) at a pH of 7.2 and 36.5°C.[*] The reaction led to the formation of a silica-rich layer at the interface. Calcium citrate complexes formed that did not allow the Ca,P–rich layer to form, which subsequently crystallized into apatite. In the absence of the citric acid (i.e., in *tris*-HCl buffer solution) the Na and P were less than proportional, whereas the Ca was more than proportional to that of the solution. In this solution apatite formed only after 8 hrs. The accumulation of Ca and P at the interface was attributed to the formation of bonds with non-bridging silica (of a silica-rich layer formed by the leaching) by complexation that stabilized the surface and decreased the leaching.

Hench et al. 1991 developed a more detailed series of bioactive reactions, which are listed below:

1. exchange of Na^+ or K^+ with H^+ or H_3O^+,
2. dissolution of silica forming Si–OH at interface,
3. condensation and repolymerization of SiO_2-rich layer,
4. diffusion of Ca^{2+} and PO_4^{3-} through SiO_2-rich layer forming a CP^+-rich layer on top of the SiO_2-rich layer,
5. crystallization of the amorphous CP-rich layer by incorporating OH^- and $CO_3^=$ or F^- from solution forming a mixed HCA or HCFA[‡] layer, and

---

[*] A pH of 7.2 and a temperature of 36.5°C are very close to accepted human body values of pH 7.4 and 37°C.

[†] CP = calcium phosphate.

[‡] HCA = hydroxyl-carbonate-apatite and HCFA = hydroxyl-carbonate-fluoro-apatite.

6. chemical bonding of biological moieties with growing HCA layer leading to incorporation of collagen fibrils.

The HCA layer was on the order of 30 μm thick and the SiO_2-rich layer was between 100 and 120 μm thick, depending upon the composition of the implant, time, and fit of the implant.

For glasses with greater than 60 wt% SiO_2, the reactions slowed and the CP-rich film did not form and no bonding took place. With SiO_2 contents ranging from 45 to 53 wt%, the surface silanol formation slowed, increasing the time to form the silica-rich layer. Thus the time to form the CP layer increased and the crystallization of HCA were reduced. This happened even though sufficient CaO and P_2O_5 were available for CP film formation. *This indicates that the kinetics of the condensation and polymerization steps are the critical steps in formation of the CP-rich film and subsequent crystallization of HCA.* Any composition changes that decrease the rate of reaction #3 above decreases the rate of HCA layer formation and eventual tissue bonding.

Bioactive glass (53 wt% SiO_2) spheres in a matrix of bioabsorbable polymer was investigated by Niemelä et al. 2005 After immersion in a phosphate buffered solution at $pH = 7.4$ and 37°C for times ranging up to 12 days, they found that a calcium phosphate precipitation formed on the surface of the glass spheres. Although Niemelä et al. 2005 did not determine the particular phase of calcium phosphate present, it appeared from their data to be close to tetracalcium phosphate, TTCP.

The chemical durability of calcium phosphate glasses was examined in SBF* at 37°C by Kasuga 2005. Glasses of varying CaO and P_2O_5 contents were prepared along with several containing TiO_2 and/or Na_2O. In addition several compositions were heated to ≈800°C to prepare glass-ceramic materials. The glass-ceramics contained crystalline phases of β-$Ca_3(PO_4)_2$ and β-$Ca_2P_2O_7$. A glass with a CaO/P_2O_5 ratio of 50/50 and others with an $x/90–x$ ratio containing 10 mol% TiO_2 and particle sizes ranging between 45 and 150 μm were investigated. The Ca^{2+} and P^{5+} concentrations in the SBF were an order of magnitude less for the glass containing TiO_2. The Ti^{4+} concentration was negligible. The pH of the SBF remained at 7.4 after 10 dys of soaking for the glass containing TiO_2 whereas the pH for the 50/50 mixture had dropped to 6.2. A calcium-deficient apatite layer about 5–10 μm thick formed on the surface of the compositions containing TiO_2 after soaking in SBF for 10 days. An intermediate gel layer 0.5–2.0 μm thick also formed. With a partial substitution of Na_2O for TiO_2, apatite formed within several days and the particle surfaces were completely covered in 7 days. Glass samples containing a CaO content ≤ 55 mol% contained no orthophosphate groups, $(PO_4)^{3-}$, and formed no apatite. The other compositions contained only pyrophosphate groups, $(P_2O_7)^{4-}$. By autoclaving[†] the glass-ceramics, Kasuga was able to enhance the

---

* SBF—simulated body fluid.
† Heated to 140°C for 1 hr in distilled water.

bioactivity of these compositions. Autoclaving formed TiO_2 (anatase) crystals on the surface. It was suggested that the negatively charged hydrated titania and anatase induced apatite nucleation.

The antimicrobial activity of glasses in the system $Na_2O-CaO-P_2O_5-SiO_2$ was studied by Waltimo et al. 2007. They found that the over 10-fold increase in surface area by using nano-sized bioactive glass[*] released more alkali into SBF at 37°C than micron-sized glasses and exhibited an enhanced antimicrobial effect. In addition a 10-fold increase in silica release was noted. Weber et al. 2004 have suggested that silica solutions act as food disinfectants.

In contrast to the silicate-based bioactive glasses, bioactive borate glass did not form the initial thin silica-rich gel layer, Huang et al. 2006a, Huang et al. 2006b, and Li et al. 2008, when reacted at $pH = 7.0$ and at 37°C. The borate glass/aqueous phosphate solution reaction initially formed $CaHPO_4 \cdot 2H_2O$ (for reaction times <5 hr). Formation of the amorphous calcium phosphate followed and then finally the HA material. Li et al. 2008 stated that dissolution rates and conversion to HA could be tailored between the silicate (more durable) and borate (less durable) bioactive glasses.

Several glass-ceramics have been developed including A-W (an apatite-wollastonite material), Nakamura 1985, and BioVerit I (mica-apatite), II (mica-cordierite), and III (silica free, apatite, AlPO4, and complex phosphates), Holland and Vogel 1993. Sol-gel prepared calcium silicate glass-ceramics containing phosphorous have been shown to exhibit a faster rate of apatite layer formation when compared to those without phosphorous, Padilla et al. 2005. The crystalline phases present in phosphorous-free samples were a combination of wollastonite and pseudo-wollastonite[†] with no wollastonite and a little cristobalite being present at the highest heat treatments (1300 and 1400°C). Samples containing phosphorous exhibited a combination of wollastonite, apatite, and quartz when heat treated at temperatures below 1200°C. At higher temperatures pseudo-wollastonite, cristobalite, and tricalcium phosphate appeared. The various combinations of these crystalline phases affected the bioactivity with large cristobalite amounts causing the lowest reactivity. In addition the work of Vallet-Regi and others found that sol-gel derived calcium silicate glass-ceramics were more bioactive in simulated body fluid than melt derived materials, Vallet-Regi 2001 and Vallet-Regi et al. 2003.

Bioactive Hydroxyapatite

Hydroxyapatite (HA) containing from 4 to 6% carbonate with multiple other substitutions makes up about 70 wt% of human bones and about 96 wt% of human teeth, Leventouri et al. 2009. *Of all the calcium phosphate materials, the apatites are the most thermodynamically stable, Mathew and Takagi 2001,*

---

[*] Bioactive glass 45S5.
[†] Wollastonite and pseudo-wollastonite are polymorphs of $CaSiO_3$.

and of all the apatites, fluorapatite is the most stable. In addition the biological apatites (bioapatites) are nonstoichiometric and generally calcium deficient. Synthetic HAs are invariably phase impure. One of the more common secondary phases present is tricalcium phosphate (TCP[*]). This results from the thermal decomposition of HA that begins about 1200°C during sintering, Chen and Miao 2005. The thermal decomposition of HA is given by:

$$Ca_{10}(PO_4)_6(OH)_2 \rightarrow 3Ca_3(PO_4)_2 + CaO + H_2O \quad (14.1)$$

The incorporation of TiO_2 into the HA was found to inhibit the decomposition of HA, Li et al. 2002. Incorporation of TiO_2 was thought to decrease the thermal expansion mismatch between HA and a titanium substrate during various coating processes, Li et al. 2002. The thermal expansion coefficients for HA, TiO_2 (anatase), and titanium are 16, 10.2, and $8.9 \times 10^{-6}/K$, respectively. The dissolution behavior of carbonated hydroxyapatite (CHA) was reported by Hankermeyer et al. 2002 to be highly dependent upon the *p*H of the solution. Five other variables were also studied—solution flow rate, agitation of the solution, temperature, ionic strength of the solution, and surface area of the CHA, all of which were found to increase the dissolution rate. The mechanism of dissolution given by Hankermeyer et al. for a calcium-deficient CHA (prepared to simulate a calcified occluded artery) was one of protonation of both the carbonate and phosphate groups forming carbonic and phosphoric acids, respectively. The source of the protons was HCl in concentrations ranging from 0.1 to 1.0 N. The equation that represents this reaction is:

$$Ca_{8.8}(HPO_4)_{0.7}(PO_4)_{4.5}(CO_3)_{0.7}(OH)_{1.3} + 17.6H^+ \rightarrow$$
$$8.8Ca^{2+} + 5.2H_3PO_4 + 0.7H_2CO_3 + 1.3H_2O \quad (14.2)$$

The assumption here is that the carbonate group is substituted onto the hydroxyl site, although the phosphate site is preferred by a factor of 5:1 as discussed in Chapter 4. The difference may relate to comparing a synthetic CHA with a natural one.

According to Mavropoulos et al. 2003 the Ca/P ratio of HA determines the dissolution behavior in water at 37°C and *p*H ranging from 6.1 to 6.7. The Ca/P ratio ranged from 1.49 to 1.64. *Calcium-deficient HAs in general are more soluble than stoichiometric HAs.* Dissolution equilibrium was not obtained even after 7 days.

Seo and Lee 2007 reported the variation of dissolution of different crystallographic faces of hydroxyapatite (HA). They found that the (300) face remained stable during immersion into distilled water (*p*H = 7.4 and 37°C) for 14 days,

---

[*] A phase pure β-TCP is manufactured by Curasan, AG under the name Cerasorb®.

whereas the (211) and (112) faces[*] both exhibited dissolution as evidenced by XRD intensity measurements. In a study on the effects of fluoride ions upon the dissolution of HA Kwon et al. 2011 observed the dissolution of the (100)[†] face with and without a 10 mM NaF buffer solution at pH 6.0. When compared to a NaF-free solution, the 10 mM solution exhibited surface step retraction velocities that decreased by about a factor of five. At higher levels (50 mM) there was a decrease in dissolution taking about 8 hrs to remove one monolayer of atoms. They reported that a fluoride substituted HA surface[‡] was several hundred times less soluble than HA in aqueous solution.

Seo and Lee 2007 concluded that a nonstoichiometric HA phase was present at the grain boundaries of their synthetic material due to the grain boundaries being dissolved preferentially. This in turn could lead to grain separation and eventual mechanical failure. In a previous study, Seo et al. 2007, it was determined that the grain boundaries of hydroxyapatite contained α-tricalcium phosphate (α-TCP), which is more soluble than HA. In a much earlier study Nonami and Wakai 1995 found that the Ca/P ratio at the grain boundaries of HA was lower than the bulk grains. The Ca/P ratios of α-TCP and HA are 1.5 and 1.667, respectively. Thus the lower Ca/P ratio is compatible with a more soluble nonstoichiometric HA.

The solubility in *tris*-buffer solution at $pH = 7.4$ at 37°C of three apatite sources was studied by Fulmer et al. 2002. Two of the apatite sources were calcium-deficient carbonate HAs (BoneSource[§] and Norian CRS[¶]) and the other was a sintered stoichiometric HA (Calcitite[**]). Ground samples of each apatite were placed into the 0.03 M *tris*-buffer solution. At various time intervals ranging from 10 min to 120 hrs samples were taken from the solution for evaluation. Equilibrium values for calcium concentration in the solutions were at 120 hrs 7.39, 7.51, and 1.38 ppm, respectively for Norian RS, BoneSource, and Calcitite. The much better dissolution resistance of the Calcitite was attributed to the fact that it was sintered and therefore more crystalline than the others.

Ribeiro and Barbosa 1991 reported increased dissolution of HA with decreasing pH for hydroxyapatite in a 0.9% NaCl saline physiological solution at 37°C agitated for times ranging from 0.5 hr to 10 days. The Ca/P ratio in the liquid varied with a maximum at pH = 6. Below pH = 5.3 and above pH = 6.3 the Ca/P ratio was lower than that in the solid. At pH = 6 it was higher. This ratio was generally less than that in the solid indicating incongruent dissolution. Similar results were also reported by Levinskas and Neuman 1955.

---

[*] The (211), (300), and (112) are the three strongest XRD reflections for hydroxyapatite.
[†] The (100) faces are the sides of an elongated hexagonal prism.
[‡] Human bone contains <1.0% fluorine, and pure fluorapatite contains 3.77 wt%, Chen and Miao 2005.
[§] Bonesource—Howmedica, Rutherford, NJ.
[¶] Norian CRS—Norian Corp., Mountain View, CA.
[**] Calcitite—Calcitek, Carlsbad, CA.

Addition of Al, Ti, or V to the solution caused variation in the dissolution of Ca and P as listed below:

1. Al caused increased Ca and decreased P dissolution,
2. Ti caused decreased Ca and decreased P dissolution, and
3. V caused increased Ca and decreased or no P dissolution.

Velayudhan et al. 2005 developed a composite[*] of bioactive HAP[†] particles in a matrix of ethylene vinyl acetate copolymer (EVA) for use in cranioplasty. In this case the polymer was non-resorbable and ductile and the HAP provided an osteoconductive[‡] ceramic. The polymer provided the composite with sufficient malleability for surgeons to sculpt the composite to fit the repair site. A 50/50 mixture exhibited the best results (i.e., noncytotoxic to mouse fibroblast cells).

Leonor et al. 2003 developed a composite composed of a biodegradable starch thermoplastic matrix and the bioactive hydroxyapatite for implantation into the human body. The degradation of the composite implant must be controlled to allow the gradual transfer of load to the healing bone. Hydroxyapatite of 30 wt% is required to cause the formation of calcium phosphate on the surface of the composite for adhesion to the bone. Samples immersed into a simulated body fluid at $pH = 7.35$ showed no change after 8 hrs. With increased immersion time calcium phosphate nuclei formed, grew in number and size, and coalesced fully covering the surface of the composite within 24 hrs. A dense uniform calcium phosphate layer was formed after 126 hrs. Hydroxyapatite-reinforced polyhydroxybutyrate was investigated by Ni and Wang 2002. They found that the bioactivity in SBF increased as the HA content increased, starting after 7 days. With increased times the apatite surface layer increased in connectivity and thickness.

In a study of the stability of apatite in glassy matrices, Berger et al. 1996 found that chloroapatite dissolved more slowly than hydroxyapatite at elevated temperatures. They attributed the greater dissolution rate of the HA to the reaction of OH forming water and oxygen compared to the chloroapatite dissolution being a diffusion-controlled process.

Bioresorbable Materials

Resorption of synthetic materials is quite different from that of natural bone. Natural bone is essentially composed of loosely bonded nanocrystals with a very high surface area that have grown in an organic matrix. Synthetic calcium phosphate bioceramics have much stronger crystal-to-crystal bonds

---

[*] The first bioceramic/polymer composite was a hydroxyapatite-reinforced polyethylene.
[†] Some authors use HAP rather than HA to represent hydroxyapatite.
[‡] Osteoconduction occurs when a bone graft material acts as a scaffold for new bone growth.

and much lower surface areas. According to Bandyopadhyay et al. 2006 resorption is caused by the following three factors:

1. physiologic dissolution (dependent upon pH and the material composition),
2. physical disintegration due to either grain boundary attack or high porosity, and
3. biological factors—phagocytosis[*].

It has been reported that the *in vivo* resorption of β-tricalcium phosphate was incomplete even after 9.5 minutes in a human mandible (29% of the TCP remained), Zerbo et al. 2001. Arinzeh et al. 2005 have shown that the HA/β-TCP ratio in biphasic materials influenced the rate of mesenchymal stem cell (MSC)[†] induced bone formation in ectopic sites. A 20 wt% HA/80 wt% β-TCP ratio scaffold seeded with MSC exhibited the greatest rate of bone formation when compared to the two end members or other ratios. In a separate study Jensen et al. 2008 showed that degradation and bone formation was inversely proportional to the HA/β-TCP ratio (20/80, 60/40, and 80/20). A porous tri-phasic (HA, TCP, and CS) ceramic-coated HA scaffold was investigated by John et al. 2008. They found that in a phosphate-buffered saline ($pH = 7.4$) solution silica was released and that the TCP completely disappeared after 60 days. Single-phase HA showed no change after immersion in the same solution for 60 days.

Since natural bone contains a significant number of trace metal ions, Bandyopadhyay et al. 2006 investigated the effects that the ions Na^+, Ag^+, Mg^{2+}, Zn^{2+}, Si^{4+}, Ti^{4+}, and F^- had upon the resorption of TCP. They found that Mg^{2+}, Zn^{2+}, and Si^{4+} had the largest effect. It is thought that Mg^{2+} stabilizes the TCP structure by forming shorter and therefore stronger metal-oxygen bonds than does calcium. Zinc, like magnesium, is smaller than calcium and should form stronger metal-oxygen bonds. However, the increased bioactivity of the zinc-doped TCP was attributed to a distorted structure by Yin et al. 2002. The incorporation of Mg and Zn into β-TCP reduced the dissolution process both *in vitro* and *in vivo* as reported by LeGeros 1991. This result should not be unexpected since more compact structures are more durable.

Ahmed et al. 2004(a) studied the solubilities of sodium-calcium-phosphate glasses as potential degradable scaffolds for tissue regeneration. They found that the glasses with constant phosphate content (45, 50, and 55 mol%) and increasing calcium content (30, 35, and 40 mol%) exhibited decreased solubilities. They also discussed the solubilities with relation to the glass structure. *More dense structures were more durable.* In a second study Ahmed et al. 2004(b) investigated the solubility of sodium-calcium-phosphate glass fibers

---

[*] Phagocytosis is the mechanism used to remove pathogens and cell debris including bacteria, dead tissue cells, and small mineral particles.

[†] MSC are stem cells from bone marrow.

as potential cell delivery vehicles for muscle stem cells. Fiber solubilities were similar to the bulk glass except that the fibers were more soluble due to their increased surface area. In an additional study, sodium-calcium-phosphate glass fibers containing Fe_2O_3 were examined by Ahmed et al. 2004c for tissue engineering applications.[*] Increasing calcium content related to more durable fibers, as did increasing iron content. They also reported data that indicated that glass T_g values increased with increasing calcium or iron content. (See the discussion in Chapter 16 on the relationship among glass properties and dissolution.) Compositions with no iron were too soluble for cell attachment. It was found that 4–5 mol% Fe_2O_3 was sufficient for cell attachment. Ahmed et al. 2004(c) attributed the improved performance to the replacement of P-O-P bonds with Fe-O-P bonds and to the phosphate chains being cross-linked by iron ions.

Calcium phosphate cements, commonly used as moldable bone substitute materials and as a filling for bone defects, containing Mg, Sr, and Zn were reviewed by Pina and Ferreira 2010. Two types of cements are currently used, one being apatite, the other being dicalcium phosphate dehydrate (DCPD). Although the name DCPD would indicate that the chemical formula would contain two calcium atoms, it is generally written as $CaHPO_4 \cdot 2H_2O$, which is the mineral brushite.[†] This mineral is the most soluble of all the calcium phosphates. Due to its fast dissolution rate (0.25 mm/wk, Ohura et al. 1996) brushite may cause the formation of immature bone.

Dental Materials

The process of degradation of ceramics in the human mouth is quite complex, involving dissolution in salvia and the various physicochemical reactions caused by food, chewing, and bacteria. Within a short period of implantation, restorations are covered with a biofilm called the pellicle.[‡] It is for this reason that no *in vitro* test can duplicate this process, Øilo 1992.

Acidic foods (e.g., soft drinks with pH of 4–5) and bacteria provide the most important form of chemical attack upon tooth enamel, Sollböhmer et al. 1995, at temperatures between 0 and 80°C. Gastric acid[§] (in the form of gastric reflux) can also provide a significant source for enamel dissolution, Meurman et al. 1994. Material loss in acidic beverages (there was no difference between regular or diet sodas) was 0.1 nm/s (3 mm/yr), whereas in distilled water ($pH = 7$) it was 0.013 nm/s (0.4 mm/yr). These values correspond

---

[*] Since the morphology of glass fibers mimics that of muscle fibers, glass fibers are excellent candidates for tissue engineering of muscle.
[†] Brushite is one of the most common cave minerals and is also found in guano deposits.
[‡] The pellicle is a thin layer of salivary glycoproteins.
[§] Gastric acid can have a pH as low as 1.5.

to 3 and 0.4 mm/yr, respectively. Sollböhmer et al. also noted that the maximum material loss was mostly from grain boundaries of the HA. This is the case for all polycrystalline ceramics since grain boundaries are regions of higher energy compared to the bulk grain and also because grain boundaries often contain a more soluble second phase.

As stated in the section above on bioactive hydroxyapatite, natural dental enamel is composed of about 96 wt% HA and contains about 4 to 6% carbonate, Leventouri et al. 2009. Sollböhmer et al. 1995, however, reported that mature enamels contain up to 86% HA. Leventouri et al. found that carbonate content increased with age. This increased carbonate decreased the a-lattice parameter with no significant change to the c-lattice parameter. In addition HA crystallinity decreased with age. Although Leventouri et al. did not discuss the effects of carbonate content and crystallinity changes upon dissolution, it would appear that dissolution would slow with age due to a more compact structure. However, a decrease in crystallinity may have an opposite effect. The effects of fluoride substitution for hydroxyl ions in the HA structure were discussed in the section above on bioactive hydroxyapatite. A condition called *fluorosis* can develop if too much fluorine is ingested during the tooth-forming years, Bronckers et al. 2009. This causes a more porous structure that is more susceptible to dissolution and decay.

Modern dental ceramics are generally of excellent durability. Exposure to acidulated fluoride may cause surface degradation for individuals who receive large doses of radiation for neck and/or throat cancer, Anusavice 1992.

Dicor®* along with several other glass-ceramics were tested in 4% HAc solution at pH of 1, 9, and 11 at 80°C for 15 days by Anusavice and Zhang 1997. These weight loss tests along with some others are shown in Table 14.2. The weight loss determinations were performed over a period of up to 15 days in a shaker bath using 30 cycles/min. At pH ≥ 9 the leaching rates decreased in the order $Si^{4+} > Li^+ > Ca^{2+} > Al^{3+}$. At pH < 5 the leaching rates decreased in the order $Li^+ > Si^{4+} > Ca^{2+} > Al^{3+}$. The allowable daily intake of Li^+ is equal to 2 mg.† A surface area of a full set of teeth is about 74 cm². Using lithium alumino calcium silicate (LACS) as a full set of teeth at a leach rate of 0.25 mg/cm² would yield 1.2 mg/dy of Li^+. This value is less than the allowable amount and thus would be acceptable as discussed by Anusavice 1995.

Anusavice and Zhang 1998 studied the effects of Al_2O_3 additions upon the dissolution of fluorocanasite-based glass-ceramics,‡ which they found to be less durable than Dicor. They found that alumina additions decreased the durability of fluorocanasite-based glass-ceramics. Anusavice and Zhang concluded that these materials could be used as prosthodontic core materials even though they exhibited durabilities less than Dicor due to the fact that

---

* Dicor® is a tetrasilicic fluoromica glass–ceramic manufactured by Corning ($K_2Mg_5Si_8O_{20}F_4$). Similar products are Empress® I and II and Procera®Allceram.
† For manic-depressive patients.
‡ Fluorocanasite is composed of Al_2O_3-CaO-F-K_2O-Na_2O-SiO_2.

TABLE 14.2

Weight Loss Data for Several Glass-Ceramics

Material	Mean Wt. Loss (mg/cm$^2 \cdot$ day)	Ref.
Tetrasilicic fluormica	0.0042	Grossman & Waters 1984
Ceramco	0.0057	DeRijk et al. 1985
Ceramco	0.0095	Grossman & Waters 1984
Vitar porcelain	0.0165	Grossman & Waters 1984
Vitar aluminous core	0.0200	Grossman & Waters 1984
LACS Ag[b]	0.25 ± 0.06[a]	Anusavice and Zhang 1997
LACS	0.25 ± 0.01[a]	Anusavice and Zhang 1997
Dicor®	0.27 ± 0.01[a]	Anusavice and Zhang 1997
Neph-Sye porcelain	0.92	DeRijk et al. 1985
LACSP[c]	6.80 ± 1.3[a]	Anusavice and Zhang 1997

[a] Tested in 4% HAc solution at 80°C and pH = 1, 9, and 11 for 15 days.
[b] Li_2O (27.8 mol%), Al_2O_3 (2.5%), CaO (5.9%), SiO_2 (63.8%), $AgNO_3$ (0.78%).
[c] Contains 1 mol% P_2O_5 as a nucleating agent.

these tests were done at a *p*H less than that found in the mouth. *One must be careful when doing experimental tests that do not duplicate the actual environmental conditions of the application. Mechanisms of dissolution can change drastically with pH changes.*

The dissolution of a nepheline syenite–based porcelain was determined to be 0.92 mg/cm$^2 \cdot$ day at 80°C in 4% glacial acetic acid by DeRijk et al. 1985. In a similar study the dissolution of a conventional feldspathic porcelain (Ceramco[*]) was 0.57 × 10^{-2} mg/cm$^2 \cdot$ day. (Compare these values with others in Table 14.2.) Interestingly DeRijk et al. calculated that it would require 22 years of immersion into artificial salvia at 22°C to be equivalent to the exposure in 4% acetic acid at 80°C for 1 week.

The release rates of various toxic elements from dental ceramics is of great concern; however, they are in general quite low. Lithium has been discussed above. Fluorine released from a set of 32 Dicor crowns in 4% acetic acid at 80°C has been estimated to be 0.1 mg/day by Anusavice 1992. This is well below that obtained from food and drink consumption, which ranges from 0.3 to 2.0 mg/day. Uranium oxide was the first material used to produce fluorescence that matched natural tooth fluorescence.[†] The hazard of radioactive uranium has forced the dental manufacturers to find other means of producing fluorescence, Mackert 1992. Most of the rare earth oxides and some transition element oxides (e.g., Mn) have been investigated. The release of some components of restorative composites is beneficial. Recent research has been

---

[*] Ceramco contains about 0.3 wt% ZrO_2.
[†] The fluorescence of teeth seems to be a problem only for those people who are in contact with ultraviolet light sources in darkened areas (i.e., discos).

TABLE 14.3

Chemistry of Artificial Saliva

Compound	Amount
K_2HPO_4	0.1 L of 25 mM
Na_2HPO_4	0.1 L of 24 mM
$KHCO_3$	0.1 L of 150 mM
NaCl	0.1 L of 100 mM
$MgCl_2$	0.1 L of 1.5 mM
$CaCl_2$	0.1 L of 15 mM
Citric acid	0.006 L of 25 mM
NaOH or HCl	Sufficient to adjust pH to 6.7
Distilled H_2O	Sufficient to adjust volume to 1 L
Thymol	0.05 wt% (to avoid bacteria growth)

Source: Söderholm, K.-J.M., Mukherjee, R. and Longmate, J. 1996. Filler Leachability of Composites Stored in Distilled Water or Artificial Saliva, *J. Dental Res.*, 75(9), 1692–99.

directed toward nanocomposites that release Ca, PO_4 (Xu et al. 2007), and F, Xu et al. 2010. The Ca and PO_4 released effectively remineralizes the tooth lesion by reprecipitating as HA. Any F released can diffuse into surrounding teeth helping to minimize decay in those teeth.

Söderholm et al. 1996 studied the leachability of dental composites containing ceramic fillers (either quartz or barium-alumino borosilicate glass) in distilled water compared to artificial saliva (see Table 14.3 for the chemical composition of the saliva). They confirmed that artificial saliva is the more aggressive leachant, indicating that results from *in vitro* tests in distilled water must be considered carefully. Based on the differences in elemental concentration gradients one would think the opposite to be true. This can all be explained by the work of Charles 1958 and others, as discussed in Chapter 18. One should also remember that human saliva contains organic components, such as proteins, that contribute to pellicle and plaque coatings. The differences between artificial and human saliva, which also varies among patients, most likely will contribute to contradictory results between *in vitro* and *in vivo* tests. Leaching rates have also been shown to be enhanced by incorporation of elements such as sodium, barium, and strontium into the filler glass, Söderholm 1981. In addition Söderholm et al. 1994 found that specimens stored in distilled water lost more weight than those stored in artificial saliva. This would indicate water to be the more aggressive leachant; however, they found that ions from the saliva diffused into the composite, offsetting some weight loss. *All this indicates that researchers must design their experiments carefully and use more than one analytical technique to evaluate the results.*

The method of choice for applying HA coatings onto metallic implants is plasma spraying, Campbell 2003. Gottschling et al. 1995 reported that

TABLE 14.4

Materials Formed during Plasma Spraying of HA onto Dental Implants

Material	Abbreviation	Min T (°C) for Formation	Solubility Relative to HA
$Ca_x(PO_4)_y$	ACP	Amorphous	>
$Ca_4P_2O_9$	TTCP	1450	>>
$Ca_3(PO_4)_2$	α-TCP	1300	>
	β-TCP	800–1000	>>
$Ca_{10}(PO_4)_6(OH)_2$	HA	<1000	

plasma-sprayed HA consisting of 40% crystals onto titanium tooth root implants exhibited 15% dissolution after 140 days at room temperature in doubly deionized water. Other materials may form during plasma spraying depending upon the temperature of application. These are shown in Table 14.4.

As reported by LeGeros et al. 1995, the dissolution (in pH 5 and 6 acetate and lactate acid buffer solutions and pH 7.3 *tris* buffer solutions) is dependent upon the composition and percent crystallinity. Plasma-sprayed coatings on metal substrates actually contain all the phases listed in Table 14.4. Amorphous CP and the more soluble CP phases dissolve preferentially to HA. Bioreactivity is related to CO_3–apatite formation on the surface of the CP phases. Therefore highly crystalline HA may not be desirable for implant coatings on metals.

Mineral Dusts

Asbestiform Materials

The diseases related to mineral dusts that have received the most attention are asbestos-related—asbestosis, mesothelioma, and lung cancer—and thus the minerals that have received the most attention are asbestos and the asbestiform amphiboles. Silica and coal dusts should also be added to that list. Guthrie 1992 reviewed epidemiological studies of many minerals including oxides, hydroxides, 1:1 and 2:1 layered silicates, and zeolites. In some cases the results were conflicting. One of the main things that he found was that many studies were not well defined mineralogically. Guthrie stated that the mineral content of the samples tested was "potentially suspect" due to various varietal names being used. The mineral species were incorrectly identified in many cases. Therefore it was extremely difficult to draw any conclusions from the data. Samples of unknown purity or mixed minerals were tested at various dose levels and for various durations, all leading to extreme difficulty in drawing any meaningful conclusions. A major lack in these studies is any information concerning the surface characteristics of the minerals (e.g., surface area). Most experimental tests evaluate the toxicity of a particular material rather than the solubility in some body fluid. In some cases the solubility data may be buried within the test data, but evaluating all the available literature and extracting the solubility data is a monumental task. According to May et al. 1977 the simulated body fluid compositions are oversimplified and may not include a sufficient array of organic ligands. This results in *in vitro* tests understating the true metal mobility from minerals. *Probably the most important parameter for obtaining accurate laboratory results is using a test fluid that truly represents the real situation.* Even performing tests with animals doesn't always result in accurate data since an animal's response may be different than that of a human. Even the animal response can vary with the type of animal—for example, mice versus hamsters. In addition, the clearance rates for insoluble particles are much greater for rats than for humans, McClellan 2000. A recent study by Favero-Longo et al. 2009 reported that most experiments to evaluate the toxicity of asbestos were performed on laboratory fibers and not naturally occurring fibers. They stated that naturally occurring fibers are less toxic than laboratory ones due to the effects of weathering. Weathering causes the release of cations, which in turn alters the surface reactivity. Thus performing a meaningful experiment involves many parameters, some of which are often overlooked.

A particular mineral may vary in biodurability or bioaccessibility based upon the route that it travels through the body. The most bioreactive minerals are those that contain substantial alkali constituents (e.g., cement and concrete) or acidic components (e.g., salts from mine wastes). Amphibole asbestos and erionite[*] are less readily dissolved in lung, interstitial, and phagolysosomal fluids[†] than chrysotile asbestos. This creates a greater potential for fibrosis and cancer from these minerals, Sébastien et al. 1989. Carbonate minerals are generally more stable in near-neutral bicarbonate-rich lung and interstitial fluids, whereas phosphates dissolve more readily in the phosphate-poor gastric fluids than in lung, interstitial, or lysosomal fluids that are either less acidic or contain some phosphorous, Plumlee et al. 2006.

The zero point of charge of chrysotile is close to pH 11 due to chrysotile's essentially basic surface character (i.e., exposed Mg ions on surface). At the lower pH of the various body fluids chrysotile would exhibit a positive surface charge and therefore would interact with anionic species, Jolicoeur and Poisson 1987. Ménard et al. 1984 in a study of the binding of various aromatic hydrocarbons to chrysotile reported that their affinity increases with molecular weight.

Hume and Rimstidt 1992 have studied the *in vitro* dissolution rate of silicate minerals in simulated lung fluids (SLF) and lysosomal fluids. Results

---

[*] Erionite, $(Na_2,K_2,Ca)_2Al_4Si_{14}O_{36} \cdot 15H_2O$, is a fibrous zeolite often found in volcanic ash.
[†] Phagolysosomal fluids are those found in the deep lung.

showed that the dissolution rate of chrysotile was high in SLF and higher than amphibole asbestos, erionite, talc, and quartz. Several studies have shown that magnesium was selectively leached from talc, chrysotile, and crocidolite fibers. If organic iron–chelators were present, Werner et al. 1995 found iron to be selectively leached from crocidolite. Mg–O and Fe–O bonds are relatively easy to break compared to Si–O bonds, and therefore selective leaching of magnesium and/or iron should be expected.

There is some evidence that the chrysotile/tremolite ratio decreases while in the human lung. Wood et al. 2006 have reported that chrysotile dissolved much more rapidly than tremolite in lung fluid. They also indicated that both dolomite and hydroxyapatite formed, dolomite forming at the expense of the chrysotile but the apatite forming at the expense of none of the dissolving materials. There was conflicting evidence as to whether the tremolite was present along with the chrysotile when inhaled or whether the chrysotile converted to tremolite in the lung.

Hume and Rimstidt 1992 have studied the dissolution of chrysotile in an effort to develop a general test for mineral dust biodurability. At $pH < 9$ the reaction:

$$Mg_3Si_2O_5(OH)_4 + 6H^+ \rightarrow 3Mg^{2+} + 2H_4SiO_4 + H_2O \qquad (14.3)$$

described the dissolution. Based upon some reported concentration levels of Mg^{2+}, H^+, and silica in lung tissue fluids, Hume and Rimstidt determined that chrysotile would be in equilibrium with these fluids at a pH of 8. However, body fluids never reach $pH = 8$, thus creating an environment for continuous dissolution. Its persistence was due to a very slow dissolution rate. Dissolution occured in two steps; first Mg was leached and then the silica matrix dissolved. Thus the lifetime of chrysotile was determined by the silica dissolution. Hume and Rimstidt gave the following equation for calculating the lifetime of a chrysotile fiber:

$$t = (3/4)/(d/V_m k) \qquad (14.4)$$

where:
   d = fiber diameter in meters,
   k = rate constant for silica dissolution (mol/m^2s), and
   V_m = volume of one mol of silica in chrysotile (equal to 5.4×10^{-5} m^3/mol).

A 1 µm diameter[*] chrysotile fiber will take approximately 9 months to dissolve (based upon the silica concentration of the fluid at a pH of 4–7 and 37°C). This lifetime is an order of magnitude less than the time required for

---

[*] Fibers greater in diameter than 3 µm cannot be inhaled into the lower lung, although fibers with diameters <3 µm and lengths greater than 20 µm can penetrate deeply into the lung, Hadley, 2000.

TABLE 14.5

Comparison of Dissolution (Si Release) Lifetimes for 1 μm
Diameter Fibers

Mineral	SGF[a] (pH ≈ 1.2)	SLF[b] (pH ≈ 7.4)	References
Chrysotile	33 hrs	19 mns	Oze & Solt 2010
		7 mns	Jurinski & Rimstidt 2001
		9 mns	Hume & Rimstidt 1992
Tremolite	9 mns	4 yrs	Oze & Solt 2010
Asbestos		14 yrs	Hadley 2000
Fused silica		438 yrs	Rimstidt & Barnes 1980
Fiberglass		50 dys	Hadley 2000
Aluminosilicate		14 yrs[c]	Bauer 2000
Quartz		≈ 2 M yrs	Rimstidt & Barnes 1980
Talc		8 yrs	Jurinski & Rimstidt 2001

[a] SGF = Simulated gastric fluid
[b] SLF = Simulated lung fluid
[c] Dissolution based on fiber diameter decrease

the onset of disease symptoms. These data were based upon a silica release
rate of about 6×10^{-10} mol Si/m^2 from a shrinking cylinder model. Thus any
biological model must explain this difference. By comparison a fused silica
fiber of 1 μm diameter under the same conditions takes 438 years to dissolve completely. Asbestos, on the other hand, takes about 14 years for a 1
μm fiber to completely dissolve (see Table 14.5). One should also remember
that chrysotile, being a sheet silicate with a misfit between the octahedrally
coordinated magnesium-containing layer and the tetrahedrally coordinated
silica layer, curls up into a tube forming fibrils with the octahedral layer on
the exterior (see Chapter 8 on Surface Effects and Figure 8.2). Thus the dissolution of a mineral of this structure can be quite different from one that
forms a uniform chemistry across a solid fiber.

Others have also reported that chrysotile dissolves more rapidly than tremolite in simulated lung fluids, Plumlee et al. 2006. Actual comparisons must
be done very carefully, since the surface area of chrysotile, on an equal weight
basis, is more than tremolite due to their differences in morphology. An excellent discussion of these differences between chrysotile and tremolite was
given by Oze and Solt 2010. They related the faster dissolution rates in gastric
fluids (pH ≈ 1.2) when compared to lung fluids (pH ≈ 7.4) to the enhanced
action of protons breaking Si–O–Mg connecting oxygen atoms along with the
production of silonal and hydrated Mg^{2+} ions, as earlier reported by Rosso
and Rimstidt 2000 for forsterite. Forsterite is constructed of only Si–O–Mg
bonds and thus this is the only mechanism for dissolution. In contrast Hume
and Rimstidt 1992 reported the Si release rate to be pH independent. These
results were obtained in dilute HCl solutions as opposed to simulated gastric

fluid (SGF) or SLF. If one considers equivalent surface areas, Oze and Solt 2010 reported that chrysotile released Si about twice as fast as tremolite in SGF but in SLF tremolite released Si about five times as fast as chrysotile. However, if one considered equivalent masses, the results were very different. Chrysotile released Si about 1.3 times faster than tremolite in SLF and about 9 times faster in SGF. But one must be cautioned that the surface area of chrysotile is 8.3 times larger than tremolite on a per mass basis.

Comparing this to the dissolution of silicate glasses, one finds that silica release is independent of pH below about $pH = 8$ and above $pH = 9$ becomes increasingly more rapid (see Chapter 16).

The rate of asbestos dissolution was given by Oze and Solt 2010 as:

$$d(Si)/dt = k_{SBF} \{asbestos\}'' \tag{14.5}$$

where $d(Si)/dt$ is the rate in $\mu mol/h$ of Si released into solution, k_{SBF} is the rate constant in $\mu mol\ m^{-2 \times n}/h$ (n is the reaction order) of Si released into simulated body fluid, and $\{asbestos\}''$ is the surface area in m^2. This equation ignores any back reactions of secondary precipitation or adsorption, which was confirmed by the results. This equation was used to obtain the results reported above. They also calculated saturation index values for both chrysotile and tremolite using the Visual MINTEQ program and the Lindsay database described by Gustafsson 2006. All values were negative, indicating undersaturation in both lung and gastric fluids. Oze and Solt 2010 calculated fiber lifetimes for 1 μm diameter by 10 μm long and treating the fibers as cylinders with chrysotile having a surface area 10 times greater than tremolite and each fiber containing either 2 moles Si per mole of chrysotile or 8 moles Si per mole of tremolite. Their results indicated that a fiber of chrysotile would last about 19 months in SLF and about 33 hrs in SGF, whereas tremolite would last about 4 years and about 9 months, respectively. These numbers are quite a bit different than that reported by Hume and Rimstidt 1992 of about 9 months at $pH = 4–7$. But then the method of calculation was quite different. Oze and Solt also pointed out that their calculations did not take into account the fact that fibers shrink during dissolution, thus causing the surface area to shrink. As can be seen, comparison of different minerals is a very complex situation requiring a complete understanding of the minerals themselves (morphologies, surface areas, etc.) and the fluids performing the dissolution (dilute acids, SBF, or actual body fluids). By comparison a 1 μm diameter particle of quartz would last almost 2 million years in lung tissue, Hume and Rimstidt 1992.

The reason for the greater toxicity of crocidolite compared to chrysotile was attributed by Holland and Smith 2001 to the greater content of iron in crocidolite than chrysotile, up to 36 wt% and 2–3 wt%, respectively. Iron caused the formation of reactive oxygen species (ROS). The body uses some ROS beneficially;[*] however, they have been reported to cause oxidative dam-

---

[*] Macrophages of the lung can form ROS to aid the attack of engulfed particles or microbes.

age to lipids, proteins, and DNA, which in turn leads to various forms of cancer, Plumlee 2006.

SiC

A study by Birchall et al. 1988 has indicated that SiC whiskers are on the same order of toxicity as crocidolite asbestos. The samples they tested were from the carbo-thermic reduction of silica and consisted predominantly of the cubic (beta) form of silicon carbide with minor amounts of the hexagonal alpha form and traces of corundum and mullite. Birchall et al. also tested submicron-sized nonfibrous alpha silicon carbide that exhibited a negative response to their tests.

Talc

The dissolution of talc,[*] Jurinski and Rimstidt 2001, is similar to chrysotile in that both are incongruent with an initial release of magnesium and then silica. Thus the silica release is the rate-limiting step and is independent of pH between 2 and 8. The solvent chemistry (i.e., pH, buffering capacity, chelators, or protein) has no effect upon the dissolution. Exposure to inhaled talc dusts has been shown to cause a relatively benign lung pneumoconiosis, Gamble 1950, talcosis. Because talc dusts are so prevalent[†] in the workplace talc dusts are also regulated.

Jurinski and Rimstidt 2001 used a shrinking sphere model to describe the dissolution of talc particles in several fluids—deionized water adjusted to pH = 2, 4, and 7.4; phosphate-buffered saline solution (pH = 7.4); and two types of Gamble's solutions,[‡] one containing protein. This model assumes uniform dissolution over the surface of the sphere. The time required to dissolve a sphere is given by:

$$t = d/2kV_m \qquad (14.6)$$

where t is the time (sec), d is particle diameter (m), k is the rate constant (mol/ m^2s), and V_m is the molar volume (m^3/mol). Using this equation Jurinski and Rimstidt 2001 estimated the lifetime of a talc particle that started out at 1 µm in diameter to be 8 years. This model may overestimate the lifetime since talc most likely does not dissolve evenly across the basal plane.

---

[*] $Mg_3Si_4O_{10}(OH)_2$—a 2:1 layered silicate (chrysotile is a 1:1 layered silicate) containing an octahedral layer of Mg on either side of the Si tetrahedral layer and thus remains flat or platy as opposed to chrysotile, which curls up into fibrils.

[†] Talc is a component in many industrial products including pharmaceuticals, paints, and coatings.

[‡] Gamble's solution is a simulated extracellular lung fluid and contains various inorganic salts mostly NaCl, and HCO_3^-, H_2CO_3, $HPO_4^=$, $SO_4^=$, organic acid, and protein with a pH of 7.4.

Nanoparticles

Although a large amount of literature is available concerning the toxicity of nanomaterials, very little is available concerning their solubility in body fluids. Their extremely small size makes their chemical and physical properties different fundamentally from larger particles of the same composition. Carbon nanotubes, copper oxide, iron oxide, titanium dioxide, and zinc oxide have all been investigated, with copper oxide being the worst health risk, Anon. 2008. Carbon nanotubes have been shown to exhibit the same effects as asbestos fibers, Poland et al. 2008.

Below the concentration level of 1 mg/L, SiC exhibited no cytotoxicity against macrophages, Guevara-Lora et al. 2009.

An S-shaped curve for the generation of ROS was observed by Saliner et al. 2009 for nine different particle sizes for TiO_2. Above 30 and below 10 nm the activity was constant. The activity decreased from 30 to 10 nm. ROS generation was greatest for amorphous TiO_2, least for rutile, and intermediate for anatase. Using density functional theory (DFT),[*] Vittadini et al. 2007 studied the reactivity of various crystallographic faces of anatase and found that the (001) face was more reactive than the (101) face. They attributed this to the development of tensile stresses due to the unnatural geometric configuration of bridging oxygens.

Other ceramic nanoparticles that have all been investigated in various drug delivery applications are alumina, calcium phosphate, calcium carbonate, iron oxides, silica, and titania. The literature in general discusses the production of the nanoparticles, the incorporation of the drugs, and the release of the drugs, but not the solubility of the ceramic nanoparticles in the human body.[†] For these data one must rely on the solubility data developed for other applications.

Synthetic Mineral Fibers

The dissolution behavior in simulated body fluid of fibers[‡] of β-dicalcium silicate (β-Ca_2SiO_4) and the natural mineral hillebrandite ($Ca_2(SiO_3)(OH)_2$), both prepared by hydrothermal methods, were investigated by Gou et al. 2004. Within 2 weeks the two fiber types had dissolved by 30.42% and 30.91%,

---

[*] DFT is an *ab initio* simplified quantum mechanical–based calculation method used for systems of atoms less than several hundred.

[†] Paul and Sharma reported, "One issue that still needs to be documented is how quickly the CaP particles break down," Paul and Sharma 2009.

[‡] Fibers of 1.0–1.5 μm diameter and 15–25 μm in length were tested in SBF at 37°C for 1, 7, and 14 days.

TABLE 14.6

Names and Compositional Ranges
of Synthetic Glass Fibers

Oxide	Rock Wool	Glass Wool
SiO_2	43–50	60–70
Al_2O_3	6–15	3–7
FeO	3–8	<0.5
CaO+MgO	23–33	8–12
Na_2O+K_2O	1–3.5	13–18
B_2O_3	0.00	3–7
R_2O/RO	<1	>1

respectively. Both hillebrandite and β-Ca_2SiO_4 exhibited an initial surface layer formation of $Ca_3Si_2O_7 \cdot H_2O$ after 1 day. At longer times hydroxyapatite formed. Gou et al. 2004 concluded that these characteristics indicated that these fibers were excellent candidates as reinforcement for biocomposites.

In addition to the beneficial bioactive glasses there is the extremely important area of hazardous health effects from manufactured glasses. One such case is that of inhalation of glass fibers. The dissolution of these fibers is very critical in determining their health risk. According to OSHA there is insufficient evidence to support respiratory disease in humans, although experiments with animals indicate there may be limited evidence of carcinogenicity from synthetic mineral fibers. Synthetic mineral fibers vary in composition depending upon their application. They also vary in names (see Table 14.6). Three major categories exist—fiberglass, mineral wool, and refractory ceramic fiber. Fiberglass is further broken down into glass wool and continuous or textile fibers. Mineral wool can be either rock/stone wool or slag wool. The names are related to the source of the raw materials. Glass fibers tend to break transversely, forming many shorter fibers. Because of this they tend to be expelled easier than asbestos, which tends to split longitudinally, forming more long fibers not as easily expelled.

Bauer 2000 reported the work of Eastes and Hadley that glass fibers greater than 20 μm, if inhaled, have been correlated to respiratory disease in laboratory animals. The dissolution was dependent upon the fiber surface chemistry and physical nature. The continuous movement of fluids in the human lung can increase the dissolution rate and also transport the dissolved species to other parts of the body via the bloodstream. Mattson 1994 stressed that flow rates and various analytical methods can contribute substantial error to experimental dissolution rates. Aluminosilicate fibers were the most durable, while the dissolution rate of borosilicate fibers (i.e., home insulation) was 1000 times greater. The biopersistence of 1 μm diameter fibers varied from several days to as long as 14 years depending upon their chemistry. Annealing fibers at temperatures below the transition temperature decreased the dissolution rate in simulated extracellular fluid (pH = 7.4) by 2 to 3 times. The fact that they have not shown any major adverse reaction in human lungs was attributed by Bauer to the high dissolution rate of glass fibers.

Several investigators, Scholze and Conradt 1987, Bauer et al. 1988, Potter and Mattson 1999, have determined the dissolution rates of various glass fibers under *in vitro* conditions believed to simulate the extracellular space of the lung. Considerable variation of the magnitude of the dissolution rates and the relative ranking of the different compositions was present. These studies used different modifications of Gamble's solution and different flow rates and also used different analytical methods to determine the dissolution rates.

Implantation of fibers into laboratory animals was done by artificial means that bypassed the normal body defenses and thus the results are questionable as to their relevance to humans. Hesterberg et al. 2001 actually stated that "implantation studies (into rodents) are not valid for risk assessment or for concluding anything about the human health hazard associated with the inhalation of airborne SVFs (synthetic vitreous fibers)."

Therapeutic Materials

A ^{90}Y-containing radiotherapeutic REAS is sold under the trade name TheraSphere™.* White and Day 1994 reported no detectable weight loss of a 1 × 1 × 0.2 cm glass sample before 6 weeks in 100 mL of distilled water ($pH = 7$) or saline ($pH = 7.4$) at 37, 50, or 70°C. Dissolution rates of $\leq 3 \times 10^{-9}$ g/cm^2 · min were determined after 6 weeks. In a comparison study of fused silica, a Corning glass (CGW-1723™) and yttria aluminosilicate (YAS), Oda and Yoshino 1991 showed that YAS was significantly more durable than fused silica in saturated steam at 300°C and 8.6 MPa. The dissolution mechanism is very important for applications in the human body; however, it is very difficult to determine whether these glasses exhibit congruent or incongruent dissolution. Surface analyses of microspheres and bulk glasses indicated that the mechanism was congruent, White and Day 1994. Using inductively coupled plasma (ICP) and atomic adsorption (AA) spectroscopy it has been determined that the yttrium release from YAS microspheres in distilled water or saline at 37 or 50°C was below detectable limits, Erbe and Day 1991.

More recently Conzone et al. 2002 have reported the development of borate glasses for use in treatment of rheumatoid arthritis, since these glasses are potentially more reactive with physiological liquids. Borate glasses containing only alkali ions dissolved uniformly (i.e., congruently) in simulated physiological liquids at temperatures ranging from 22 to 75°C. When the borate glasses contained other cations (such as Ca, Mg, Fe, Dy, Ho, Sm, and Y) in

---

* TheraSphere™ is manufactured by MDS Nordion located in Ottawa, Ontario, Canada.

amounts ranging from 2 to 30 wt% dissolution was nonuniform (i.e., incongruent) with the formation of new compounds. Day 2002/03 gave an example of Dy_2O_3-containing borate solid glass microspheres that reacted to form hollow spheres, shells of concentric layers, or microspheres filled with homogeneous gel-like material depending upon the Dy_2O_3 content. The dissolution mechanism involved the selective leaching of lithium and boron, allowing the rare earth (i.e., Dy) to react and form an insoluble phosphate.[*] When calcium-containing borate glasses were reacted a semicrystalline or gel calcium phosphate formed that had a composition very similar to hydroxyapatite. *Although early work by Hench and colleagues has indicated the need for the formation of a silica gel surface layer for silicate glasses to be bioactive, the work of Day and colleagues has indicated that a silica gel is not always necessary for bioactivity.*

Professor Delbert Day and his son Ted along with a former student, Dr. Steve Jung, developed a bioactive calcium-containing borate glass nanofiber (called DermaFuse[†]) for wound-healing applications, Wray 2011. They determined that borate glasses reacted up to five times faster with human body fluids than did silicate glasses. They also found that calcium-containing borate glasses helped the body regulate the wound-healing process. Apparently calcium is a factor in wound healing since it is expected that calcium is required for migration of epidermal cells. They also tested a lithium-containing borate glass for its antibacterial function. It was thought that bacteria were killed when lithium was leached[‡] from the borate glass, causing a local increase in the *p*H.

Avent et al. 2003 studied the dissolution of Na-Ca-phosphate glasses containing small amounts of silver in an attempt to develop biocompatible controlled-release glasses for application in medical equipment such as urological catheters. It has been known for a long time that traces of silver have bactericidal properties.[§] With that in mind, Avent et al. investigated the dissolution of several glass compositions in distilled water and two different simulated urine solutions at 25 and 35°C. They found that silver release was dependent upon the Na/Ca ratio of the glass and that silver release was double in simulated urine compared to distilled water. They concluded that these glass compositions dissolved by destruction of the links between polyphosphate species with the dominant polyphosphate species being *cyclo*-hexophosphate.

The dissolution of sol-gel derived silica fibers along with the release of the drug dexmedetomidine hydrochloride was studied by Czuryszkiewicz et al. 2002. The release of the drug and fiber dissolution were studied in 0.9 wt% NaCl solution and in pure ethanol at 37°C. The dissolution was found

---

[*] The phosphorus is from a phosphate-buffered saline simulated physiological liquid.

[†] DermaFuse is designed to mimic the microstructure of fibrin, which is one of the main components of blood clots. Like fibrin, the glass fibers trap blood platelets and provide a scaffolding across which the wound covering forms.

[‡] The wound fluids are rich in enzymes.

[§] Other elements with antibacterial effects include Cu, F, Hg, and Zn.

to be dependent upon the sol-gel pH during manufacture of the fibers. The drug release from these fibers occurred in three steps. An initial burst was followed by a diffusion-controlled step and then a final slower release rate. This three-step process was attributed to the surface structure of the fibers.

Phosphate-based glass fibers containing copper in amounts up to 10 mol% were investigated as a localized antibacterial delivery system by Neel et al. 2005. They found that increasing copper content caused a decrease in dissolution of the glass fibers in deionized water at 37°C; however, increased copper content caused an increased copper release. The most effective composition for killing the pathogen *Staphylococcus epidermidis* contained 10 mol% copper.

Queiroz et al. 2003 investigated the reinforcement of HA with glass containing $15CaO$, $75P_2O_5$, and $10CaF_2$ mol% for potential drug delivery applications. They found that dissolution increased as the glass content increased from 4.0 to 7.5 wt%, both composites being more soluble than HA. The control of dissolution is extremely important for drug release applications. Due to the sintering conditions during preparation, the composites may contain some tricalcium phosphate (TCP). This is an advantage for drug delivery since TCP is more soluble than HA. Dissolution studies were performed in deionized water and simulated body fluid for times up to 60 days. As the composite dissolved, a precipitate of calcium phosphate formed on the surface, being a homogeneous layer on the higher glass content composite.

PerOssal®,* a composite of nanocrystalline hydroxyapatite and calcium sulfate (approximately a 50/50 mixture), not only is a bone replacement material but also can be inserted into infected bone defects with a simultaneous delivery of antibiotics. Calcium sulfate (plaster of Paris) has been used for many years to treat bone defects and as an antibiotic carrier, Rauschmann et al. 2005. However, there is a short-term cytotoxic effect with the use of calcium sulfate. Dissolution of the calcium sulfate caused a local acidic environment with subsequent inflammation. Combining the calcium sulfate with hydroxyapatite apparently solved this problem.

Additional Recommended Reading

Brown, P.W. and Constantz, B. 1994. *Hydroxyapatite and Related Materials*, CRC Press, Boca Raton, FL.

Hench, L.L. and Wilson, J. 1993. *An Introduction to Bioceramics*, Advanced Series in Ceramics, Vol. 1, World Scientific.

Weigand, A., Buchalla, W. and Attin, T. 2007. Review on Fluoride-Releasing Restorative Materials—Fluoride Release and Uptake Characteristics, Antibacterial Activity and Influence on Caries Formation, *Dent. Mater.*, 23, 343–62.

---

* Manufactured by Coripharm, Dieburg, Germany.

Exercises, Questions, and Problems

1. Describe the four types of ceramic biomaterials.
2. What is the general chemistry of the bioactive crystalline materials?
3. Discuss the chemical characteristics of bioactive apatite.
4. Discuss the effects of pH upon the dissolution of characteristics of hydroxyapatite.
5. What is the difference between a bioactive and a bioresorbable material?
6. What property of brushite limits its use as a bone substitute material?
7. Why is fluorescence important in dental ceramics? Discuss.
8. What are asbestiform materials?
9. Discuss the dissolution of chrysotile in simulated lung fluid.
10. What are therapeutic materials and what characteristics must they possess?

References

Ahmed, I., Lewis, M., Olsen, I. and Knowles, J.C. 2004a. Phosphate Glasses for Tissue Engineering: Part 1. Processing and Characterisation of a Ternary-Based P_2O_5-CaO-Na_2O Glass System, *Biomaterials, 25,* 491–99.

Ahmed, I., Lewis, M., Olsen, I. and Knowles, J.C. 2004b. Phosphate Glasses for Tissue Engineering: Part 2. Processing and Characterisation of a Ternary-Based P_2O_5-CaO-Na_2O Glass Fibre System, *Biomaterials, 25,* 501–7.

Ahmed, I., Collins, C.A., Lewis, M.P., Olsen, I. and Knowles, J.C. 2004c. Processing, Characterisation and Biocompatibility of Iron-Phosphate Glass Fibres for Tissue Engineering, *Biomaterials, 25,* 3223–32.

Andersson, O.H., Yrjas, K.P. and Karlsson, K.H. 1991. Reactions in and at the Surface of Bioactive Glasses in Aqueous Solutions, pp. 127–133 in *Bioceramics, Vol. 4,* Proc. 4th Int. Sym. Ceram. in Med., Bonfield, W., Hastings, G.W. and Tanner, K.E. (eds.), Butterworth-Heinmann, London, UK.

Anonymous. 2008. Study Sizes Up Nanomaterial Toxicity, *Chem. & Engr. News, 86*(35), 1 Sep.

Anusavice, K.J. 1992. Degradability of Dental Ceramics, *Advances in Dental Res., 6,* 82–89, Sep.

Anusavice, K.J. 1995. Development and Testing of Ceramics for Dental Restorations, pp. 101–24 in *Bioceramics: Materials & Applications,* Fischman, G., Clare, A. and Hench, L.L. (eds.), *Ceramic Transactions Vol. 48,* Amer. Ceram. Soc., Westerville, OH, p. 329.

Anusavice, K.J. and Zhang, N-Z. 1997. Chemical Durability of Dicor® and Lithia-Based Glass-Ceramics, *Dental Materials, 13*(1), 13–19.

Anusavice, K.J. and Zhang, N-Z. 1998. Chemical Durability of Dicor and Fluorocanasite-Based Glass-Ceramics, *J. Dent. Res., 77*(7), 1553–59.

Arinzeh, T.L., Tran, T., Mcalary, J. and Daculsi, G. 2005. A Comparative Study of Biphasic Calcium Phosphate Ceramics for Human Mesenchymal Stem-Cell-Induced Bone Formation, *Biomaterials, 26*, 3631–38.

Avent, A.G., Carpenter, C.N., Smith, J.D., Healy, D.M. and Gilchrist, T. 2003. The Dissolution of Silver-Sodium-Calcium-Phosphate Glasses for the Control of Urinary Tract Infections, *J. Non-Cryst. Solids, 328*, 31–39.

Bandyopadhyay, A., Bernard, S., Xue, W. and Bose, S. 2006. Calcium phosphate-based resorbable ceramics: Influence of MgO, ZnO and SiO_2 Dopants, *J. Amer. Cer. Soc.,* 89(9), 2675–88.

Bauer, J.F. 2000. Corrosion and Surface Effects of Glass Fiber in Biological Fluids, *Glass Res.* (2), 4–5.

Bauer, J.F., Law, B.D., Roberts, K.A. 1988. A Study of the Solubility and Durability of Man-Made Mineral Fibers in a Synthetic Physiological Fluid, Presented at the Environmental Chemistry Symposium, 30th Rocky Mountain Conference, 1 August, Denver, CO.

Berger, C., Breitinger, D., Grellner, F. and Greil, P. 1996. Stability of Hydroxyapatite in Glassy Matrices, in *Bioceramics: Materials and* Applications II, Ceram. Trans., 63, Rusin, R.P. and Fischman, G.S. (eds.), Amer. Cer. Soc., Westerville, OH, pp. 129–33.

Birchall, J.D., Stanley, D.R., Mockford, M.J., Pigott, G.H., and Pinto, P.J. 1988. Toxicity of Silicon Carbide Whiskers, *J. Mat. Sci. Lets., 7*(4), 350.

Bronckers, A.L.J.J., Lyaruu, D.M. and denBesten, P.K. 2009. The Impact of Fluoride on Ameloblasts and the Mechanisms of Enamel Fluorosis, *J. Dental Res., 88*(10), 877–93.

Campbell, A. 2003. Bioceramics for Implant Coatings, *Materialstoday*, Nov., 26–30.

Charles, R.J. 1958. Static Fatigue of Glass, *J. Appl. Phys., 29*(11), 159–63.

Chen, Y. and Miao, X. 2005. Thermal and Chemical Stability of Fluoro-Hydroxyapatite Ceramics with Different Fluorine Contents, *Biomaterials, 26*(11), 1205–10.

Chow, A.Y. 1993. Electrical Stimulation of the Rabbit Retina with Subretinal Electrodes and High Density Microphotodiode Array Implants, *Invest. Ophtalmol. Vis. Sci. (Suppl.), 34*, 835.

Conzone, C.D., Brown, R.F., Day, D.E. and Ehrhardt, G.J. 2002. *In Vitro* and *In Vivo* Dissolution Behavior of a Dysprosium Lithium Borate Glass Designed for the Radiation Synovectomy Treatment of Rheumatoid Arthritis, *J. Biomedical Materials Research, 60*(2), 260–68.

Czuryszkiewicz, T., Ahvenlammi, J., Kortesuo, P., Ahola, M., Kleitz, F., Jokinen, M., Lindén, M. and Rosenholm, J.B. 2002. Drug Release from Biodegradable Silica Fibers, *J. Non-Cryst. Solids, 306*, 1–10.

Day, D.E. 2002/03. Reactions of Bioactive Borate Glasses with Physiological Liquids, *Glass Researcher, 12*(1&2), 21–22.

DeRijk, W.G., Jennings, K.A. and Menis, D.L. 1985. A Comparison of Chemical Durability Test Solutions for Dental Porcelains, in *Biomedical Engineering—Recent Developments*, Sauer, B.W. (ed.), Proceedings, Southern Biomedical Engineering Conf., New York, Pergamon, pp. 152–55.

Erbe, E.M. and Day, D.E. 1991. *Proceedings: Science & Tech of New Glasses*, Sakka, S. and Soga, N. (eds.), Ceram. Soc. Japan, Tokyo, p. 105.

Favero-Longo, S.E., Turci, F., Tomatis, M., Compagnoni, R., Piervittori, R. and Fubini, B. 2009. The Effect of Weathering on Ecopersistence, Reactivity, and Potential Toxicity of Naturally Occurring Asbestos and Asbestiform Minerals, *Journal of Toxicology and Environmental Health, Part A, 72,* 305–14.

Fulmer, M.T., Ison, I.C., Hankermayer, C.R., Constantz, B.R. and Ross, J. 2002. Measurements of the Solubilities and Dissolution Rates of Several Hydroxyapatites, *Biomaterials, 23*(3), 751–55.

Gamble, J.L. 1950. *Chemical Anatomy, Physiology, and Pathology of Extracellular Fluid,* Harvard Univ. Press, Cambridge, MA.

Gomez, P.F. and Morcuende, J.A. 2005. Early Attempts at Hip Arthroplasty—1700s to 1950s. *Iowa Orthop. J., 25,* 25–29.

Gottschling, S, R., Engel, A. and Oel, H.J. 1995. Characterization of Plasma Sprayed HA Coatings on Titanium Toothroot Implants, pp. 201–13 in *Bioceramics: Materials & Applications,* Fischman, G., Clare, A. and Hench, L.L. (eds.), *Ceramic Transactions Vol. 48,* Amer. Ceram. Soc., Westerville, OH, 329 pp.

Gou, Z., Chang, J., Gao, J. and Wang, Z. 2004. In Vitro Bioactivity and Dissolution of $Ca_2(SiO_3)(OH)_2$ and β–Ca_2SiO_4 Fibers, *J. Eur. Ceram. Soc., 24,* 3491–97.

Grossman, D.G. and Waters, H.V. 1984. The Chemical Durability of Dental Ceramics (abstract), *J. Dent. Res., 63,* 234.

Guevara-Lora, I., Czosnek, C., Smycz, A., Janik, J.F. and Kozik, A. 2009. SiC Nanoparticles as Potential Carriers for Biological Active Substances, *J. Physics: Conference Series, 146*(1), 012022.

Gustafsson, J.P. 2006. Visual MINTEQ 2.5.1, Department of Land and Water Resources Engineering, KTH, Stockholm, Sweden.

Guthrie Jr., G.D. 1992. Biological Effects of Inhaled Minerals, *Am. Miner., 77,* 225–43.

Hadley, J.G. 2000. A New Understanding of Fiber Safety, Insulation Outlook, National Insulation Assoc., http://fibrox.com/art02.htm.

Hämmerle, H., Kobuch, K., Kohler, K., Nisch, W., Sachs, H. and Stelzle, M. 2002. Biostability of Micro-photodiode Arrays for Subretinal Implantation, *Biomaterials, 23,* 797–804.

Hankermeyer, C.R., Ohashi, K.L., Delaney, D.C., Ross, J. and Constantz, B.R. 2002. Dissolution Rates of Carbonated Hydroxyapatite in Hydrochloric Acid, *Biomaterials, 23,* 743–50.

Hench, L.L. 1998. Bioceramics, *J. Amer. Cer. Soc., 81,* 1705–1728.

Hench, L.L. 2002/03. Bioactive Glasses Help Heal, Repair and Build Human Tissue, *Glass Researcher, 12*(1&2), 18.

Hench, L.L. and Wilson, J. 1993. Introduction, pp. 1–24 in *An Introduction to Bioceramics,* Advanced Series in Ceramics, Vol. 1, World Scientific.

Hench, L.L., Andersson, O.H. and Latome, G.P. 1991. The Kinetics of Bioactive Ceramics, Part III: Surface Reactions for Bioactive Glasses Compared with an Inactive Glass, pp. 155–62 in *Bioceramics, Vol. 4,* Proc. 4th Int. Sym. Ceram. in Med., Bonfield, W., Hastings, G.W. and Tanner, K.E. (eds.), Butterworth-Heinmann, London, UK.

Hesterberg, T.W., Anderson, R., Bunn III, W.B., Chase, G.R. and Hart, G.A. 2001. Man-Made Mineral Fibers, pp. 1227–40, Chap. 112 in *Clinical Environmental Health and Toxic Exposures,* 2nd ed., Sullivan Jr., J.B. and Krieger, G.R. (eds.), Lippincott Williams & Wilkins, a W. Kluwer Company, Philadelphia, PA, 1323 pp.

Holland, J.P. and Smith, D.D. 2001. Asbestos, pp. 1214–27, Chap. 111 in *Clinical Environmental Health and Toxic Exposures*, 2nd ed., Sullivan Jr., J.B. and Krieger, G.R. (eds.), Lippincott Williams & Wilkins, a W. Kluwer Company, Philadelphia, PA, 1323 pp.

Huang, W., Day, D.E., Kittiratanapiboon, K. and Rahaman, M.N. 2006a. Kinetics and the Mechanism of the Conversion of Silicate (45S5), Borate, and Borosilicate Glasses to Hydroxyapatite in Dilute Phosphate Solutions, *J. Mater. Sci. Mater. Med.*, 17, 583–96.

Huang, W., Rahaman, M.N., Day, D.E. and Li, Y. 2006b. Mechanisms for Converting Bioactive Silicate, Borate, and Borosilicate Glasses to Hydroxyapatite in Dilute Phosphate Solutions, *Phys. Chem. Glass.: Europ. J. Glass Sci. Technol. Part B*, 47(6), 647–58.

Hulbert, S.F., Bokros, J.C., Hench, L.L., Wilson, J. and Heimke, G. 1987. pp. 189–213 in *High Tech Ceramics*, Vincenzini, P. (ed.), Elsevier Science B.V., Amsterdam.

Hume, L.A. and Rimstidt, J.D. 1992. The Biodurability of Chrysotile Asbestos, *Amer. Mineral.*, 77(9/10), 1125–28.

http://www.osha.gov/SLTC/syntheticmineralfibers/table2.html.15.36.

Jensen, S.S., Bornstein, M.M., Dard, M., Bosshardt, D.D. and Buser, D. 2009. Comparative Study of Biphasic Calcium Phosphates with Different HA/TCP Ratios in Mandibular Bone Defects: A Long-Term Histomorphometric Study in Minipigs, *J. Biomed. Mater. Res., Part B, Appl. Biomaterials*, 90B(1), 171–81.

John, A., Nair, M.B., Varma, H.K., Bernhardt, A. and Gelinsky, M. 2008. Biodegradation and Cytocompatibility Studies of a Triphasic Ceramic-Coated Porous Hydroxyapatite for Bone Substitute Applications, *Int. J. Appl. Ceram. Tech.*, 5(1), 11–19.

Jolicoeur, C. and Poisson, D. 1987. Surface Physico-Chemical Studies of Chrysotile Asbestos and Related Minerals, *Drug and Chemical Toxicology*, 10(1&2), 1–47.

Jurinski, J.B. and Rimstidt, J.D. 2001. Biodurability of Talc, *Amer. Mineral.*, 86, 392–99.

Kasuga, T. 2005. Bioactive Calcium Pyrophosphate Glasses and Glass-Ceramics, *Acta Biomaterialia*, 1, 55–64.

Kokubo, T., Hata, K., Nakamura, T. and Yamamuro, T. 1991. Apatite Formation on Ceramics, Metals and Polymers Induced by a CaO-SiO$_2$ Based Glass in a Simulated Body Fluid, pp. 113–20 in *Bioceramics, Vol. 4*, Bonfield, W., Hastings, G.W. and Tanner, K.E. (eds.), Butterworth-Heinmann, London, UK.

LeGeros, R.Z. Calcium Phosphates in Oral Biology and Medicine, in *Monographs in Oral Science*, Vol. 15, Myers, H. (ed.), Karger AG, Basel, 200 pp.

LeGeros, R.Z., LeGeros, J.P., Kim, Y., Kijkowska, R., Zheng, R., Bautista, C. and Wang, J.L. 1995. Calcium Phosphate in Plasma-Sprayed HA Coating, pp. 173–89 in *Bioceramics: Materials & Applications*, Fischman, G., Clare, A. and Hench, L.L. (eds.), *Ceramic Transactions Vol. 48*, Amer. Ceram. Soc., Westerville, OH, 329 pp., 15.14.

Leonor, I.B., Ito, A., Onuma, K., Kanzaki, N. and Reis, R.L. 2003. In Vitro Bioactivity of Starch Thermoplastic/Hydroxyapatite Composite Biomaterials: An In-Situ Study Using Atomic Force Microscopy, *Biomaterials*, 24, 579–85.

Leventouri, Th., Antonakos, A., Kyriacou, A., Venturelli, R., Liarokapis, E. and Perdikatsis, V. 2009. Crystal Structure of Human Dental Apatite as a Function of Age, *Int. J. Biomaterials*, Vol. 2009, Article ID 698547, 6 pp., doi: 10.1155/2009/698547.

Levinskas, G.J. and Neuman, W.F., 1955. The Solubility of Bone Mineral. Solubility Studies of Synthetic Hydroxyapatite, *J. Chem. Phys., 59,* 164–68.

Li, H., Khor, K.A. and Cheang, P. 2002. Titanium Dioxide Reinforced Hydroxyapatite Coatings Deposited by High Velocity Oxy-Fuel (HVOF) Spray, *Biomaterials, 23*(1), 85–91.

Li, Y., Rahaman, M.N., Bal, B.S., Day, D.E. and Fu, Q. 2008. Early Stages of Calcium Phosphate Formation on Bioactive Borosilicate Glass in Aqueous Phosphate Solution, *J. Am. Ceram. Soc., 91*(5), 1528–33.

Mackert Jr., J.R. 1992. Side-Effects of Dental Ceramics, *Advances in Dental Res., 6,* 90–93, Sep.

Mathew, M. and Takagi, S. 2001. Structures of Biological Minerals in Dental Research, *J. Res. Natl. Inst. Stand. Technol., 106*(6), 1035–44.

Mattson, S.M. 1994. Glass Fiber Dissolution in Simulated Lung Fluid and Measures Needed to Improve Consistency and Correspondence to In-Vivo Dissolution, in *Environmental Health Perspectives, Vol. 102, Supplement 5: Biopersistence of Respirable Synthetic Fibers and Minerals* (Oct.), pp. 87–90.

Mavropoulos, E., Rossi, A.H., da Rocha, N.C.C., Soares, G.A., Moreira, J.C. and Moure, G.T. 2003. Dissolution of Calcium-Deficient Hydroxyapatite Synthesized at Different Conditions, *Materials Characterization, 50,* 203–7.

May, P.M., Linder, P.W. and Williams, D.R. 1977. Computer Simulation of Metal-Ion Equilibria in Biofluids: Models for the Low-Molecular-Weight Complex Distribution of Calcium (II), Magnesium (II), Iron (III), Copper (II), Zinc (II), and Lead (II) Ions in Human Blood Plasma, *J. Chem. Soc. Dalton Trans.,* 588–95.

McClellan, R.O. 2000. Particle Interaction with the Respiratory Tract, pp. 3–66 in *Particle-Lung Interactions*, Gehr, P. and Heyder, J. (eds.), Marcel-Dekker.

Ménard, H., Noël, L., Kimmerle, F.M., Tousignant, L. and Lambert, M. 1984. Adsorption Isotherms of Polycyclic Aromatic Hydrocarbons on Asbestos Chrysotile by High-Pressure Liquid Chromatography, *Anal. Chem., 56*(8), 1240–42.

Meurman, J.H., Toskala, J., Nuutinen, P. and Klemetti, E. 1994. Oral and Dental manifestations in Gastroesophageal Reflux Disease, *Oral Surg., Oral Med., Oral Pathol., 78,* 583–89.

Nakamura, T., Yamamuro, T., Higashi, S., Kokubo, T. and Ho, S. 1985. A new glass-ceramic for bone replacement: Evaluation of its bonding to bone tissue, *J. Biomed. Mater. Res., 19,* 685–98.

Neel, E.A.A., Ahmed, I., Pratten, J., Nazhat, S.N. and Knowles, J.C. 2005. Characterisation of Antibacterial Copper Releasing Degradable Phosphate Glass Fibres, *Biomaterials, 26,* 2247–54.

Nel, A. 2006. Toxic Potential of Materials at the Nanolevel, *Science, 311*(5761), 622–27, 3 Feb.

Ni, J. and Wang, M. 2002. In Vitro Evaluation of Hydroxyapatite Reinforced Polyhydroxybutyrate Composite, *Mat. Sci. and Engr., C20,* 101–9.

Niemelä, T., Niiranen, H., Kellomäki, M., et al. 2005. Self-Reinforced Composites of Bioabsorbable Polymer and Bioactive Glass with Different Bioactive Glass Contents. Part I: Initial Mechanical Properties and Bioactivity, *Acta Biomaterialia, 1*(2), 235–42.

Nonami, T. and Wakai, F. 1995. Evaluation of Crack Propagation in Hydroxyapatite by Double-Torsion Method in Air, Water and Toluene, *J. Ceram. Soc. Jpn., 103*(1198), 648–52.

Oberdörster, G., Oberdörster, E., Oberdörster, J. 2005. Nanotoxicology: An Emerging Discipline Evolving from Studies of Ultrafine Particles, *Environmental Health Perspectives*, 113(7), 823–39, Jul.

Oda, K. and Yoshino, T. 1991. Properties of Y_2O_3-Al_2O_3-SiO_2 Glasses as a Model System of Grain Boundary Phase of Si_3N_4 Ceramics, Part 2: Leaching Characteristics, *J. Ceram. Soc. Japan*, 99(11), 1150–52.

Ohura, K., Bohner, M., Hardouin, P., Lemaitre, J., Pasquier, G. and Flautre, B. 1996. Resorption of, and Bone Formation from, New β-Tricalcium Phosphate Cements: An In Vivo Study, *J. Biomed. Mater. Res., B*, 30, 193–200.

Øilo, G. 1992. Biodegradation of Dental Composites/Glass-Ionomer Cements, *Advances in Dental Res.*, 50, 50–54.

Oze, C. and Solt, K. 2010. Biodurability of Chrysotile and Tremolite Asbestos in Simulated Lung and Gastric Fluids, *Am. Mineral.*, 95, 825–31.

Padilla, S., Román, J., Carnas, A. and Vallet-Regi, M. 2005. The Influence of the Phosphorous Content on the Bioactivity of Sol-Gel Glass-Ceramics, *Biomaterials*, 26, 475–83.

Pantano, C.G., Clark, A.E. and Hench, L.L. 1974. Multilayer Corrosion Films on Bioglass Surfaces, *J. Amer. Cer. Soc.*, 57, 412–13.

Paul, W. and Sharma, C.P. 2010. Fatty Acid Conjugated Calcium Phosphate Nanoparticles for Protein Delivery, *International J. App. Ceramic Tech.*, 7(2), 129–34.

Pina, S. and Ferreira, J.M.F. 2010. Brushite-Forming Mg-, Zn- and Sr-Substituted Bone Cements for Clinical Applications, *Materials*, 3, 519–35.

Plumlee, G.S., Morman, S.A. and Ziegler, T.L. 2006. The Toxicological Geochemistry of Earth Materials: An Overview of Processes and the Interdisciplinary Methods Used to Understand Them, *Reviews in Mineralogy & Geochemistry, Vol. 64, Medical Mineralogy and Geochemistry*, Sahai, N. and Schoonen, M.A.A. (eds.), Min. Soc. Am., Chantilly, VA, 5–57.

Poland, C.A., Duffin, R., Kinloch, I., Maynard, A., Wallace, W.A.H., Seaton, A., Stone, V., Brown, S., MacNee, W., and Donaldson, K. 2008. Carbon Nanotubes Introduced into the Abdominal Cavity of Mice Show Asbestos-Like Pathogenicity in a Pilot Study, *Nature Nanotechnology*, 3(7), 423.

Potter, R.M. and Mattson, S.M. 1991. Glass Fiber Dissolution in a Physiological Saline Solution, *Glastech. Ber.*, 64, 16–28.

Queiroz, A.C., Santos, J.D., Monteiro, F.J. and Prado da Silva, M.H. 2003. Dissolution Studies of Hydroxyapatite and Glass-Reinforced Hydroxyapatite Ceramics, *Materials Characterization*, 50, 197–202.

Rauschmann, M.A., Wichelhaus, T.A., Stirnal, V., Dingeldein, E., Zichner, L., Schnettler, R. and Alt, V. 2005. Nanocrystalline Hydroxyapatite and Calcium Sulphate as Biodegradable Composite Carrier Material for Local Delivery of Antibiotics in Bone Infections, *Biomaterials*, 26, 2677–84.

Ribeiro, C.C. and Barbosa, M.A. 1991. Influence of Metal Ions on the Dissolution Behavior of Hydroxyapatite, pp. 145–53 in *Bioceramics, Vol. 4*, Proc. 4th Int. Sym. Ceram. in Med., Bonfield, W., Hastings, G.W. and Tanner, K.E. (eds.), Butterworth-Heinmann, London, UK.

Rimstidt, J.D. and Barnes, H.L. 1980. The Kinetics of Silica-Water Reactions, *Geochimica et Cosmochimica Acta*, 44, 1683–99.

Rosso, J.J. and Rimstidt, J.D. 2000. A High Resolution Study of Forsterite Dissolution Rates, *Geochimica et Cosmochimica Acta*, 64, 797–811.

Saliner, A.G., Burello, E. and Worth, A. 2009. Review of Computational Approaches for Predicting the Physicochemical and Biological Properties of Nanoparticles, *JRC Scientific & Technical Reports*, EUR 23974 EN, ISSN 1018-5593, Luxembourg.

Scholze, H. and Conradt, R. 1987. An In Vitro Study of the Chemical Durability of Siliceous Fibers. *Ann. Occup. Hyg.*, *31*, 683–92.

Sébastien, P., McDonald, J.C., McDonald, A.D., Case, B. and Harley, R. 1989. Respiratory Cancer in Chrysotile Textile and Mining Industries: Exposure Inferences from Lung Analysis, *Br. J. Indust. Med.*, *46*, 180–87.

Seo, D.S., and Lee, J.K. 2007. The Crystallographic Influence on the Dissolution and Aligned Crystallites of Hydroxyapatite, *J. Ceram. Processing Res.*, *8*(4), 233–37.

Seo, D.S., Kim, H. and Lee, J.K. 2007. Mechanism of Surface Dissolution in Dense Hydroxyapatite, *Solid State Phenom.*, *121–23*, 1241–44.

Shikha, D., Jha, U., Sinha, S.K., Barhai, P.K., Kalavathy, S., Nair, K.G.M., Dash, S., Tyagi, A.K. and Kothari, D.C. 2008. Improvement in Corrosion Resistance of Biomaterial Alumina after 60keV Nitrogen Ion Implantation, *Int. J. Appl. Ceram. Tech.*, *5*(1), 44–48.

Söderholm, K.-J.M. 1981. Degradation of Glass Filler in Experimental Composites, *J. Dental Res.*, 1867–75.

Söderholm, K.-J.M., Achanta, S. and Olsson, S. 1994. Variables Affecting the Water Diffusing Coefficient of Experimental Composites (abstract), *J. Dental Res.*, *73* (Special Issue), 106.

Söderholm, K.-J.M., Mukherjee, R. and Longmate, J. 1996. Filler Leachability of Composites Stored in Distilled Water or Artificial Saliva, *J. Dental Res.*, *75*(9), 1692–99.

Sollböhmer, O., May, K-P. and Anders, M. 1995. Force Microscopical Investigation of Human Teeth in Liquids, *Thin Solid Films*, *264*(2), 176–83.

Taunton, A.E., Wood, S.A. and Gunter, M.E. 2002. The Thermodynamics of Asbestos Mineral Dissolution and Conversion in the Human Lung, 12th Annual Goldschmidt Conference, Davos, Switzerland.

Vallet-Regi, M. 2001. Ceramics for Medical Applications, *J. Chem. Soc. Dalton Trans.*, 97–108.

Vallet-Regi, M., Ragel, C.V. and Salinas, A.J. 2003. Glasses with Medical Applications, *Eur. J. Inorg. Chem.*, *6*, 1029–42.

Velayudhan, S., Anilkumar, T.V., Kumary, T.V., et al. 2005. Biological Evaluation of Pliable Hydroxyapatite-Ethylene Vinyl Acetate Co-polymer Composites Intended for Cranioplasty, *Acta Biomaterialia*, *1*(2), 201–9.

Vittadini, A., Casarin, M. and Selloni, A. 2007. Chemistry of and on TiO_2 Anatase Surfaces by DFT Calculations: A Partial Review, *Theoretical Chemistry Accounts*, *117*, 663–71.

Waltimo, T., Brunner, T.J., Vollenweider, M., Stark, W.J. and Zehnder, M. 2007. Antimicrobial effect of nanometric bioactive glass 45S5, *J. Dent. Res.*, *85*, 754–7.

Weber, G.H., O'Brien, J.K. and Bender, F.G. 2004. Control of *Escherichia coli* O157:H7 with Sodium Metasilicate, *J. Food Prot.*, *67*, 1501–6.

Werner, A.J., Hochella Jr., M.F., Guthrie Jr., G.D., Hardy, J.A., Aust, A.E. and Rimstidt, J.D. 1995. Asbestiform Riebeckite (Crocidolite) Dissolution in the Presence of Fe-Chelators: Implications for Mineral-Induced Disease, *Am. Mineral.*, *80*, 1093–1103.

White, J.E. and Day, D.E. 1994. Rare Earth Aluminosilicate Glasses for In Vivo Radiation Delivery, pp. 181–208 in *Rare Elements in Glasses*, Shelby, J.E. (ed.), Key Engineering Materials, Vols. 94–95, Trans Tech, Switzerland.

Wood, S.A., Taunton, A.E., Normand, C. and Gunter, M.E. 2006. Mineral-Fluid Interaction in the Lungs: Insights from Reaction-Path Modeling, *Inhalation Toxicology, 18*, 975–84.

Wray, P. 2011. Cotton Candy That Heals?, *Am. Ceram. Soc. Bull., 90*(4), 25–29.

Yin, X., Calderin, L., Stott, M.J. and Sayer, M. 2002. Density Functional Study of Structural, Electronic and Vibrational Properties of Mg- and Zn-Doped Tricalcium Phosphate Biomaterials, *Biomaterials, 23*(20), 4155–63.

Zerbo, I.R., Bronckers, A.L., de Lange, G.L., van Beek, G.J. and Burger, E.H. 2001. Histology of Human Alveolar Bone Regeneration with a Porous Tricalcium Phosphate—A Report of Two Cases, *Clin. Oral Implants Res., 12*(4), 379–84.

Xu, H.H.K., Weir, M.D., Sun, L., Takagi, S. and Chow, L.C. 2007. Effects of Calcium Phosphate Nanoparticles on Ca-PO$_4$ Composite, *J. Dental Res., 86*(4), 378–83.

Xu, H.H.K., Weir, M.D., Sun, L., Moreau, J.L., Takagi, S., Chow, L.C. and Antonucci, J.M. 2010. Strong Nanocomposites with Ca, PO$_4$, and F Release for Caries Inhibition, *J. Dental Res., 89*(1), 19–28.

15

Specific Crystalline Materials

You can observe a lot by just watching.

Yogi Berra

In an effort to classify corrosion along chemical lines, the following sections are separated into attack by various environments and then specific materials. The area of thermal barrier coatings and the newer one of environmental barrier coatings contain a wealth of information concerning the gaseous corrosion of ceramics. Even though a special section on coatings may be appropriate, these materials are discussed here based upon their chemistry. A strong attempt has been made to follow the original plan of the first edition to follow a classification based upon chemistry.

Attack by Liquids

Attack by Molten Glasses

In the indirect corrosion of oxides by glasses the crystalline phase that forms at the interface is dependent upon the glass composition and the temperature. Various interface phases that form in some silicate melts are listed in Table 15.1. Whether the system is under forced convection or not will also play an important role in the formation of a crystalline interface phase. An excellent study of the effects of forced convection is that by Sandhage and Yurek 1991, who in their studies of the indirect dissolution of chrome-alumina crystalline solution materials in $CaO-MgO-Al_2O_3-SiO_2$ melts at 1550°C reported that the reaction layer thickness of the spinel that formed decreased with increasing rotational rpm but did not change with time at constant rpm. The reaction layer was an order of magnitude thinner (30 vs. 300 µm) at 1200 rpm when compared to the case with no forced convection. *The investigator must be careful in his interpretation of the crystalline phases present after an experiment has been completed so that he does not confuse phases that precipitate during cooling with those that were present during the experiment.* The reader, if interested in a particular system, should examine the original articles of those listed in Table 15.1 to determine the exact experimental conditions. The following sections describe some of the more important

Corrosion of Ceramic Materials

TABLE 15.1

Interfacial Reaction Products Caused by Molten Liquid Attack

Oxide	Liquid[a]	Interface[a]	Ref
Al_2O_3	CAS	CA_2 & CA_6	Oishi et al. 1965
Al_2O_3	Coal Slag	Mixed spinel	Bonar et al. 1980
Al_2O_3	CMAS	$MgAl_2O_4$	Sandhage & Yurek 1988; 1990
Al_2O_3	S	A_3S_2	McCauley 1975
Al_2O_3-Cr_2O_3	CMAS	Mixed spinel	Sandhage & Yurek 1988; 1991
Al_2O_3-Cr_2O_3	Coal Slag	Mixed spinel	Bonar et al. 1980
AZS	Coal Slag	CA_6 & C_2AS	Bonar et al. 1980
AZS	NCS	NAS_2 & Z	McCauley 1975
AZS	KPS	KAS_2 & Z	Hilger et al. 1984
CaO	CAS	C_2S & C_3S	Matsushima et al. 1977
CaO	CFS	C_2S & C_3S	Matsushima et al. 1977
Cr_2O_3-spinel	Coal Slag	Mixed spinel	Bonar et al. 1980
Fused SiO_2	CAS	Cristobalite	Bates 1987
MgO	CAS	C_2MS_2 & M	Bates 1987
MgO	CFS	MF solution	Umakoshi et al. 1980
$MgAl_2O_4$	CAS	C_2AS or CAS_2	Bates 1987
$Al_6Si_2O_{13}$	NCS	NAS_2 & A	McCauley 1975
$Y_3Al_5O_{12}$	CAS	C_2AS	Bates 1987
$ZrSiO_4$	KPS	KZS_3 & Z	Bonetti et al. 1969

[a] A = Al_2O_3, C = CaO, F = FeO, K = K_2O, M = MgO, N = Na_2O, P = PbO, S = SiO_2, Z = ZrO_2.

systems that have been investigated, but no attempt has been made for an exhaustive survey.

Alumina-Containing Materials

The corrosion of multicomponent materials proceeds through the path of least resistance. Thus, those components with the lowest resistance are corroded first. This is really a form of selective corrosion and may proceed through either the direct or indirect corrosion process. The corrosion of a fusion-cast alumina-zirconia-silica (AZS) refractory will be used as an example of a case when selective direct corrosion is operative. This particular material is manufactured by fusing the oxides, casting into a mold, and then allowing crystallization to occur under controlled conditions. The final microstructure is composed of primary zirconia, alumina, alumina with included zirconia, and a glassy phase that surrounds all the other phases (Figure 15.1). The glassy phase (about 15% by volume) is necessary for this material to provide a cushion for the polymorphic transformation of zirconia during cooling and subsequent use. This material is widely used as a basin-wall material in soda-lime silica glass furnaces. The corrosion proceeds by the diffusion of

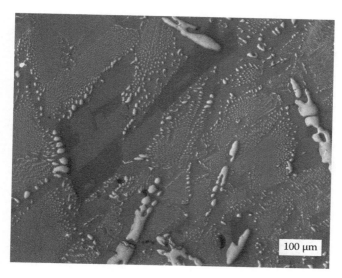

FIGURE 15.1

Microstructure of an unused fusion cast alumina-zirconia-silica refractory. Reflected oblique illumination (magnification 200×). Brightest areas are ZrO_2, next darker areas are Al_2O_3, next darker areas are silicate glass, and the few darkest spots are pores.

sodium ions from the bulk glass into the glassy phase of the refractory. As sodium ions are added to this glass, its viscosity is lowered and it becomes corrosive toward the refractory. The corrosion next proceeds by solution of the alumina and finally by partial solution of the zirconia. Under stagnant conditions, an interface of zirconia embedded in a high-viscosity alumina-rich glass is formed (Figure 15.2). If the diffusion of sodium ions into the glassy phase is sufficient, the glassy phase may contain sufficient sodium so that upon cooling, nepheline ($Na_2Al_2Si_2O_8$) crystals precipitate, or if the temperature is proper, the nepheline may form in service. The presence of nepheline has been reported by several investigators: McCauley 1975, Clauss and Salge 1974, and Derobert 1975. In actual service conditions, however, the convective flow of the bulk glass erodes this interface, allowing continuous corrosion to take place until the refractory is consumed. This type of corrosion can take place in any multicomponent material where the corroding liquid diffuses into a material that contains several phases of varying corrosion resistance.

Hilger et al. 1984 reported the corrosion of an AZS refractory by a potassium-lead-silicate glass at 1200°C to be very similar to that discussed above. In this case, the potassium ions diffuse into the glassy phase of the refractory, dissolving the alumina of the refractory and forming a glassy phase with a composition very similar to leucite ($K_2Al_2Si_4O_{12}$). Actual crystals of leucite were found upon examination of used blocks. It is interesting that very little lead diffuses into the refractory.

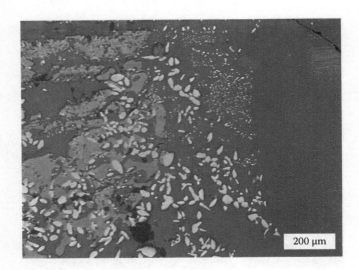

FIGURE 15.2
Refractory/glass interface of the refractory in Figure 15.1 corroded by a soda-lime-silica glass at 1450°C for 7 days showing the interface of ZrO_2. Reflected light illumination. (Sample courtesy of Corning, Inc.)

In these refractory materials containing ZrO_2, one should note that the ZrO_2 is very insoluble in soda-lime-silicate and potassium-lead-silicate glasses. Thus the corrosion of AZS refractories in these glasses is very similar to that which occurs in alumino-silicate (e.g., mullite) refractories. The major difference is the skeletal interface layer of undissolved ZrO_2 that forms on the AZS materials. The presence of lead in the corroding glass acts predominantly to lower the viscosity, with increasing lead contents producing more severe corrosion, Bonetti et al. 1969.

Lakatos and Simmingskold 1967 studied the effects of various glass constituents upon the corrosion of two pot clays, one with 21% alumina and one with 37% alumina. Their silicate glasses contained K_2O, Na_2O, CaO, and PbO in varying amounts. They found that PbO had no significant effect upon corrosion, that Na_2O was two to three times more corrosive than K_2O, and that CaO followed a cubic function. Since their tests were conducted at 1400°C, it should be obvious that the glass viscosities varied considerably. They concluded that 95–96% of the total variance in corrosion was due to viscosity differences and that the specific chemical effects existed only to a small extent.

Lakatos and Simmingskold 1971 later found in isoviscosity tests that the corrosion of alumina depended upon the lime and magnesia content of the glass, whereas the corrosion of silica depended upon the alkali content.

During the testing of refractories for resistance toward coal-ash slags, Bonar et al. 1980 determined that AZS type refractories exhibited complete dissolution at the slag line, alumina exhibited significant corrosion, and a chrome-spinel refractory exhibited negligible attack at 1500°C and 10^{-3} Pa oxygen pressure for 532 hours. These results were consistent with the determined acid/base ratios of the slags and what one would predict knowing the acid or base character of the refractories.

Figure 15.3 shows the results of a mullite refractory that was removed from the regenerator division wall of a soda-lime-silicate container glass furnace. The sample was in service for 1 year at a temperature of approximately 1480°C. The attacking glass was from batch particulate carryover and condensation of volatiles. A small amount of convective flow down the vertical face of the wall was present due only to gravity. The alteration of the refractory due to corrosion occurred to a depth of about 25 mm. As can be seen from Figure 15.3 the mullite has completely converted to predominantly corundum and nepheline. Figure 15.3a is the XRD pattern supporting the presence of only corundum and nepheline. The optical micrograph shown in Figure 15.3b indicated the presence of an additional phase. Upon examination of elemental maps via SEM/EDS shown in Figure 15.3c, it was determined that the nepheline contained a reasonable amount of dissolved calcium and that the crystalline nepheline was embedded in a matrix of vitreous potassium-titanium-silicate. The potassium diffused into the refractory from glass batch impurities and the titanium was present in the original refractory in minor amounts.

Orlova et al. 1979 reported that divalent cation additions to alkaline silicate melts decreased the dissolution temperature of mullite while increasing the intensity of dissolution in the order CdO > ZnO > BaO. Additional crystalline phases form depending upon the divalent cation added: corundum (CdO), spinel (ZnO), or celsian (BaO).

There are times when the microstructure of the resultant corrosion product can offer information useful in determining the cause of the deterioration. Figure 15.4 shows a sample taken from surface runnage on a corundum refractory that was attacked by silica in a glass furnace. The dendritic and fibrous nature of the mullite formed is indicative of crystallization from a mullite melt containing a slight excess of silica and a variable cooling rate. The sample shown had apparently been at a temperature near or slightly in excess of 1850°C (the melting point of mullite). This temperature was approximately 300°C above the normal operating temperature for this furnace. The mullite identification for the dendrites was confirmed by SEM/EDS.

Carbides and Nitrides

Silicon carbide and nitride are relatively inert to most silicate liquids as long as they do not contain significant amounts of iron oxide. The reaction:

$$SiC + 3FeO \rightarrow SiO_2 + 3Fe + CO \tag{15.1}$$

can occur and becomes destructive at temperatures above about 1100°C, Muan 1984.

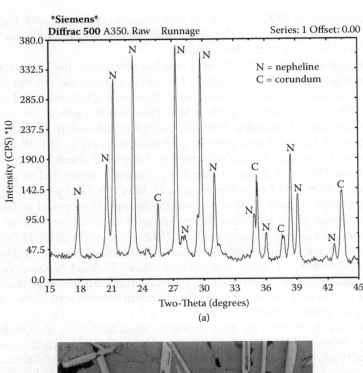

(a)

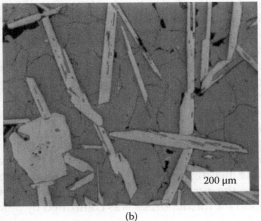

(b)

FIGURE 15.3
Corrosion of a mullite refractory: (a) XRD pattern, (b) reflected light optical micrograph (magnification 100×; lighter areas are corundum and darker areas are nepheline), and (c) EDS elemental spot maps.   *Continued*

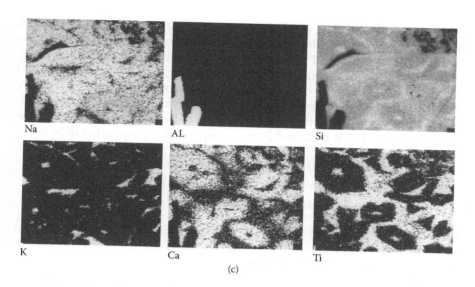

(c)

FIGURE 15.3 (*Continued*)
Corrosion of a mullite refractory: (a) XRD pattern, (b) reflected light optical micrograph (magnification 100×; lighter areas are corundum and darker areas are nepheline), and (c) EDS elemental spot maps.

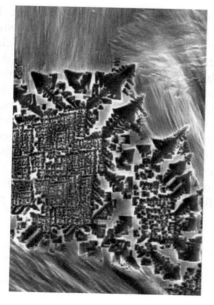

FIGURE 15.4
Dendritic and fibrous mullite formation caused by silica attack of a corundum refractory.

The dissolution of Si_3N_4 by glass is important not only in evaluating attack by various environments but also for gaining an understanding of the operative mechanism in liquid phase sintering and solution/precipitation creep phenomena occurring in materials that contain a glassy bonding phase. Tsai and Raj 1982 studied the dissolution of beta-silicon nitride in a Mg-Si-O-N glass, which they reported separated into SiO_2-rich and MgO- and N-rich regions. They concluded that the dissolution of beta-silicon nitride into the glass at 1573 to 1723 K occurred in three steps, with precipitation of Si_2N_2O:

1. beta-Si_3N_4 dissolves into the melt as Si and N,
2. this Si and N diffuses through the melt toward Si_2N_2O, and
3. the Si and N then attach to the growing Si_2N_2O.

Ferber et al. 1985 reported that the corrosion of α-SiC at 1175 to 1250°C when coated with a static layer of a basic coal slag involved at least three reaction mechanisms. These were as follows:

1. oxidation of SiC with the formation of silica between the slag and SiC,
2. dissolution of the silica by the slag, and
3. localized formation of Fe-Ni silicides at the SiC surface due to reaction of the SiC with the slag.

Which of these predominated was dependent upon the thickness of the slag layer, which in turn determined the local partial pressure of oxygen available by diffusion through the layer. When the slag thickness was <100 μm, passive oxidation occurred with the formation of SiO_2. As the slag thickness increased, the available oxygen at the surface was insufficient for SiO_2 formation, causing SiO to form instead. This active oxidation, forming the gaseous phases of SiO and CO, disrupted the silica layer allowing the slag to come in contact with the SiC, thus forming iron and nickel silicides.

McKee and Chatterji 1976 have reported a similar effect upon the corrosion mechanism of SiC where a molten salt layer provided a barrier to oxygen diffusion promoting the formation of SiO gas.

Based upon the work of Deal and Grove 1965, Ferber et al. 1985 gave the following equation for calculating the oxygen partial pressure at the SiC/slag interface:

$$pO_2 = RTC^*/(1 + 2X_1/A) \tag{15.2}$$

where:
   R = gas constant,
   T = temperature,
   C^* = equilibrium concentration of oxygen in slag,
   X_1 = slag layer thickness, and
   A = constant determined from kinetics.

For oxidation in air only, estimating A as 0.31 μm at 1250°C and taking C^* as 0.086 mol/m³ (oxygen concentration in silica), the critical slag layer thickness for the change from passive to active oxidation was calculated as 155 μm. Based upon the various assumptions involved in the calculation, this is very close to the experimentally determined value of 100 μm.

Reaction-bonded SiC (RBSiC) is produced by the reaction of either liquid or gaseous silicon or SiO with carbon in a silicon carbide/carbon compact. This results in a porous body with continuous silicon carbide phase; however, these pores can be filled with nonreacted Si (2–10%) yielding a dense product that results in excellent mechanical properties. The excellent wettability between Si and SiC allows this to be done for RBSiC but not for RBSN (silicon nitride). This interpenetrating grain boundary phase of silicon metal limits the high temperature mechanical properties to the melting point of Si (1410°C). When exposed to aggressive environments, the silicon may be attacked relatively easily, leading to degradation of properties.

Zirconia-Containing Materials

The attack of zircon by soda-lime-silicate glasses is similar to that of AZS materials in that an interface of zirconia crystals embedded in a highly viscous glass is formed. The difference is the lack of alumina, which keeps nepheline from forming, and the viscous glass is now a silicious glass as opposed to an alumina-rich glass. Thomas and Brock 1974 reported that as the sodium content of the attacking glass decreased, the thickness of the zirconia layer decreased. The attack of zircon by E-glass that contains only about 0.5% Na_2O exhibits no observable alteration. Zircon has been successfully used in contact with high temperature lithium-alumino-silicate glasses. A protective layer of zirconia crystals suspended in a very viscous glass is formed by the leaching of silica from the zircon. Since these glasses are melted at temperatures above 1700°C it is quite possible that the zircon dissociates into zirconia and silica,* with the silica then going into solution rather than the silica being leached from the zircon. As long as thermal cycling does not occur, this protective layer remains intact.

Because of the polymorphic transitions associated with zirconia it has not been widely used as material for furnace linings except when combined with other materials, such as the AZS refractories described above, or when stabilized in the cubic phase. Stabilization of the cubic phase by incorporation of CaO, MgO, or Y_2O_3 has been successful. These oxides, however, are susceptible to leaching when in contact with liquids. Chung and Schlesinger 1994 studied the effects of molten calcia-ferro-silicate slags upon the stability of zirconia stabilized by magnesia and calcia. Three slags were chosen: a basic, an acid, and a high-iron slag. They found that low basicity slag attacked magnesia-stabilized zirconia by leaching the magnesia. This subsequently allowed the more soluble monoclinic phase to form. Higher iron content

---

* Zircon dissociates at about 1675°C into ZrO_2 and SiO_2, Butterman and Foster 1967.

slags had lower viscosities and thus were more corrosive. High-basicity slag was found to attack magnesia-stabilized zirconia by direct dissolution of the cubic phase. Calcia-stabilized zirconia samples were completely destroyed by these slags even after short time periods.

Attack by Aqueous Solutions

The resistance to attack by aqueous solutions can be very important for many applications and especially where the shape-forming step involves slip casting of powders suspended in slurries. In Chapter 1 it was pointed out that a tremendous amount of literature is available concerning the dissolution in aqueous media of soil minerals. Some of these are mentioned below; however, the area of soils dissolution is too extensive to warrant an exhaustive review as is the area of cement/concrete chemistry and the dissolution of the various cement phases. Both of these areas are important ones to consider for those interested in hazardous waste disposal. Anyone interested in the hydrous and anhydrous cement phase chemistry should see Lea 1970. Calcia-silica-water chemistry has been discussed by Taylor 1979 and Jennings 1986. Hydration of dicalcium silicate has been discussed by McConnell 1955. The area of dissolution studies related to nuclear waste disposal for at least the last 30 years has appeared in a series of symposia proceedings published by the Materials Research Society under the series title *Scientific Basis for Nuclear Waste Management* 1982 through 2009 and by the American Ceramic Society under the series title *Nuclear Waste Management*, Advances in Ceramics 1984 and 1986, and Ceramic Transactions 1990 and 1991.

Alumina

Alumina has been shown by Sato et al. 1991 to dissolve into aqueous solution at 150 to 200°C containing NaOH by the following reactions:

$$Al_2O_3 + OH^- + 2H_2O \rightarrow Al(OH)_4^- + AlOOH \tag{15.3}$$

$$AlOOH + OH^- + H_2O \rightarrow Al(OH)_4^- \tag{15.4}$$

with the second reaction being the faster of the two. Although no surface interfacial layers were reported, AlOOH solid formed as part of the overall reaction as shown above. The rate of dissolution was linearly proportional to the NaOH concentration. Since the samples of Sato et al. were impure, containing a silicate grain boundary phase (7 and 0.5%), the grain boundaries exhibited enhanced corrosion.

The ligand-enhanced dissolution of alumina was discussed by Brady and House 1996. They reported that the dissolution was dependent upon the structure of the ligand; the oxalate five-membered ring was the most corrosive, with the malonate six-membered ring being intermediate, and the succinate seven-membered ring being the least corrosive. These three ligands are all adsorbed through two functional groups (i.e., bidentate adsorption) to a single site. Benzoate, which is adsorbed through only one functional group (i.e., unidentate adsorption), was even less corrosive than those adsorbed through two functional groups.

Carbides and Nitrides

The transition metal carbides and nitrides are chemically stable at room temperature but exhibit some attack by concentrated acid solutions. The one exception to this is VC, which slowly oxidizes at room temperature.

Bowen et al. 1990 reported the formation of $Al(OH)_3$ (bayerite) on AlN powder after 16 hours in contact with deionized water at 25°C. In the first 8 hours, growth of an amorphous hydrated layer occurred with a chemistry very close to AlOOH, while the pH of the solution drifted from 7 to 10 after 5.5 hours. The kinetics indicated a linear rate controlled by the surface reaction. Anyone involved in the aqueous processing of materials like AlN should be aware of the potential reactions that can take place with the incorporation of oxygen into their product through the formation of hydrated surface layers.

The behavior of sintered SiC in 0.045 M Na_2SO_4 + 0.005 M H_2SO_4 and 0.1 M LiOH aqueous solutions at 290°C was studied by Hirayama et al. 1989. They examined weight losses for up to 200 hours in both oxygenated and deoxygenated solutions. Weight losses increased with increasing pH and were greater for oxygenated solutions. No surface silica layers were found, with dissolution progressing through SiC hydrolysis. The proposed reaction follows:

$$SiC + 4H_2O \rightarrow Si(OH)_4 + CH_4 \tag{15.5}$$

and

$$Si(OH)_4 \rightarrow H_3SiO_4^- + H^+ \rightarrow H_2SiO_4^{2-} + 2H^+ \tag{15.6}$$

where the $Si(OH)_4$ sol that formed immediately dissolved. The dissolution of the $Si(OH)_4$ in acidic solutions ($pH = 4$) was slower than that in alkaline solutions and provided a small degree of protection, leading to a rate law that was approximately parabolic. In alkaline solutions ($pH = 10$) the rate law was linear.

The corrosive effect of HCl aqueous solutions at 70°C upon Si_3N_4 has been shown by Sato et al. 1988a to be dependent upon the sintering aid used or more specifically the grain boundary phase present in hot isostatically pressed materials. In solutions of <1 M HCl the corrosion was surface reaction controlled, whereas in solutions of >5 M HCl the corrosion was controlled by diffusion through the interfacial reaction layer that formed (assumed to be silica). Corrosion occurred through dissolution of the Al and Y ions (Y ion dissolution was about twice that of Al) contained in the grain boundary phase, with dissolution decreasing as the degree of crystallinity increased for this phase. Negligible dissolution of silicon ions was reported. In contrast to the above dissolution, the corrosive effect of 0.1–10 M aqueous HF solutions between 50 and 80°C for HIP or hot-pressed Si_3N_4 containing Y_2O_3, Al_2O_3, and AlN additives has been shown by Sato et al. 1988b to involve the selective dissolution of Si and Al ions but not Y ions. The Y ions instead formed insoluble YF_3.

TiN, CrN, and (TiAl)N coatings are widely used for protection of steels toward wear and corrosion. These coatings are generally applied by physical vapor deposition (PVD). The major problem with these hard coatings in aggressive environments is that they are prone to forming pinholes and open porosity during the deposition process. Ibrahim et al. 2002 studied the corrosion of these materials in 0.05 M H_3BO_3, 0.05 M $Na_2B_4O_7$, and 0.5 M NaCl aqueous solutions by open circuit potential, potentiodynamic polarization, and cyclic polarization techniques. Samples of 304 stainless steel were coated by PVD with a 3 μm thick layer of the three nitrides. Both TiN and (TiAl)N exhibited higher free corrosion potentials than either CrN or 304 stainless steel. The NaCl solution was more corrosive than the borate solutions for all three nitrides. The resistance to corrosion in borate solution decreased in the order (TiAl)N > TiN > CrN > 304SS and in the NaCl solution it decreased in the order CrN > (TiAl)N > TiN > 304SS. The susceptibility to pitting was highest in the NaCl solution. The presence of pinholes allowed penetration of the solution to the substrate and subsequent galvanic corrosion between the coating and the substrate.

Seshadri and Srinivasan 1988 investigated the corrosion of a titanium diboride particulate reinforced silicon carbide at room temperature in several aqueous solutions (aqua regia, NaOH, and HF/HNO_3) for up to 500 hours. Aqua regia was the most corrosive and a 50% NaOH solution was the least. Preferential leaching of the TiB_2 from the surface was reported to be the cause of decreasing weight loss with time. After about 100 hours weight loss was stopped for the aqua regia and HF/HNO_3 solutions, whereas it took approximately 250 hours in the 50% NaOH solution.

Silica and Silicates

The dissolution at room temperature and $pH = 7$ of the various forms of silica has been reported to be a function of the silica tetrahedra packing density by Wilding et al. 1977. Thus the dissolution increases in the following sequence: quartz, cristobalite, opal, amorphous silica. A wide variation in the solubility data has been reported in the literature, which has been attributed to the various investigators using different test conditions—pH, temperature, particle size, silica surface condition, and various components dissolved in the water. Quartz is not attacked by HCl, HNO_3, or H_2SO_4 at room temperature; however, it is slowly attacked by alkaline solutions. At elevated temperatures, quartz is readily attacked by NaOH, KOH, Na_2CO_3, Na_2SiO_3, and $Na_2B_4O_7$. The presence of organics dissolved in the water has been shown to increase greatly the solubility of silica with the formation of

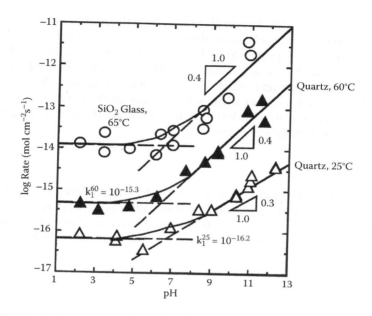

FIGURE 15.5
Dissolution rates of quartz at 25° and 60°C and SiO$_2$ glass at 65°C. (Brady and House 1996, Copyright 1996, CRC Press, Boca Raton, FL, reprinted with permission of CRC Press.)

Si-organic molecular complexes (see Schnitzer and Kodama 1977 for comparison to silicates). Acetic and oxalic acids have been reported to increase the rate of mineral breakdown, accelerating their solubility, Graustein et al. 1977. Various chemisorbed metallic ions (especially Al$_3^+$) have been reported to inhibit dissolution with the formation of relatively insoluble silicates.

Figure 15.5 shows a comparison of the dissolution of quartz at 25 and 60°C in 0.1 M NaCl solution with that of fused silica at 65°C.[*] These data exhibit several important points—first that dissolution rates increase with temperature, second that the dissolution rates are relatively constant below pH = 7 and increase after that, and third that the crystalline form of silica dissolves at a slower rate than the amorphous form. The difference in rates between crystalline and amorphous silica is confirmation of the effects of tetrahedra packing density. Although the temperature of the tests between crystalline and amorphous silica were different, Brady and House 1996 concluded that this difference was probably insignificant.

The dissolution of quartz in 49% HF solutions has been reported to vary depending upon the crystallographic face being attacked. Liang and Readey 1987 reported that the rate of dissolution of X-cut quartz was about twice as high for the positive end versus the negative end, whereas for Y- and Z-cut

---

[*] The fused silica was tested in a buffered solution of distilled, deionized water. Different inorganic and organic buffers were used for each pH range.

samples both ends exhibited similar rates. Both the X- and Y-cut ends exhibited dissolution rates much lower than the rate for Z-cut samples. The rates of dissolution were reported to be HF concentration dependent and surface reaction controlled. The substitution of HF molecules in solution with a surface complex ion was suggested as the surface reaction. The differences among the various X-, Y-, or Z-cut samples was attributed to a difference in the number of reactive or kink sites, rather than a difference in structure for the various crystallographic faces.

Superconductors

Murphy et al. 1988 reported that reaction with water liberates oxygen, forms Y_2BaCuO_5 and CuO, in addition to barium hydroxide, and is a function of temperature and surface area. This is similar to the leaching of barium from perovskites in aqueous solutions when the pH is less than 11.5 reported by Myhra et al. 1984.

Titanates and Titania

A crystalline titanate mineral assemblage called *SYNROC* has been under investigation for many years as a possible encapsulant for high-level radioactive wastes. The titanates are commonly a mixture of perovskites, $CaTiO_3$ and $BaTiO_3$, zirconolite, $CaZrTi_2O_7$, and hollandite, $BaAl_2Ti_6O_{16}$. In a study of the dissolution of these titanates in CO_2-enriched (4 ppm) deionized water ($pH = 5$–6) at 300 and 350°C and 500 bars, Myhra et al. 1984 reported the following reactions:

$$CaTiO_3 + CO_2 \rightarrow TiO_2 + CaCO_3 \tag{15.7}$$

$$BaTiO_3 + CO \rightarrow TiO_2 + BaCO_3 \tag{15.8}$$

$$CaZrTi_2O_7 + CO_2 \rightarrow 2TiO_2 + ZrO_2 + CaCO_3 \tag{15.9}$$

$$BaAl_2Ti_6O_{16} + CO_2 + H_2O \rightarrow 6TiO_2 + BaCO_3 + 2AlO(OH) \tag{15.10}$$

$$CaTiO_3 + H_2O \rightarrow TiO_2 + Ca(OH)_2 \tag{15.11}$$

$$BaTiO_3 + H_2O \rightarrow TiO_2 + Ba(OH)_2 \tag{15.12}$$

The dissolution mechanism proposed for these titanates was one involving initial selective leaching of the alkaline earth ions along with hydration of the titanate surface. This first step was rather rapid, but then overall dissolution slowed as the solution became saturated. When the solubility product was exceeded, precipitation and equilibration with CO_2 occurred. As the precipitate concentration increased, the dissolution rate decreased. Thus the overall dissolution of these titanates was dependent upon the solubility of the alteration products in the solution. In contrast, Kastrissios et al. 1987 proposed that the calcium was not selectively leached from the perovskite but instead the perovskite dissolved congruently, forming an amorphous titanium-rich surface layer from which TiO_2 precipitated. This titania layer was not continuous and therefore did not protect the underlying material from continued corrosion.

Buykx et al. 1988 gave a diagram of relative phase stability for various titanium-containing compounds, among others, for dissolution in water at 150°C for 3 days. No alteration was found for zirconolite-zirkelite ($CaZrTi_2O_7$). Some alteration and precipitation of TiO_2 was found for hollandite ($BaAl_2Ti_6O_{16}$), loveringite-landauite ($FeTi_3O_7$), pseudobrookite (Fe_2TiO_5), and rutile (TiO_2). Extensive replacement by TiO_2 was found for perovskite ($CaTiO_3$) and freudenbergite ($Na_2Ti_6Fe_2O_{14}(OH)_4$). Complete and rapid dissolution was found for any glassy phases. The stoichiometries given above are only approximate; the complete analyzed stoichiometries for the compounds investigated are given in the original paper.

Titania was investigated by Bright and Readey 1987 as the least complex titanate to evaluate the quantitative dependence of kinetics upon ambient conditions. Powdered anatase (~0.54 μm agglomerate size and ~0.13 μm crystallite size) was added to acid solutions of HF-HCl and stirred for several hours at temperatures ranging from 37.5 to 95.0°C. Although very little is known about the titanium species in HF-HCl solutions, it was believed that the most predominant complex was (TiF_6)$^{2-}$. The rate-controlling step in the kinetics of dissolution was concluded to be the removal of the highly charged cations from kink sites on the surface. The average calculated initial (for the first hour) dissolution rate was 59.0 wt% TiO_2 dissolved per hour.

Slightly reduced titania has been investigated for its use in electricity generation and for water decomposition, Harris et al. 1977. In these applications *n*-type semiconducting titania was used as a photoanode in an aqueous solution of 0.5 M H_2SO_4. The photogenerated positive holes in the valence band of illuminated *n*-type titania reacted with the solution according to the following equation:

$$SO_4^{2-} + p^+ \rightarrow SO_4^- \tag{15.13}$$

The (SO_4)$^-$ that formed was an active species that reacted with titania forming etch pits. This phenomenon is called *photoelectrochemical aging*.

Transition Metal Oxides

The use of transition metal oxides (RuO_2, NiO, MoO_2, Mo_4O_{11}, Mo_8O_{23}, Mo_9O_{26}, and WO_2) as fuel cell electrocatalysts requires that they be stable in aqueous solution of 1 N H_2SO_4. These oxides are relatively stable in acid solutions but undergo redox reactions in the region of *p*H = 7. Horkans and Shafer 1977a reported that Mo_4O_{11} exhibited anodic dissolution but that WO_2 did not; however, it did form a layer of WO_3 on its surface. They reported that MoO_3 was more soluble than WO_3 in acid solutions, whereas MoO_2 was more stable.

Horkans and Shafer 1977b reported that the oxidized surface layers that formed were generally less conductive than the bulk reduced phase, that they were generally of a wide range of compositions, and that the actual composition of the reaction surface layer was highly dependent upon the electrode potential. They also found that MO_2 (M = Mo, Ru, W, Re, Os, and Ir) was substantially more stable in acid solutions than was indicated by their Pourbaix diagrams.

The anodic dissolution of RuO_2-coated Ti electrode in 0.5 mol/L H_2SO_4 solution at 25°C and high anodic potentials was investigated by Gajić-Krstajić et al. 2004. At high anodic potentials the major reaction was oxygen evolution. Dissolution was dependent on the current density, the pH of the solution, and the composition of the coated layer. The mechanism of dissolution was proposed to be through hydrolysis of RuO_2, then oxidation to RuO_3, and then reduction to RuO_2 releasing oxygen.

In a study of the corrosion of nuclear fuels, Clayton 1988 investigated the effects of 633 K flowing water upon various thoria- and urania-containing materials. He found the following order of corrosion resistance: ThO_2 > ThO_2-UO_2 > ZrO_2-UO_2 > ZrO_2-CaO-UO_2 > UO_2. Corrosion in highly oxidative conditions (caused by the fission fragment radiolysis of the water) was attributed to the oxidation of uranium from the four-valent to the six-valent state. UO_2 corroded easily even in low-oxygen (5 ppm) water. The test conditions of pH, sample preparation, and compact attributes had no effect upon corrosion resistance. He et al. 2009 demonstrated that the rate of corrosion of UO_{2+x} varied over a large range that was dependent upon the degree of nonstoichiometry and the diversity of surface structures. Higher nonstoichiometry led to higher corrosion rates and thus greater release of radionuclides at those sites. A review of the literature by He et al. revealed that UO_2 was quite insoluble (in groundwater) under reducing conditions but increased about 5 orders of magnitude when oxidized. Evaluation of the corrosion of spent fuel in a deep geological repository under the unfortunate case of container failure indicated a dual front of corrosion, Broczkowski et al. 2010. Reaction of the steel container would release Fe^{2+} and H_2 and produce some Fe_3O_4. Although much has been written about the presence of Fe^{2+} and how it suppresses fuel corrosion, other studies have indicated that H_2 has a much greater effect. The lowered corrosion rates were attributed to the presence of the H^+ radical (from H_2 dissolved in water) produced by various activation steps. The interested reader should review the article by Broczkowski et al. since the references are too numerous to list here.

Zirconia-Containing Materials

The hydrothermal effect of water upon the dissolution of yttria (14 mol%) stabilized zirconia (YSZ) single crystals was investigated by Yoshimura et al. 1986. They found four regimes of behavior for YSZ treated at 600°C and 100 MPa for 24 hrs, depending upon the *p*H of the solution. In alkali solutions (those containing LiOH, KOH, NaOH, or K_2CO_3), partial decomposition and dissolution/precipitation were found, with yttria being the more soluble component. In acidic solutions (those containing HCl or H_2SO_4) rapid dissolution of yttria occurred, forming an interface of polycrystalline monoclinic ZrO_2. In reactions with H_3PO_4 solution the interface layer formed was ZrP_2O_7. In neutral solutions the dissolution was minimal.

A thin coating of several phases containing zirconia was prepared by the micro-plasma oxidation* (MPO) process of a Ti-6Al-4V alloy in an aqueous electrolyte containing K_2ZrF_6 and H_3PO_4 (added to adjust the *p*H to 5–6) with and without additions of Na_2SO_4, Yao et al. 2006. The ceramic coating is composed of both monoclinic and tetragonal ZrO_2 and $KZr_2(PO_4)_3$. The coating thickness was about 206 µm without the Na_2SO_4 addition and about 94 µm with the addition. Corrosion of these coatings was evaluated in an aerated 3.5% NaCl solution. Both uniform corrosion and localized pitting corrosion were reduced by use of these coatings with the one containing Na_2SO_4 being somewhat better. Even though the coating containing Na_2SO_4 was thinner, it exhibited better corrosion resistance, most likely because this coating contained a higher proportion of the two ZrO_2 phases compared to $KZr_2(PO_4)_3$.

Attack by Molten Salts

Carbides and Nitrides

The normally protective layer of SiO_2 that forms on SiC and Si_3N_4 can exhibit accelerated corrosion when various molten salts are present. McKee and Chatterji 1976 described several different modes of behavior of SiC in contact with gas-salt mixed environments relating to the formation of various interfacial reaction layers. Salt mixtures containing Na_2SO_4 and Na_2CO_3, Na_2O, $NaNO_3$, Na_2S, or graphite were tested. McKee and Chatterji found that a SiO_2 protective layer corroded in a basic salt solution but not in an acid salt solution. With low oxygen pressures, active corrosion took place by formation of SiO gas.

The activity of Na_2O has been shown to be an important parameter in the action of molten sodium salts by Jacobson and coworkers, Jacobson and Smialek 1985, Smialek and Jacobson 1986, Jacobson 1986, Fox and Jacobson 1988, Jacobson and Fox 1988, and Jacobson et al. 1986. The higher this activity, the greater the potential reaction with silica. The relationship of soda activity and SO_3 partial pressure can be obtained from the following equation:

---

* MPO uses a high-voltage alternating current that is applied to a metal part submerged into an electrolytic bath. Micro arc discharges at the surface of the metal create an intense plasma that oxidizes the surface of the part and grows a nanostructured oxide coating from the parent metal surface. The oxide film is produced by subsurface oxidation. The MPO process can grow coatings up to 250 microns. http://www.ibccoatings.com/microplasma-oxidation.html.

$$Na_2SO_4 \leftrightarrow Na_2O + SO_3 \qquad (15.14)$$

where the equilibrium constant k (which can be written in terms of concentrations, activities, or partial pressures) is given by:

$$k = \frac{[Na_2O][SO_3]}{[Na_2SO_4]} = pSO_3 \qquad (15.15)$$

Therefore, the highest Na_2O activity is related to the lowest partial pressure of SO_3. Jacobson 1986 reported that at partial pressures of SO_3 greater than 0.1 Pa no reaction occurred between SiC and Na_2SO_4 at 1000°C for at least up to 20 hours. It is assumed as always that Na_2O and Na_2SO_4 are chemically pure stoichiometric compounds and that SO_3 acts as an ideal gas. Experimentally the Na_2O activity can be set by the appropriate partial pressure of SO_3.

The decomposition of sodium sulfate by reaction 15.14 is not something that takes place readily. Sodium sulfate melts at 884°C and is relatively nonreactive toward silica, even at temperatures as high as 1400°C. To increase the reactivity, the sulfate must be reduced to some lower oxide. This has been known by the manufacturers of soda-lime-silica glass for many years. Sodium sulfate has been used not only as a source of sodium but also as a fining agent to remove the bubbles from the glass melt during processing. If the sulfate is not reduced, it either floats on the surface or forms lenticular immiscible inclusions in the finished product. References in the old glass literature refer to *blocking the furnace,* a term used to describe the process of adding wooden blocks (i.e., carbon) to pools of nonreactive sodium sulfate floating upon the surface of the molten glass.[*] The carbon from the burning wood reduced the sulfate to a form reactive or at least miscible with the molten glass. This reaction, shown below:

$$2Na_2SO_4 + C \rightarrow 2Na_2SO_3 + CO + 1/2\ O_2 \qquad (15.16)$$

is controlled more scientifically in modern glass manufacture through the use of coal as a batch ingredient and precise control of the combustion system to control the partial pressure of oxygen above the melt, which in turn controls the SO_3 equilibrium through:

$$SO_2 + 1/2\ O_2 \leftrightarrow SO_3 \qquad (15.17)$$

and subsequently the soda (or some sodium sulfur containing compound) activity. The reaction of Na_2SO_3 with silica according to:

$$Na_2SO_3 + xSiO_2 \rightarrow Na_2O \cdot xSiO_2 + SO_2 \qquad (15.18)$$

---

[*] There have been stories told by glass workers that potatoes also work well for this purpose.

is the one of importance in the dissolution of silica in the manufacture of glass and is most likely the one of major importance in the corrosion of the silica layer formed on SiC or Si_3N_4. Continued reduction of the Na_2SO_3 to Na_2S, although still reactive with silica, is not necessary for excessive dissolution of carbides and nitrides. The more reduced forms of the sodium-sulfur compounds are the basis of the amber color formed in the manufacture of brown bottles.

Jacobson and Smialek 1985 found that the partial decomposition of Na_2SO_4 enhanced the oxidation of SiC, forming a layer of tridymite, a sodium silicate glass, and some Na_2SO_4. Any free carbon in the SiC enhanced the corrosion, since it aided in the reduction of the sulfate. This enhanced corrosion was due to the ease of diffusion of oxygen through the predominantly sodium silicate amorphous layer compared to that of a crystalline silica layer. A somewhat different mechanism has been proposed for the corrosion of SiC by potassium sulfate, although details of the behavior have not yet been reported, Sato et al. 1991. In the case of the potassium salt, no silica layer is formed on the carbide, since it immediately is dissolved by the sulfate according to:

$$16SiC + 13K_2SO_4 \rightarrow 4(K_2O \cdot 4SiO_2) + 16CO + 9K_2S_{1.44} \qquad (15.19)$$

presumably due to dissolution being faster than oxidation.

Cree and Amateau 1987 in their studies of the behavior of SiC in molten lithium salts found that fracture strengths were reduced by a factor of two when tested above 600°C. In the mixed salts of LiS/Li/LiF and LiCl/Li/LiF (i.e., nonoxide) the corrosion was via grain boundary penetration, whereas in $LiSO_4$/Li/LiF the penetration was uniform. The decrease in fracture strength above 600°C was attributed to the large volume increase when lithia reacted with the silica surface coating on the SiC to form Li_2SiO_3.

A composite consisting of a combination of crystalline phases (41.6 wt% Cr_2AlC, 38.9 wt% $AlCr_2$, 5.8 wt% Al_8Cr_5, and 13.7 wt% Cr_7C_3) was evaluated for resistance to molten Na_2SO_4 by Lin et al. 2007. Samples coated with 4 mg/ cm^2 of a saturated Na_2SO_4 solution were heated to 900 and 1000°C for up to 20 hrs. The weight gain for 900°C was 1.2×10^{-3} mg/cm^2 · hr and for 1000°C it was 4.4×10^{-3} mg/cm^2 · hr. After 900°C for 20 hrs the oxide surface scale was about 2 μm thick. Selective oxidation of Al from the composite formed an oxide scale rich in Al_2O_3 that effectively provided protection from excessive corrosion.

The corrosion of Ti_3SiC_2[*] with respect to its resistance to molten salts (Na_2SO_4, Li_2CO_3-K_2CO_3, and Na_2SO_4-NaCl) has been studied by Liu and coworkers, Liu et al. 2002a, Liu et al. 2002b, Liu et al. 2003, and Liu et al. 2005. Severe corrosion was exhibited in Na_2SO_4 at 900 and 1000°C, in the

---

[*] Ti_3SiC_2 and Cr_2AlC of the previous paragraph are members of what are called MAX phases. These phases are stoichiometrically $M_{n+1}AX_n$, where n = 1, 2, or 3; M = early transition metal; A = an A-group element; and X = either C or N.

K_2CO_3-Li_2CO_3 eutectic at 700–850°C and in a Na_2SO_4-NaCl mixture with Na_2SO_4 content >35% at 850°C. The Ti_3SiC_2 was oxidized by dissolved oxygen in the melts, forming a double surface layer of SiO_2 near the substrate and an outer layer of TiO_2. In addition the CO produced in this reaction might be sufficient to reduce the Na_2SO_4 to Na_2S. It was speculated that this low melting Na_2SO_4/Na_2S mixture could react with the SiO_2 and TiO_2 surface layers causing the excessive corrosion. Compare this with equations 15.16 and 15.18.

The corrosion of hot-pressed silicon nitride (HPSN), reaction-bonded silicon nitride (RBSN), and silicon carbide by molten sodium sulfate, sodium chloride, and the eutectic composition between these two salts at temperatures from 670 to 1000°C for up to 120 hours was reported by Tressler et al. 1976. Molten sodium sulfate was the most corrosive, the eutectic composition was intermediate, and sodium chloride was the least effective in dissolving the silica surface layer present on these materials. HPSN was the most resistant, whereas silicon carbide completely dissolved in sodium sulfate at 1000°C within 20 minutes. The lower reactivity of Si_3N_4 compared to SiC with molten Na_2SO_4 was reported by Fox and Jacobson 1988 to be due to the formation of an inner protective layer of silica that stops the continued reaction of Si_3N_4. The formation of this inner protective layer was highly dependent upon whether oxidation or dissolution was the faster mechanism. Sato et al. 1991 reported that this inner protective layer of silica formed on pressureless sintered Si_3N_4 containing 5 wt% Y_2O_3 and Al_2O_3 in contact with molten potassium sulfate at 1200°C when the tests were conducted in air but not when conducted in nitrogen. This same situation was not true for attack by molten potassium carbonate at 1013°C. In this case, attack occurred in both nitrogen and air, with air causing a greater degree of reaction. Compared to studies performed in molten sodium and lithium sulfate and carbonates, Sato et al. found that the corrosion rate, in a nitrogen atmosphere, was independent of the alkali present, with the sulfates yielding an activation energy of 430 kJ/mol compared to that of the carbonates of 106 kJ/mol.

Wang et al. 2009 examined the corrosion of a Si_3N_4-bonded SiC[*] removed from the sidelining of an aluminum electrolysis cell after 270 days of operation. Even though these samples were in contact with a crust of cryolite and various gases above the molten zone, the main mode of corrosion was oxidation. The phases present in these materials before use in the electrolysis cell were 3C-, 6H-, and 4H-SiC[†] and α(fibers)- and β(grains)-Si_3N_4 along with a trace of elemental silicon. Since the Si_3N_4 bonding phase was considerably smaller in size than the SiC and also because the alpha phase existed as fibers, the Si_3N_4 had a considerably higher surface area than the SiC and thus was the first to corrode. The major products of corrosion were Si_2ON_2 in the gas zone and Na_2SiO_3 in the electrolyte zone. The fact that Si_2ON_2 was present indicated that a low partial pressure of oxygen was present, since Si_2ON_2

---

[*] Si_3N_4-bonded SiC typically contains 20-28wt% Si_3N_4.
[†] 3C, 6H and 4H are either cubic or hexagonal polymorphs of differing stacking sequences.

is stable only under reducing conditions. The Si_2ON_2 formed by oxidation of the Si_3N_4 (mostly alpha) with air or CO/CO_2. Na_2SiO_3 can form from both Si_3N_4 and SiC through the action of gaseous sodium, which diffused through the carbon bottom block. Another degradation reaction was that of HF in the gas zone reacting with silica to produce SiF_4 gas. In addition the reaction of HF with Si_3N_4 could also produce SiF_4 along with H_2 and N_2.

Oxides

The importance of molten salt reactions is well known in alumina reduction cells for the production of aluminum metal (Hall-Heroult Process). In this process, the electrolyte consists of a solution of alumina (<10 wt%) dissolved in molten cryolite (Na_3AlF_6), Grjotheim et al. 1966. Pure molten cryolite contains AlF_6^{3-}, AlF_4^-, F^-, and Na^+ ions. When alumina is added the complex ion $AlOF_x^{(1-x)}$ ($x = 3–5$) forms in addition to the others. In a study of the cryolite-mullite and cryolite/sodium fluoride-mullite systems, Siljan and Seltveit 1991 reported that materials with high Si/Al ratios experience high weight losses when in contact with NaF-cryolite eutectic melts due to the formation of gaseous SiF_4. They reported that mullite dissolved readily in cryolite and cryolite-NaF melts and that NaF reacted with alumina to form beta-alumina. Allaire 1992 studied the resistance of 15 different commercial refractories with varying Al_2O_3/SiO_2 ratios in a simulated alumina reduction cell and found that the resistance to attack by molten metallic sodium and NaF increased with alumina content. He reported that above 700°C an Al_2O_3/SiO_2 ratio greater than 0.90 should be used.

The corrosion of fused silica by molten sodium sulfate in atmospheres containing either 1% SO_2/O_2 or pure oxygen at 700 and 1000°C has been described by Lawson et al. 1990 to take place according to the ease of sodium diffusion in the various phases that form. Sodium diffused into the fused silica, leading to the nucleation of cristobalite. Once a continuous layer of cristobalite formed, sodium diffusion was minimized. The sodium at the cristobalite/fused silica interface then diffused further into the fused silica. The basicity of the reaction determined whether a cristobalite layer formed, with less cristobalite forming as the reaction became more acidic. Cristobalite globules, however, were reported to precipitate from the salt solution. Low partial pressures of SO_3 were reported to promote the fluxing action of the molten sulfate by increasing the activity of Na_2O. The type and extent of the reactions was determined by the activity of Na_2O or the partial pressure of the SO_3 in the molten salt.

The corrosion of yttria-stabilized zirconia (YSZ) and rare-earth zirconates ($La_2Zr_2O_7$ and $La_2(Zr_{0.7}Ce_{0.3})_2O_7$) in a molten mixture of V_2O_5 and Na_2SO_4 at 900°C for 100 hrs in air was studied by Xu et al. 2010. The samples were deposited as thermal barrier coatings upon a Ni-based alloy using electron beam–physical vapor deposition (EB-PVD). YSZ has been the material of choice for applications in aircraft engine turbines and land-based industrial engines;

however, it is limited to about 1200°C for long-term use due to polymorphic transitions that cause spalling. The rare-earth zirconates have attracted attention as an alternative to YSZ. A major problem with the EB-PVD process is that it creates a unique columnar microstructure that provides paths for penetration of molten salts. At elevated temperatures Na_2SO_4 dissociates into Na_2O liquid and SO_3 gas. The Na_2O liquid will react with any V_2O_5 present forming $NaVO_3$ liquid. The reaction of this $NaVO_3$ liquid with YSZ yields YVO_4, monoclinic ZrO_2, and Na_2O liquid. It is the depletion of the yttrium from YSZ that causes the ZrO_2 to transform to the monoclinic phase. This transition is accompanied by a large volume expansion, which mechanically disrupts the coating. There was no chemical effect of the Na_2SO_4 upon the YSZ. Mohan et al. 2007 also found similar results when the liquid was P_2O_5. In this case, the YSZ formed ZrP_2O_7. If excess La_2O_3 existed in the $La_2Zr_2O_7$, rapid degradation occurred by reaction with $NaVO_3$ and/or Na_2SO_4, forming $LaVO_4$, $La_2O_2SO_4$, and $La_2(SO_4)_3$. The cerium-containing zirconate exhibited no significant degradation under these test conditions, most likely due to the highly acidic nature of CeO_2.

Several investigators have shown that alumina coatings on various metals performed quite well in molten Na_2SO_4, Chen et al. 1992, Lawson et al. 1993, and Tang et al. 2000, and molten NaCl, Ramaswamy et al. 1997, indicating that alumina might be an excellent choice for gas turbine applications. Wu et al. 2005 tested this idea by depositing a layer of Al_2O_3 (by EB-PVD)[*] upon YSZ and then exposing it to a mixture of Na_2SO_4 + 0–15 wt% V_2O_5 at 950°C for 10 hrs. The substrate metal was a Ni-based superalloy (Inconel 610). The as-deposited alumina was determined to be the gamma phase. The gamma alumina began to convert to the alpha phase as the amount of V_2O_5 in the molten salt increased, with complete conversion at 15 wt% V_2O_5. In addition, as the V_2O_5 content increased the amount of YSZ destabilization increased, indicating that the alumina coating was no longer protective at the higher levels of V_2O_5. When compared to YSZ without the alumina coating, destabilization was much less (8% with Al_2O_3 compared to 65% without at the 5 wt% V_2O_5 level). Wu et al. observed that the morphology of the alumina coating after exposure was different than prior to exposure. This led them to determine that the conversion from gamma to alpha alumina was through a solution/precipitation mechanism. Gamma alumina was dissolved by molten $NaVO_3$ and then when the temperature decreased to below the $NaVO_3$-Al_2O_3 eutectic (610°C) the alpha phase precipitated. At the lower levels of V_2O_5 (<5 wt%), the alumina coating acted as a barrier to molten salt infiltration into the YSZ, even though it was partially degraded by dissolution. According to Jose et al. 1985 α-alumina will dissolve in molten sulfate by basic fluxing, forming $NaAlO_2$ liquid if the Na_2O activity is high. When the Na_2O is low the dissolution is by acidic fluxing, forming $Al_2(SO_4)_3$ liquid. Negligible dissolution will occur at intermediate levels of Na_2O activity, which was apparently the case in the work of Wu et al.

---

[*] EB-PVD = electron beam physical vapor deposition.

Ueno et al. 2006 evaluated alumina, mullite, Lu_2O_3, and $Lu_2Si_2O_7$ as environmental barrier coatings (EBC) for silicon nitride. An EBC is necessary to protect nonoxide materials against oxidation and corrosion associated with water vapor and alkalies in gas turbine applications. The four materials listed above were coated with $NaHCO_3$ and heated to 1500°C for 5 hrs. Alumina was easily attacked as was mullite. The $Lu_2Si_2O_7$ resisted attack well although impurity phases of Lu_2SiO_5 and SiO_2 confused the matter. Very little, if any, corrosion was noted for the Lu_2O_3 samples, although substantial grain growth occurred.

In the evaluation of cathode materials for molten carbonate fuel cells, Baumgartner 1984 reported solubility data for NiO, CuO, ZnO, $LiFeO_2$, and $LaNiO_3$ in a molten binary carbonate of Li/K (62/38 molar ratio) between 823 and 1223 K. Both NiO and CuO exhibited dissolution (CuO being more soluble than NiO) into the molten carbonate and diffusion toward the anode until the local partial pressure of oxygen was sufficiently low for metal precipitation. At temperatures exceeding 1123 K $LaNiO_3$ decomposed to La_2NiO_4 and NiO, which dissolved and reduced to metallic Ni. A similar situation was found for $LaCoO_3$, which decomposed to La_2CoO_4 and CoO at temperatures exceeding 1073 K. Dissolution of $LiFeO_2$ into the molten carbonate resulted in reduction at the anode to $LiFe_5O_8$, while ZnO at the anode became nonstoichiometric. The solubilities of these oxides were in the order $LaNiO_3$ < NiO < $LiFeO_2$ < CuO < ZnO below 1023 K. Above this temperature the relative solubilities of CuO and ZnO reversed.

In the evaluation of $LiNi_{1/2}Mn_{1/2}O_2$[*] soaked in 1M $LiPF_6$/ethylene carbonate/dimethyl carbonate electrolyte for various times Dupré et al. 2009 followed the formation of lithium-containing species on the surface of the $LiNi_{1/2}Mn_{1/2}O_2$ grains. The reaction, which was initially rapid, slowed during longer exposure times and produced a heterogeneously covered surface. $LiCO_3$ initially formed that further reacted, producing various fluorinated species. At longer exposure times organic products from the decomposition of the electrolyte appeared. These interphases that form between the positive electrode and the electrolyte are currently not fully understood.

An example of when corrosion is beneficial is the removal of ceramic cores in the investment casting process. The new process of directional solidification and the new alloys involved (NiTaC) require contact between the molten metal and the core material for times up to 20 hours at temperatures as high as 1800°C, Huseby and Klug 1979. The requirement that the core material must withstand these conditions and then be chemically removed is a contradiction in stability. Core removal requires high dissolution rates at low temperatures. Potential core materials are Al_2O_3, Y_2O_3, $Y_3Al_5O_{12}$, $LaAlO_3$, and $MgAl_2O_4$, which all possess satisfactory resistance to the casting conditions as reported by Huseby and Klug. These materials, except for Y_2O_3, are insoluble in aqueous acids or bases. The solvents used must be aggressive toward

---

[*] $LiNi_{1/2}Mn_{1/2}O_2$ is one of the positive electrode materials used in Li-ion batteries.

the core material but not toward the alloy. Borom et al. 1981 reported that the weakly basic or amphoteric oxides of Al_2O_3, Y_2O_3, and La_2O_3 can be dissolved by molten M_3AlF_6, $M_3AlF_6 + MF$, $M_3AlF_6 + M'F_2$, or $M_3AlF_6 + MCl$, where M = Li, Na, or K and M' = Mg, Ca, Ba, or Sr. The more acidic core materials such as ZrO_2 or ThO_2 required alkali or alkaline earth oxide additions to make the molten salt more basic.

Another field of study where the solution in molten salts is beneficial is that of crystal growth. The solubility of Be_2SiO_4 and $ZrSiO_4$ in various solvents was studied by Ballman and Laudise 1965. Solvents studied included alkali vanadates and molybdates. Due to solvent volatility problems (more important for molybdates than vanadates) most of their data contain substantial error at higher temperatures. The reported ion solubilities were greater for Be_2SiO_4 ranging from 3 to 5 1/2 mol% in the 900–1000°C range than for $ZrSiO_4$, which ranged around 1 mol% in the 900–1400°C region. Except for the solution by $Na_2O \cdot 3V_2O_5$, which was as much as 6 mol% at 1400°C, the vanadates were more corrosive than the corresponding molybdates of those studied. The greater solubility of Be_2SiO_4 over that of $ZrSiO_4$ can also be predicted from acid/base theory, since BeO is a stronger base than ZrO_2.

Superconductors

In an investigation of the molten-salt synthesis of $YBa_2Cu_3O_{7-x}$ *(123)*, Raeder and Knorr 1990 reported the stability of *123* against decomposition toward several molten salts at 1173 K. They concluded that *123* was not stable in molten LiCl or the dichlorides of Cu, Ca, Mg, or Ba or their mixtures. However, minimum decomposition was found in the NaCl-KCl system. The mechanism of decomposition was postulated as being one of selective dissolution of the barium in the *123* forming $BaCl_2$ and causing the *123* to decompose into several oxide phases that were consistent with the phase diagram reported by Lee and Lee 1989. These oxide phases were generally CuO and $Y_2Cu_2O_5$ or CuO and Y_2BaCuO_5 depending upon the amount of barium in the initial mixture.

Attack by Molten Metals

The application of ceramics to withstand the attack by molten metals is a very large part of the ceramic industry. Refractories are used to line furnaces for the manufacture of steel and the nonferrous metals of which aluminum and copper are probably the most important. The steel and nonferrous metals industries consume approximately 70% of all refractories manufactured today. Thus an understanding of the potential problems one may encounter from metal attack is quite important.

The attack by molten metals generally involves mechanisms of corrosion other than those by liquids in combination with liquid attack. The actual

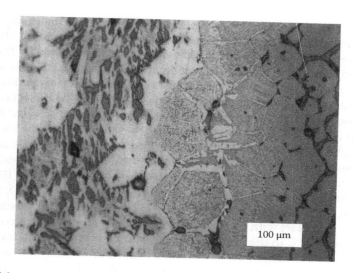

FIGURE 15.6

Corrosion interface between iron oxide-rich slag and a 60% MgO magnesite-chromite refractory (reflected light optical micrograph, magnification 150×). Brightest regions are an iron-rich mixed spinel. (Sample courtesy of Harbison-Walker.)

process that occurs in a blast furnace, for example, is truly a combination of corrosion mechanisms. In many cases small amounts of metal become oxidized and the corrosion is through essentially a molten slag process. An example of this is shown in Figure 15.6 that illustrates the corrosion of a 60% MgO magnesite-chrome refractory from an electric furnace that was in contact with a high iron oxide content slag. Diffusion of the iron oxide into the refractory and reaction with the magnesia and chrome-containing spinel formed an interface of large iron-rich mixed spinel crystals. Diffusion of iron into the magnesia caused precipitation of smaller iron-rich mixed spinel crystals within the magnesia and at the grain boundaries.

The reaction of silica-containing refractories with molten iron containing dissolved manganese has been known to be very deleterious. This reaction, however, is not only a reaction with a molten metal but also a reaction with an oxide of manganese (MnO). The initial reaction between SiO_2 and Mn forms MnO and Si metal. Although this reaction is thermodynamically unlikely, it has been reported by Kim et al. 1974 to occur at 1600°C under an argon atmosphere. The subsequent reaction of MnO and silica can form one of two intermediate compounds, but more importantly can form a eutectic liquid with a solidus temperature of 1250°C.

Many steel plant refractories are carbon-containing from manufacturing processes involving pitch or tar, organic resins, or graphite. Solid-solid corrosion through reduction by carbon takes place, or if the carbon becomes oxidized corrosion by molten metals is then very seldom a simple reaction involving only solid ceramic and molten metal.

Nitride ceramics and especially AlN are becoming of greater importance in the steel industry. AlN contains some Al_2O_3 and AlON[*] at the grain boundaries. The principal mechanism of corrosion is the attack of the AlON by oxygen dissolved in the steel, Labbe and Laïmeche 1996. This reaction produces Al_2O_3 and N_2 gas. A simultaneous reaction is the decomposition of AlN into Al dissolved in the steel and N_2 gas. This Al dissolved in the steel subsequently oxidized to Al_2O_3. To reduce the corrosion by attack of the grain boundary phases Amadeh et al. 1996 investigated the incorporation of carbon and BN into AlN and its effects upon the corrosion attack by various steels. Samples of AlN-BN-C (80-10-10, 70-10-20, and 75-5-20 wt%) were tested in pure argon up to 1625°C by Amadeh et al. 2004. The BN added reacted during the sintering process to eliminate AlON and Al_2O_3 forming more AlN, B_2O_3 gas, and N_2 gas. All samples exhibited partial decarburization when in contact with the steel.[†] The steel penetrated to a depth of 300 μm after 1 hr at 1550°C in the sample with the least carbon (this penetration was related to contact angles, surface tensions, and porosity). Since carbon has a greater affinity for oxygen than either BN or AlN, the presence of carbon helped protect the BN and AlN. In addition the presence of BN lessened the oxidation of AlN by reacting with any Al_2O_3 present to form more AlN and volatile B_2O_3.

The attack of molten aluminum upon materials containing silica follows the reaction:

$$4Al + 3SiO_2 \rightarrow 2Al_2O_3 + 3Si \tag{15.20}$$

that should be expected from examination of the free energy versus temperature data of an Ellingham diagram. The alumina that forms in many cases provides an adherent protective layer against further corrosion, Brondyke 1953. This reaction is accompanied by a volume decrease of about 26% according to Siljan et al. 2002. Although many previous investigators have attributed the spalling of refractory linings to the volume expansion of this reaction, Siljan et al. have attributed any expansive spalling that may occur to the growth of corundum along thermal gradients and/or any contraction due to the volume decrease.

At temperatures above about 800°C Al_2O_3 reacts with molten aluminum to produce the gas phase Al_2O, Shao et al. 2004. During the experiments, however, the molten aluminum was covered with a protective layer of Al_2O_3 that did not allow the Al_2O gas to escape, limiting the reaction. If the oxide layer were partially removed, the reaction would progress with continued evaporation of Al_2O.

---

[*] This amorphous grain boundary AlON is formed during the sintering process and is valuable for densification. It is formed only above 1600°C, Ado et al. 1985.

[†] Decarburization of the refractories releases carbon into the steel, lowering its melting point and forming Fe_3C (cementite) upon cooling.

The action of molten aluminum upon any beta-alumina contained in materials such as high alumina (70%) refractories produces metallic sodium, Allaire and Desclaux 1991. The metallic sodium present can then lead to reduction of silica, and if oxidized it can lead to the formation of $NaAlO_2$. The formation of $NaAlO_2$ is enhanced in the presence of a reducing atmosphere containing nitrogen by the intermediate formation of aluminum nitride according to the following reactions:

$$2Al_2O_3 + 2N_2 \rightarrow 4AlN + 3O_2 \qquad (15.21)$$

$$2AlN + Na_2O + 3/2\ O_2 \rightarrow 2NaAlO_2 + N_2 \qquad (15.22)$$

The difference in densities between alumina and sodium aluminate (3.96 vs. 2.69 g/cc) implies that a considerable volume expansion can take place during conversion of the original protective layer to a nonprotective aluminate, thus leading to continued corrosion.

According to Lindsay et al. 1964 mullite, when attacked by molten aluminum, converted to silicon metal and alumina; when attacked by molten magnesium-containing aluminum alloys mullite converted to spinel and magnesia. Brondyke 1953 has shown that molten aluminum will reduce silica in all forms contained in alumino-silicate refractories.

According to Cornie et al. 1986 the attack of SiC by molten aluminum at temperatures greater than 700°C can be represented by the following equation:

$$3\ SiC + 4\ Al \rightarrow Al_4C_3 + 3\ Si \qquad (15.23)$$

Nickel-based eutectic alloys such as NiTaC demand very severe temperature (as high as 1800°C) requirements upon their containers. Huseby and Klug 1979 studied the reactions of many oxides in contact with NiTaC-13 at 1700 and 1800°C and found that only Al_2O_3, $Y_3Al_5O_{12}$, and $LaAlO_3$ formed no interfacial reaction layers.

In a study to evaluate sialon crucible materials as candidates for containing molten silicon, Wills et al. 1980 found that the grain boundary chemistry is of utmost importance, with these phases being attacked first. Two sialon solid solutions (called O′ and β′) were evaluated at different compositions at 1450°C for 1 to 6 hrs under 0.84 MPa of argon.

A silicon nitride bonded SiC was evaluated for resistance to molten magnesium and a magnesium alloy at 750°C for 20 hrs by Chen et al. 2006. The major secondary element in the alloy was aluminum (9 wt%). An interface layer of 20–35 μm thick was formed that contained MgO and Mg_2Si in the case of the pure magnesium melt and MgO, Mg_2Si, and AlN in the case of the alloy. MgO formed by reaction of magnesium with any free silica present. The Mg_2Si could possibly form by reaction of magnesium with SiC; however, free energy calculations indicated that this reaction does not occur. Mg_2Si most likely formed by the two-step reaction proposed by Yu et al.

1999. Magnesium first reacts with silica, forming MgO and Si metal. This Si metal then reacts with Mg to form Mg_2Si. The AlN formed by reaction of the aluminum in the alloy with the Si_3N_4. A possible reaction of magnesium with Si_3N_4 would form Mg_3N_2; however, none was found.

Attack by Gases

The corrosion of a ceramic by vapor attack is generally much more severe than that by liquids or solids. The major reason for the more severe attack is related to the increased surface that is available to gases as opposed to liquids or solids. The various interfacial reaction products that may form due to attack by gases for several selected materials have been listed in Table 15.2. The interested reader should examine the original articles to determine the experimental conditions under which the various reaction products formed and also to determine the exact nature of the ceramic tested. The oxidation of

TABLE 15.2

Interfacial Reaction Products Caused by Vapor Attack

Material	Vapor	Interface[a]	Ref
Al_2O_3	Potassium	KA_1	Anderson 1979
Al_2O_3	Sodium	NA	Yamaguchi 1974
Al_2O_3	Potassium	KA_5 + Al	van Hoek et al. 1992
Al_2O_3/TiC	Oxygen	AT	Borom et al. 1988
AlN	Oxygen	A	Abid et al. 1986
B_4C	Oxygen	B	Rosner & Allendorf 1970
$MgAl_2O_4$	Sodium	NA	Yamaguchi 1974
$MgAl_2O_4$	Potassium	KA + M	Yamaguchi 1974
Mg_2SiO_4	Sodium	$N_2M_2S_3$	Yamaguchi 1974
Mg_2SiO_4	Potassium	KMS + M	Yamaguchi 1974
$Al_6Si_2O_{13}$	Sodium	$NAS_2 + NA_{SS}$	Yamaguchi 1974
$Al_6Si_2O_{13}$	Potassium	$KAS + KA_{SS}$	Yamaguchi 1974
SiC	Oxygen	S	McCauley 1975
Si_3N_4	Oxygen	$S + Si_2O_2N$	McCauley 1975
Si_3N_4/R_2O_3	Oxygen	$RS_2 + S$	Mieskowski & Sanders 1985
SiAlON	Oxygen	A_3S_2	Singhal & Lange 1977
TiB_2	Oxygen	B + T	Vedula et al. 1988
TiC	Oxygen	T	Stewart & Cutler 1967
TiN	Oxygen	T	Mukerji & Biswas 1990
$ZrSiO_4$	Potassium	$KZS_3 + Z$	Yamaguchi 1974

[a] $A = Al_2O_3$, $B = B_2O_3$, $K = K_2O$, $M = MgO$, $N = Na_2O$, $R = R_2O_3$, ($R = Y$, Ce, La, Sm), $S = SiO_2$, $T = TiO_2$, $Z = ZrO_2$, subscript SS = solid solution.

ceramics is very dependent upon the additives (and also impurities) used for such purposes as sintering aids and to the formation of liquids that promote oxygen diffusion (the lower the liquid viscosity, the greater the oxygen diffusion). In the sections below, selected materials are described in more detail.

Although oxidation is generally the most prevalent form of attack by gases, hydrogen reactions are becoming more important due to the development of miniature fuel cells for applications such as cell phones. Historically hydrogen fuel cells were the realm of the space industry. According to Nelson 1993 the corrosion of a material by hydrogen is dependent upon the following parameters:

1. ease of transport of H into the material,
2. type of reactions available between H and material, and
3. design of the structure.

Corrosion of ceramics by hydrogen is generally not a problem below about 1100°C where the molecular form exists. Thermal dissociation begins to become significant above 1100°C and transport rates to material surfaces increase rapidly.

Borides

Several of the diborides are of considerable interest because of their high melting points* and high strengths at elevated temperatures. Probably the one that has received the most attention is TiB_2; however, ZrB_2, HfB_2, NbB_2, and TaB_2 in combination with several other materials (e.g., SiC, Si_3N_4, and C) are of interest for hypersonic vehicle applications. These are the most attractive due to their high stability compared to the other diborides. Like the carbides and nitrides, the diborides possess the undesirable characteristic of oxidation. The oxidation of the diborides generally forms B_2O_3 and a metallic oxide according to:

$$M^{4+}B_2 + 5/2\ O_2 \leftrightarrow M^{4+}O_2 + B_2O_3 \qquad (15.24)$$

B_2O_3 readily vaporizes above 1100°C and therefore applications at high temperatures result in porous reaction layers. At lower temperatures, where the B_2O_3 is molten ($T_m = 490°C$), a surface layer of glassy material is formed over an inner layer of metallic oxide, Vedula et al. 1988. Above 1100°C the ZrO_2 that forms does not provide effective protection of the ZrB_2. Between 1100 and 1400°C the overall mass change is a combination of oxide formation and B_2O_3 vaporization (paralinear kinetics), Tripp and Graham 1968. Rapid linear

---

* These materials are called ultra-high-temperature ceramics, UHTC, not to be confused with a group of optical materials also called UHTC but meaning ultra-high-transmission coatings.

kinetics was exhibited above 1400°C where the rapid removal of the B_2O_3 left behind a porous nonprotective ZrO_2 scale, Fahrenholtz 2005. Oxidation kinetics at reduced pressures of oxygen have been reported but with some contradictions, Kuriakose and Margrave 1964 and Irving and Worsley 1968. Is there a relationship between oxidation rate and oxygen partial pressure?

The primary protection against continued oxidation for these materials is the barrier coating of refractory metal oxide that forms. At service temperatures in excess of 2000°C most or all of the B_2O_3 has evaporated. If SiC were added to the boride, any silica that formed would be lost due to active oxidation, Tripp et al. 1973.

The oxidation of a 20 vol% SiC/ZrB_2 composite at temperatures of 1400, 1500, and 1600°C for 2.5 hrs was studied by weight gain measurements, Carney et al. 2009. The oxide scale that formed was composed of three layers. The outer most layer was silica-rich and amorphous. An intermediate layer was composed of a ZrO_2 matrix with silica, and the innermost layer was composed of ZrO_2 enclosing partially oxidized ZrB_2 with some Si-C-B-O glass inclusions.

Although there are many aspects of the oxidation of ZrB_2 containing SiC that are similar among the various investigators, there are still several contradictions. These contradictions are generally related to the composition of the various surface layers that form. Apparently the amount of the reaction products that form (especially glasses rich in boria and/or silica and their viscosities) is important in determining the final phase analysis. Experimental parameter variation among the various investigators also plays an important role.

Fahrenholtz 2007 described a mechanism where the oxidation of ZrB_2-SiC in air at 1500°C resulted in a SiC-depleted ZrB_2 region below a ZrO_2-rich intermediate layer and an outer scale of SiO_2. This was attributed to the high vapor pressure of SiO gas under the silica oxide scale. Even though the overall reaction was one of passive oxidation of SiC (i.e., formation of a protective silica outer layer), the SiC underwent active oxidation. Oxygen partial pressures in the SiC-depleted layer were calculated to lie between 4.0×10^{-14} and 1.8×10^{-11} Pa. SiC is added to ZrB_2 to improve the oxidation resistance of pure ZrB_2 above 1100°C. The oxidation resistance is improved because of the lower volatility of SiO_2 compared to B_2O_3.

It is also well known that addition of lower valence cations into ZrO_2 (or HfO_2) forms the cubic structural modification that is oxygen ion conductive. This allows continued oxidation by oxygen transport through the outer barrier oxide scale. In an attempt to solve these problems Paul et al. 2012 investigated various additions to the base ZrB_2 (5 vol% SiC), HfB_2 (20 vol% SiC/5 wt% LaB_6), and TaC forming composites. The composite containing the LaB_6 when oxidized at 1600°C for 1 hr formed a dense surface layer of ZrO_2 and $La_2Zr_2O_7$ about 250 μm thick. With no LaB_6 and only the SiC, the surface layer was composed of porous ZrO_2 about 10 μm thick covered with a layer of amorphous silica. The HfB_2, with 20 vol% SiC composite, formed a 100 μm thick surface layer that was detached from an intermediate layer full of voids and cracks. At heat fluxes up to about 15 megawatts per square meter for 1 sec HfB_2-based composites performed better than the ZrB_2-based composites. In addition SiC and LaB_6 additions enhanced intermediate layer formation that slowed the oxidation process. At very high heat fluxes (e.g., 44 MW/m²) the composites containing SiC and LaB_6 formed volatile species that disrupted the surface protective layers. Studies with additions of carbon fibers indicated that HfB_2 carbon fiber composites exhibited the best oxidation resistance above 2500°C.

When these diborides were used as particulate reinforcement for oxide matrices, various reactions took place depending upon the oxide matrix. In a study of the oxidation of hot-pressed composites in air at 1650, 1850, and 2050°C, Vedula et al. 1988 have reported that in a zirconia matrix the titania that formed by oxidation of the TiB_2 went into solution into the zirconia and the B_2O_3 vaporized. In a yttria matrix tested in vacuum at 1600°C, they found significant reaction but were unable to determine its exact nature due to lack of published phase equilibria data. During hot pressing of a composite with an alumina matrix, the B_2O_3 that formed by oxidation of the diboride during heat-up reacted with the alumina forming a low melting liquid. Subsequent heating to 1600°C in vacuum caused reaction between the alumina and titania to form an intermediate aluminum titanate.

During sintering studies of titanium diboride Walker and Saha 1988 reported the following reactions:

$$TiB_2 + 2CO_2 \rightarrow TiC + B_2O_3 + CO \qquad (15.25)$$

$$TiB_2 + 3CO \rightarrow TiC + B_2O_3 + 2C \qquad (15.26)$$

In addition to these reactions, excess CO_2 or CO will oxidize the TiC formed to TiO_2. Davies and Phennah 1959 have shown that TiB_2 reacts with CO_2 forming titanium borate, in addition to the TiO_2 and B_2O_3 formed.

In an attempt to increase the oxidation resistance of HfB_2 by decreasing the boria loss Mattia et al. 2005 added ALN so that the overall reaction would form a stable aluminum borate, $Al_{18}B_4O_{33}$, on the surface.

Silicon hexaboride exhibits an oxidation resistance that is generally better than the above diborides due to the formation of a well-attached borosilicate film, Markarenko 1977. See also the discussion of boride composites in Chapter 17.

Carbon

The oxidation of the (100)[*] crystallographically textured CVD films of diamond was investigated by John et al. 2002. Samples were exposed to dry

---

[*] Evans and Phaal 1962 reported that the oxidation resistance of diamond crystallographic planes increased in the order (100) > (110) > (111).

O_2 at temperatures up to 800°C at partial pressures of oxygen ranging from 10 to 760 torr. To fully utilize diamond as physical and/or chemical sensors the distribution of oxygen on the surface must be controlled. The activation energy for (100) diamond oxidation was found to be constant between 23 and 800°C. The as-prepared films contained oxygen levels in the outer 12 nm at levels >5 wt%. This oxygen was present as carbonyl (C = O) or ether (COC) functional groups. Annealing under vacuum reduced this oxygen content; however, on re-exposure to the atmosphere the oxygen level returned to within 1% of the original level. This was a reversible process. All the data of John et al. supported the mechanism of oxidation of diamond as a layer-by-layer removal as carbon monoxide.

Nitrides and Carbides

None of the nitrides and carbides is thermodynamically stable in oxygen-containing environments. Under some conditions some carbides and nitrides form a protective metal oxide layer that allows them to exhibit reasonably good oxidation resistance (e.g., $Si_3Al_3O_3N_5$ forms a protective layer of mullite). As can be seen from the examples given below, the corrosion of silicon nitride and carbide materials varies considerably based upon the characteristics of the individual material and the environment. Even though many attempts have been made to determine an exact mechanism for this corrosion, there is still considerable disagreement, unless these materials are grouped according to their type and impurity level for each environment.

Silicon Nitride

Oxidation

The oxidation of Si_3N_4 is dependent upon the manufacturing process used: chemical-vapor-deposited (CVD) materials exhibit the slowest and smallest amount of oxidation due to their purity, dense structure, and in some cases larger grain size; hot-pressed (HP) and hot-isostatically-pressed (HIPed) materials exhibit an oxidation generally dependent upon the type and amount of additive used; and reaction sintered (RS) materials exhibit the most oxidation due to their large porosity. Materials containing more impurities, or at least more of those species that lower the viscosity of any silica reaction layer that may form, will provide a lower resistance to continued oxidation, since the diffusion of oxygen is easier through the lower-viscosity coating. Thus it is important to use a material that will produce a surface layer that is as close as possible to pure silica.[*]

The oxidation of Si_3N_4 has been described as occurring by either an active or a passive mechanism, Singhal 1977. The active mechanism is one where

---

[*] Compared to other oxides, silica has the lowest oxygen permeability, Klemm et al. 2003.

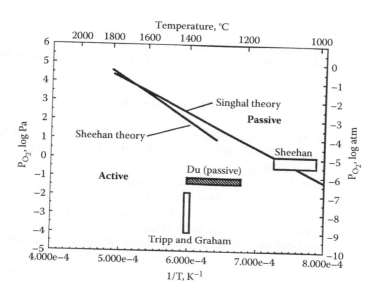

FIGURE 15.7

Literature data for active-to-passive oxidation transition for Si_3N_4. (Vaughn and Maahs 1990, reprinted with permission of the American Ceramic Society/Wiley & Sons Copyright Clearance Center.)

the fugitive SiO forms in environments with low partial pressures of oxygen by the reaction:

$$2Si_3N_4(s) + 3O_2(g) \rightarrow 6SiO(g) + 4N_2(g) \qquad (15.27)$$

Passive corrosion occurs in environments with high partial pressures of oxygen by the reaction:

$$Si_3N_4(s) + 3O_2(g) \rightarrow 3SiO_2(s) + 2N_2(g) \qquad (15.28)$$

The SiO_2 that is produced forms a protective coating and further oxidation is limited. The pressure of N_2 formed at the interface can be large enough to form cracks or pores in the protective coating, which subsequently allows additional oxidation.

The active-to-passive transition as determined by several investigators and compiled by Vaughn and Maahs 1990 is shown in Figure 15.7. The variations reported are due to differences in the silicon nitride material tested and in the experimental conditions of the test.

Thermodynamically, silicon oxynitride (Si_2N_2O) should also form; however, it is further oxidized to SiO_2 according to:

$$2Si_3N_4 + 3/2\ O_2 \rightarrow 3Si_2N_2O + N_2 \qquad (15.29)$$

$$Si_2N_2O + 3/2\, O_2 \rightarrow 2SiO_2 + N_2 \qquad (15.30)$$

The reaction of SiO_2 with Si_3N_4 can also produce a loss in weight by the reaction:

$$Si_3N_4 + 3SiO_2 \rightarrow 6SiO + 2N_2 \qquad (15.31)$$

As is quite often the case in corrosion of ceramics, the oxidation of silicon nitride initially follows a linear rate law, since the rate-limiting step is the interface reaction. When the interface reaction layer becomes sufficiently thick, the kinetics shift to one limited by diffusion through the reaction surface layer and thus obeys a parabolic rate law. This change from linear to parabolic kinetics was reported to occur at an interface layer thickness of about 50 nm for the oxidation of CVD-Si_3N_4 between 1000 and 1300°C by Choi et al. 1989. It was pointed out by Choi et al. that even though two different materials (i.e., Si metal and Si_3N_4) may exhibit the same major oxidation product (i.e., SiO_2) and follow the same oxidation kinetics (i.e., parabolic), the mechanism of oxidation may not be necessarily the same. This was based upon the very different activation energies and rates of oxidation obtained for the two materials.

Although Choi et al. reported increased refractive index determinations for the SiO_2 reaction layer, consistent with significant nitrogen concentrations, they did not mention the formation of an oxynitride layer as did Du et al. 1989. Du et al. reported that the oxidation of pure CVD alpha-Si_3N_4 formed a double reaction layer: an inner zone of Si_2N_2O and an outer zone of SiO_2. A calculated plot of the thermodynamic stability fields for SiO_2, Si_2N_2O, and Si_3N_4 is shown in Figure 15.8 that indicates a zone of Si_2N_2O separates the Si_3N_4 and SiO_2 throughout the temperature range examined (1000–2000 K). The assumption that Du et al. made in determining this was that no solutions existed among the three phases, which is not exactly correct. Indications are that the oxynitride is truly a solution of variable oxygen content with no distinct boundaries. Thus conventional thermodynamics cannot accurately predict the results. Du et al. also reported that the most probable rate-limiting step during oxidation was molecular oxygen diffusion through the inner zone of Si_2N_2O, which is structurally denser than SiO_2. In contrast to this molecular oxygen diffusion limited reaction, Luthra 1991a has suggested that the reaction is controlled by a mixed process involving nitrogen diffusion and the reaction at the Si_2N_2O/SiO_2 interface. He also stated that the rate law for this mixed controlled process can be nearly parabolic, which is consistent with all the observations reported in the literature, Luthra 1991b. Luthra's assumption was that the Si_2N_2O interface thickness was too thin to provide a significant barrier to diffusion. Ogbuji 1992, through some interesting experimentation, has given support to the diffusion barrier suggested by Du et al., although he did state that their model suffered from some inconsistencies due to incorrect assumptions. Thus it appears that an understanding of the details of oxidation of Si_3N_4 is still incomplete.

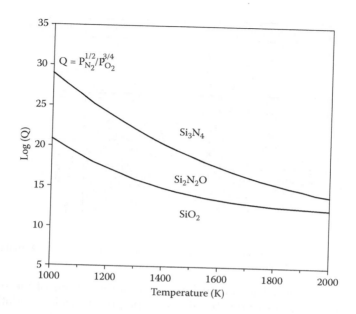

FIGURE 15.8
Thermodynamic stability fields for SiO_2, Si_2N_2O, and Si_3N_4. (Du et al. 1989, reprinted with permission of the Electrochemical Society.)

Ogbuji listed the areas of uncertainties as follows:

1. formation sequence of Si_2N_2O and SiO_2,
2. oxygen diffusion routes,
3. effects of oxide layer crystallinity upon diffusion,
4. oxidation state of product gases,
5. effects of out-diffusion of nitrogen, and
6. solubilities of O_2 and N_2 in the SiO_2 and Si_2N_2O phases.

Both de Jong 1991 and Joshi 1992 have shown through binding energy calculations obtained by XPS data that the oxidized double layer on commercially available as-received Si_3N_4 powders contained more oxygen for the inner layer than the stoichiometric Si_2N_2O. Their data matched closely that which Bergstrom and Pugh 1989 reported for Si_2ON. Joshi has suggested that the double layers consisted of a thin outer layer of amorphous SiO_2 over an inner oxynitride layer that is amorphous near the silica outer layer and crystalline near the nitride as shown in Figure 15.9. He also suggested that the oxygen content varied through the oxynitride thickness as the silazane groups of the bulk are gradually replaced by siloxane groups near the surface.

Since many silicon nitride materials contain various sintering aids and are therefore polyphase materials, their oxidation behavior is more complex

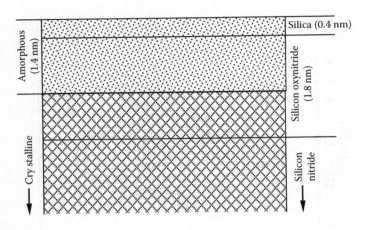

FIGURE 15.9
Typical oxide layer configuration for Si_3N_4 obtained from XPS and TEM data. Actual boundaries are diffuse. (After Joshi 1992.)

than discussed above. The outward diffusion of the cations of the sintering aids along with various impurities that may be present yields oxidation layers containing many mixed phases that depend upon what cations diffuse to the surface and how much of each is present. An indication of the equilibrium phases that should be present can be obtained from an examination of the appropriate phase diagram. The presence of any liquid phase fields greatly increases the outward diffusion of the various cations (along with the inward diffusion of oxygen) and will also provide a means for the formation of a coherent layer, since a liquid can accommodate the volume changes that occur during the reaction of the various phases. The effect that the various cations have upon the viscosity of the silica liquid phase is very important in the overall evaluation of oxidation resistance. In general, parabolic rate constants are obtained for the oxidation of these materials and these constants vary considerably from one material to the next.

Tripp and Graham 1976 reported a marked change in the oxidation rate to higher values at about 1450°C for HP-Si_3N_4. This change in rate was related to a change in mechanism from passive to active oxidation at a partial pressure of oxygen of $10^{-7.3}$ atm at 1400°C. Both mechanisms were reported to occur between 10^{-7} and 10^{-10} atm. The passive to active oxidation transition was reported to occur at 1×10^{-4} MPa at 1400°C for a HP-Si_3N_4 containing 6 wt% Y_2O_3 and 1.5 wt% Al_2O_3 by Kim and Moorehead 1990, which compares quite favorably with the calculated value of 2×10^{-4} MPa by Singhal 1976.

The rate-determining step in HP-Si_3N_4 containing MgO was reported by Cubicciotti and Lau 1978 to be the diffusion of MgO from the bulk material into the oxide surface layer. This surface layer was composed of SiO_2, $MgSiO_3$, or glass phase and some unoxidized Si_3N_4, and was porous due to released N_2. Similar results were reported by Kiehle et al. 1975 for HP-Si_3N_4

containing impurities of magnesium, iron, aluminum, manganese, and calcium (all less than 0.6 wt% each determined by emission spectroscopy). They found an amorphous silica film at temperatures as low as 750°C that converted to cristobalite at higher temperatures and/or times. Above 1000°C, sufficient migration of the impurities to the surface caused the formation of additional phases, such as enstatite ($MgSiO_3$), forsterite (Mg_2SiO_4), akermanite ($Ca_2MgSi_2O_7$), and diopside ($CaMgSi_2O_6$). Akermanite was the first silicate to crystallize, enstatite appeared only above 1350°C, and diopside appeared only after oxidation at 1450°C. Both akermanite and forsterite appeared only after longer heating times. Equilibrium phase assemblages occurred only at high temperatures and long times. It is quite interesting that Kiehle et al. reported neither a weight loss nor a weight gain (difference in weight before and after heating) for their oxidation tests. It was suggested that this was due to the simultaneous formation of both SiO_2 and the fugitive SiO. It is unfortunate that they did not report continuous weight change data, which may have shown an initial weight gain followed by a weight loss after the surface scale crystallized at about 1000°C.

Catastrophic oxidation at about 1000°C has been reported for HP-silicon nitride containing Y_2O_3 where the secondary phases were $Y_2Si_3O_3N_4$, $YSiO_2N$, or $Y_{10}Si_7O_{23}N_4$, Singhal 1977. If only Si_2N_2O and $Y_2Si_2O_7$ were present (true for materials containing less than about 6 wt% Y_2O_3) as the secondary phases, the oxidation would be very low and well behaved. Lange et al. 1977 reported that $Y_2Si_2O_7$ was the most oxidation-resistant phase in the Si_3N_4-SiO_2-Y_2O_3 system. The oxidation resistance of other rare earth silicates was found to be roughly inversely related to the eutectic temperature of the RE_2O_3-SiO_2 system by Cinibulk et al. 1992. Thus the system with the highest eutectic temperature would exhibit the lowest weight gain. Since the disilicates invariably crystallized from the liquid phase during sintering, Cinibulk et al. attributed their excellent oxidation resistance to the minimization of the amorphous phase.

In HIPed Si_3N_4 containing 5 wt% Y_2O_3 oxidized over the temperature range 1200 to 1450°C for up to 100 hours, Plucknett and Lewis 1992 found a variation in the scale microstructure from a phase mixture of small amounts of $Y_2Si_2O_7$ within larger amounts of amorphous silicate (containing some impurity cations) at short times and low temperatures, to a double layer scale of $Y_2Si_2O_7$ near the nitride and a semicontinuous SiO_2 outer layer at longer times and higher temperatures. As the oxidation times were increased, the amorphous silicate gradually converted to cristobalite.

Oxidation of RS-silicon nitride occurred through two steps: a fast internal oxidation of the open porosity until filled, and then a slower external oxidation of the surface, Singhal 1977. In reaction sintered materials with surface areas of about 0.7 and 0.3 m^2/g, Gregory and Richman 1984 reported that the surface pores were sealed by the oxidation product when the temperature of oxidation was above 1100°C. A plot of the weight gain for many RS materials

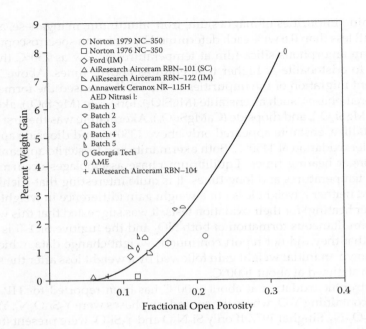

FIGURE 15.10
Weight gain–porosity relation for RS-Si$_3$N$_4$ materials exposed to static air for 1000 hrs at 1400°C. (Larsen et al, 1985, reprinted with permission of William Andrew, Inc.)

as a function of the fractional open porosity after exposure to static air at 1400°C for 1000 hours is shown in Figure 15.10, Larsen et al. 1985.

Nonporous sintered Si$_3$N$_4$ containing about 4–6 wt% of Y$_2$O$_3$, Ce$_2$O$_3$, La$_2$O$_3$, or Sm$_2$O$_3$ in addition to about 4 wt% SiO$_2$ as sintering aids was investigated for oxidation resistance at 700, 1000, and 1370°C in air for up to 200 hours by Mieskowski and Sanders 1985. All these sintering aids formed pyrosilicate and cristobalite as reaction products. The lowest oxidation rate was produced by the addition of Y$_2$O$_3$, whereas the highest rates were produced by Sm$_2$O$_3$ and Ce$_2$O$_3$.

The weight change upon oxidation for equation 15.28 is the difference between the weight of 1 mole of Si$_3$N$_4$ and 3 moles of SiO$_2$, which Horton 1969 represented by the mixed parabolic rate equation:

$$\Delta W^2 + A \, \Delta W = kt + c \tag{15.32}$$

where ΔW is the gain in weight per unit area, t is time, and A, k, and c are constants. Since linear kinetics are usually observed for the formation of very thin films of SiO$_2$, the first term in equation 15.32 would be negligibly small. Wang et al. 1991b showed that the formation of oxide layer thicknesses less than about 5 nm that form on silicon nitride powder containing 4 wt% Y$_2$O$_3$ obey a linear rate law at 900, 950, and 1000°C. For the formation of very thick films, the second term becomes negligibly small. Oxidation rates in various atmospheres for powdered samples were in the following order: dry oxygen = humid air = 2× dry air. The oxidation product was amorphous at 1065°C and tridymite between 1125 and 1340°C. Franz and Langheinrich 1972 reported the increased oxidation of amorphous CVD silicon nitride in wet oxygen over that in dry oxygen to be about four times at 1000°C for 10 hours. Horton pointed out that the exposed surface area for oxidation was actually the surface area of smaller particles that made up larger agglomerates, since solid particles of various mesh sizes all had the same BET surface area.

Reaction in Other Atmospheres

Even in environments other than pure oxygen or air silicon nitride corrodes primarily through oxidation. An example of this is the work reported by de Jong et al. 1987 in their studies of HP and RS materials at 1050°C in gas mixtures of H_2 plus 7 or 17% CO and 1300°C in a gas mixture of H_2 plus 1% CH_4 and 0.5% H_2O. Even though the partial pressure of oxygen in these experiments was on the order of 10^{-20} atm, passive oxidation occurred at the lower temperature in H_2/CO, forming a surface layer of tridymite shown in Figure 15.11. At the higher temperature in H_2/CH_4 + H_2O, active oxidation to SiO occurred followed by the formation of SiC whiskers by the VLS mechanism, which deposited onto the Si_3N_4 surface as well as the internal parts of the furnace (Figure 15.12).

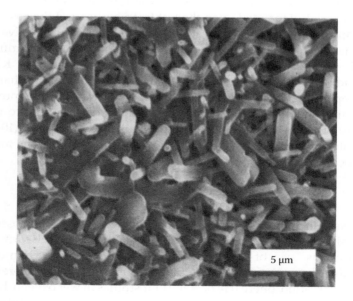

FIGURE 15.11
Tridymite surface formed on Si_3N_4 at 1050°C and 5×10^{-14} Pa oxygen for 1000 hrs (magnification 3600×).

FIGURE 15.12
Beta-SiC whiskers deposited onto Si_3N_4 at 1300°C and 10^{-15} Pa oxygen for 100 hrs (magnification 4300×).

At 1200 to 1300°C in a gas mixture of $H_2S/H_2O/H_2$, Oliveira et al. 1990 found that *HP* materials containing Y_2O_3 and Al_2O_3 (10–12 wt% total) actively corroded through the formation of gaseous SiO and SiS. The active corrosion of these materials depended upon the test temperature and the gas environment composition. In dry H_2S/H_2 at 1300°C, the outer nonprotective porous layer of the Si_3N_4 samples contained the presence of yttrium oxysulfide.

In their studies of the oxidation of CVD and HP materials, Kim and Moorehead 1990 reported that the mechanism of oxidation changed at 1400°C depending upon the partial pressure of water vapor present in H_2-H_2O mixtures. The Si_3N_4 in the CVD material decomposed to metallic silicon and nitrogen at very low partial pressures of water vapor (10^{-7} MPa). As the water vapor pressure increased up to about 10^{-5} MPa, the following reaction occurred:

$$Si_3N_4 + 3H_2O(g) \rightarrow 3SiO(g) + 2N_2(g) + 3H_2(g) \qquad (15.33)$$

with a resultant more severe weight loss. Less severe weight losses were noted at higher water vapor contents, presumably due to the formation of a discontinuous layer of SiO_2 and/or Si_2N_2O. Choi et al. 1989, in contrast, reported that the oxidation of CVD-Si_3N_4 was insensitive to the presence of H_2O in an inert atmosphere (He or Ar). Their studies in wet oxygen, however, indicated an increased oxidation rate over that of dry oxygen, exhibited by a decrease in the activation energy as the partial pressure of the water vapor increased from 2 to 80 kPa. This was thought to be due to a breaking of the

Si-O-Si bonds in the silica structure as a result of OH solution. The various gaseous reaction products that may form (e.g., NO, NH_3) can alter the counterdiffusion kinetics, thus modifying the inward diffusion of the oxidant and slightly changing the overall mechanism.

Kim and Moorehead 1990 found a similar dependence upon water vapor pressure for the oxidation of HP-Si_3N_4 containing 6 wt% Y_2O_3 and 1.5 wt% Al_2O_3. In the low-pressure region, however, the magnitude of the weight loss for the HP material was about three times greater than that of the CVD material. This difference was attributed to the greater surface area for reaction in the HP material, since it had a much smaller grain size compared to the CVD material. At the higher water vapor pressures, the grain boundaries were not preferentially attacked and thus the two materials exhibited similar weight losses.

To overcome the limitations inherent in laboratory tests, Baxter et al. 1998 studied the corrosion of low additive (0.67 wt% Al_2O_3 and 1.33 wt% Y_2O_3) Si_3N_4 in a low-velocity (0.2 m/s) burner rig. Two different sulfur-containing industrial fuels (1.0 wt% sulfur marine diesel fuel and 0.01 wt% sulfur kerosene) containing air contaminated with artificial ocean water were used. Test temperatures were 1100, 1200, and 1300°C for eight cycles of 6.25 hrs. The rate of corrosion was dependent upon the pSO_3 (i.e., fuel sulfur content). A high rate of corrosion was observed for the low pSO_3, whereas with the high pSO_3 the corrosion rate was similar to that in air. A low pSO_3 allows the rapid incorporation of sodium into the silica-rich surface scale, thus increasing the corrosion rate. See the discussion related to equations 15.14 and 15.15 concerning Na_2O activity.

Other Nitrides

The crystalline solution series of materials of alumina dissolved into beta-silicon nitride ($Si_{6-x}Al_xO_xN_{8-x}$) make up a truly interesting series of materials. The hopes have been that these materials would yield properties that are the best of the two end members. One improvement over silicon nitride is the oxidation resistance with increasing amounts of alumina while maintaining the relatively low thermal expansion characteristics of the pure silicon nitride phase. Weight gain behaviors have been reported to be parabolic with mullite being the oxide that formed on the surface. Singhal and Lange 1977 reported that mullite formed only in those compositions containing more than 20 wt% alumina and that above 40 wt% alumina, additional unidentified phases occurred. Chartier et al. 1986 prepared sialon crystalline solutions with $x = 0.4$. Since 14.05 wt% yttria was added to the original mix, the final pressureless sintered samples contained β'-$Y_2Si_2O_7$ and a glassy phase as grain boundary phases. Oxidation was parabolic and very slow below 1380°C. Above this temperature, more rapid oxidation occurred with departures from parabolic behavior. A thin aluminosilicate film formed first, but as metal cation migration occurred (predominately yttrium) reaction with this film formed more complex silicates. This film was dense below 1400°C and gradually became porous and nonprotective as the temperature was increased. Visual observation indicated a light gray zone under the surface scale that Chartier et al. reported to be due to selective oxidation of the grain boundary phase.

Wang et al. 1990 investigated the oxidation in air of rare-earth aluminum oxynitrides with the ideal formula $LnAl_{12}O_{18}N$ (Ln = La, Ce, Pr, Nd, Sm, and Gd) at temperatures ranging from 700 to 1500°C. Noticeable oxidation started at about 700°C and increased with temperature. The final reaction products depended upon the particular rare earth but progressed through several intermediate stages. At temperatures around 1000°C, the products were $LnAl_{11}O_{18}$ and alpha-alumina. $LnAlO_3$ also formed with or without the disappearance of $LnAl_{11}O_{18}$ at higher temperatures depending upon the rare earth. In cerium-containing materials, CeO_2 forms at 900°C. At temperatures as low as 700°C, lattice parameters changes (increasing c/a ratio) were noted for the oxynitrides, which was attributed to the initiation of oxidation.

Wang et al. 2005 reported that the oxidation of a polymer-derived amorphous SiAlCN was much lower (about 10 times) than the oxidation of CVD Si_3N_4 or SiC. No oxynitride layer was found as in the case of Si_3N_4 as discussed above. They proposed that the aluminum ions, which sit at the center of six-membered rings of SiO_2, blocked the diffusion of oxygen, thus lowering the oxidation rate.

AlN is an important material in the electronic ceramics industry and is an example of when a small amount of oxidation is beneficial to the application. In this case, the formation of a thin (1–2 μm) protective coating of AlON is formed and is used to improve the adhesion of copper films. Suryanarayana 1990 found the oxidation of AlN powders between 600 and 1000°C in flowing air to follow a linear rate law initially and then a parabolic law as the oxide layer thickness became sufficient to require diffusion for further growth. In contrast, Abid et al. 1986 found that the oxide layer that formed on polycrystalline AlN in air at 1200°C was α-Al_2O_3, whereas below 800°C no oxidation was observed. Dutta et al. 1992 reported that oxidation of sintered polycrystalline AlN between 20 and 200°C progressed from individual α-Al_2O_3 particles of 2–3 nm in size to a 50 nm thick film after 150 hours at 200°C. They also commented that their data were consistent with the formation of an oxynitride layer but believed α-Al_2O_3 to be the oxide formed at low temperature based upon thermodynamic calculations. Others have shown that an oxynitride formed as an intermediate preceding alumina formation at high temperatures, Katnani and Papathomas 1987, and McCauley and Corbin 1979 reported that a region of ALON stability occurred between Al_2O_3 and AlN at temperatures between 1800 and 2050°C in flowing nitrogen. Duchesne et al. 1999 reported that the oxidation of AlN between 850 and 1150°C in air followed a sequence of transition aluminas as shown by the following equation:

$$4AlN + 3O_2 \rightarrow 2(\delta + \theta)Al_2O_3 + 2N_2 \rightarrow 2\alpha\text{-}Al_2O_3 + 2N_2 \qquad (15.34)$$

TABLE 15.3

Forms of Boron Nitride

Designation	Symmetry	Structure Type
Alpha-BN	Hexagonal layered	Similar to graphite
Beta-BN	Cubic diamond-like	Sphalerite[a]
Gamma-BN	Hexagonal	Wurtzite[a]

[a] Sphalerite and wurtzite are polymorphs of ZnS, commonly called zincblend. Wurtzite is stable only above 1020°C.

Gu et al. 2006 claimed that oxidation in flowing oxygen resulted in amorphous alumina between 800 and 1000°C. At higher temperatures the amorphous phase converted to crystalline alumina.

The oxidation of BN generally forms liquid B_2O_3 and N_2 and/or NO_2. Many different oxidation rates and rate laws have been reported in the literature presumably due to porosity, impurity, and crystalline form differences, Jacobson et al. 1999. BN has also been shown to be very sensitive to moisture. Matsuda 1989 showed that BN samples processed at lower pressures and higher temperatures were less sensitive to moisture. The different forms of boron nitride are shown in Table 15.3. Oda and Yoshio 1993 have shown that α-BN was more reactive along the *a* crystallographic axis. This fact is consistent with the bonding between hexagonal planes being weaker than within the plane. After an initial release of moisture below 400°C, BN oxidized measurably by 800°C, which decreased with higher oxygen impurities levels, Podobeda et al. 1976. The vaporization of the B_2O_3 scale limits the use of BN to temperatures below 1300°C in oxidizing media, Lavrenko and Alexeev 1986.

The oxidation of TaN to Ta_2O_5 was reported to commence at about 450°C by Montintin and Desmaison-Brut 1990. As the temperature was raised, the initially powdered reaction product densified; however, the high volume expansion of Ta_2O_5 generated stresses in the coating that caused failure and spalling at high stress regions. Between 590 and 770°C in oxygen, the kinetics of the reaction was characterized by a sigmoidal rate law associated with the formation of the nonprotective Ta_2O_5.

Silicon Carbide

Oxidation

The oxidation of green hexagonal powdered SiC has been described by Ervin 1958. Ervin stated that oxidation at low oxygen pressures took place with the formation of SiO gas, while at atmospheric pressure under flowing air, SiO_2 formed. The rate-controlling step was thought to be growth of an ordered lattice of SiO_2 by solid diffusion. The following reactions are representative of the oxidation of silicon carbide:

$$SiC + 3/2\ O_2 \rightarrow SiO_2 + CO \qquad\qquad (15.35)$$

$$SiC + 3/2\ O_2 \rightarrow SiO(g) + CO_2 \qquad\qquad (15.36)$$

Jorgensen et al. 1961 proposed that the rate-controlling step in the growth of the SiO_2 layer formed on powdered SiC may be the diffusion of either oxygen ions or silicon ions. They ruled out the diffusion of molecular O_2, CO_2, and CO based upon their experimentally determined activation energies being too large for molecular diffusion. Harris 1975 studied the oxidation of crystals of 6H-α SiC and determined that the rate of oxidation on the (000$\bar{1}$) carbon face was approximately seven times greater than that on the silicon face at 1060°C for 70 hours. The thin oxide layer on the (0001) silicon face grows according to a linear rate law at all temperatures, whereas the thick oxide on the carbon face initially grows with linear kinetics but then changes to parabolic when the thickness becomes greater than 250 nm. At high temperatures and/or long times during oxidation of powdered samples, the oxidation rate changes from parabolic to linear presumably due to the fact that the growth of the linearly controlled face overtakes that of the parabolically controlled face. This change in oxidation rate at high temperatures has been attributed to a change in the oxide layer from amorphous to crystalline by Ervin 1958 and Jorgensen et al. 1959 and suggested by Costello and Tressler 1981.

The desorption of CO gas formed at the SiC/SiO_2 interface has been reported to be the rate-controlling step by Singhal 1977; however, Hinze et al. 1975 and many others have reported that it is the inward diffusion of oxygen through the surface layer of SiO_2. Spear et al. 1990 ruled out the diffusion of CO as rate-controlling based upon their experiments that exhibited a dependence of the oxidation rate upon the partial pressure of oxygen and the almost identical activation energies obtained for the oxidation of SiC and Si metal. Fergus and Worrell 1990 have concluded that the various contradictions in reported kinetics were due to a change in the diffusing species from molecular to ionic oxygen at about 1400°C. This was based upon two observations, one being that the activation energy for the growth of amorphous silica on CVD-SiC increased above 1400°C and the other being that the activation energy for the growth of cristobalite increased, but at the higher temperature of 1600°C. Decreases in oxidation rates at low temperatures have been attributed to sufficiently long times to allow crystallization of the silica scale.

In an analysis of the various possible rate-controlling steps, Luthra 1991a concluded that a mixed interface reaction/diffusion process was the limiting feature in the oxidation of SiC. This was based upon the following facts:

1. oxidation rate is lower than for pure silicon,
2. presence of gas bubbles in the oxide layer,
3. oxidation rate of single crystals dependent upon crystallographic orientation, and

4. higher activation energy than for pure silicon (although Spear et al. 1990 reported similar energies).

Since all of the above, except the presence of gas bubbles, are consistent with interface reaction control and the fact that bubbles are present, a mixed controlled process was concluded. Luthra suggested that mixed control should yield a rate law more complex than the generally observed linear or parabolic laws.

For pure monolithic CVD-SiC and Si_3N_4, Fox 1992 reported oxidation rates for 100 hours at temperatures between 1200 and 1500°C in flowing dry oxygen to be similar. In silicon nitride, any additives present will affect the oxidation rate. In general, increased levels of additives or impurities result in higher oxidation rates. These higher oxidation rates are due to the migration of the additive to the oxidized layer, thus lowering the viscosity, which increases the diffusion of the oxidant to the SiC/SiO_2 interface. Fergus and Worrell 1990 reported that 0.5 wt% boron in sintered α-SiC did not, however, significantly affect the oxidation rate.

Understandably the active oxidation of SiC has not been investigated quite as thoroughly as passive oxidation; however, it should be remembered that active oxidation to SiO gas can occur at any temperature if the oxygen partial pressure at the SiC surface falls below some critical value. Not all data reported in the literature agree. The variations reported for this transition are due to differences in the SiC materials tested and in the experimental conditions used. The partial pressure of oxygen at the transition from passive to active oxidation decreases with an increase in the total gas flow through the system, Wang et al. 1990. This is the result of a decreasing gaseous boundary layer thickness with increasing velocity. The total gas pressure of the system can also affect results as suggested by Narushima et al. 1991 since molecular gas flow exists at low pressures and viscous gas flow exists at higher pressures, thus changing the gas diffusion phenomena. Since the rate-controlling mechanism in active oxidation is the oxygen diffusion through the gaseous boundary layer, the characteristics of the gaseous boundary layer play a major role in the oxidation. If experiments are conducted at very high flow rates and very low total pressures, as was the case for the work of Rosner and Allendorf 1970, the rate-controlling step may no longer be oxygen diffusion through the gaseous boundary layer, but the kinetics of gas arrival and removal from the surface. The active-to-passive transition as determined by several investigators and compiled by Vaughn and Maahs 1990 is shown in Figure 15.13. The variations for the reported transition are due to differences in the gas flow rates of the tests and possibly differences in the SiC material tested.

A reaction that occurs in the transition region between active and passive oxidation was given by Wang et al. 2008 as:

$$SiC(s) + 2SiO_2(s) \leftrightarrow 3SiO(g) + CO(g) \tag{15.37}$$

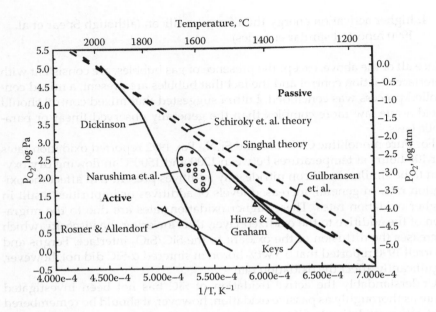

FIGURE 15.13
Literature data for active-to-passive oxidation transition for SiC. (Vaughn and Maahs 1990, reprinted with permission of the American Ceramic Society/Wiley & Sons Copyright Clearance Center.)

Sufficient oxygen is required at the transition boundary to form both SiO_2 and SiO to satisfy the equilibrium of equation 15.37. With increasing flow rates Wang et al. found that the active-to-passive boundary moved toward lower pressures. The partial pressure of oxygen at the boundary was determined by Balat et al. 1992 and is given by the following equation:

$$pO_2 = 0.734 \, K_1^{3/4} K_2^{-1/2} \tag{15.38}$$

where K_1 and K_2 are the equilibrium constants of equations 15.35 and 15.36.

The oxidation of SiC fibers and whiskers is about as diverse as it is for other forms of SiC. Not only is the corrosive degradation of fibers and whiskers complexed by their chemistry (containing impurities of C, and SiO_2) and structure (containing more than one polymorph), but their surface area to volume ratio greatly enhances reaction rates when compared to an equal weight of some other form. SiC fibers manufactured from polycarbosilane polymer precursors generally contain excess carbon, silica, and some combined nitrogen. Jaskowiak and DiCarlo 1989 reported the weight loss behavior of SiC fibers at temperatures ranging from 1000 to 2200°C at argon pressures of 0.1 and 138 MPa and under vacuum (10^{-9} MPa). Although the high external pressure delayed the onset of weight loss from about 1250 to 1550°C, active oxidation occurred through the formation of SiO. Wang et al. 1991 measured the oxide layer thickness on SiC whiskers for the low temperature (600, 700, and 800°C) linear region at times less than 4 hours to be between 2 and 10 nm as determined by XPS analysis and x-ray photoelectron spectroscopy.

The wide variation of oxidation rates and activation energies reported in the literature is due to one or a combination of many factors, including the following:

1. decrease in reactive area with advancing oxidation (taken into account by some but not all),

2. differences in materials studied (α, β, or amorphous),

3. density and porosity variations,

4. variation and amount of pre-existent surface oxide,

5. differences in oxide layer formed (crystalline, amorphous, or liquid), and

6. amounts and type of additives and impurities.

Reaction in Other Atmospheres

McKee and Chatterji 1976 reported no oxidation of SiC when exposed to gaseous environments of pure H_2, pure N_2, or H_2-10%H_2S at 900°C. No evidence of sulfide formation was found in the hydrogen-H_2S mixture. In a mixture of N_2-2%SO_2, which resulted in a partial pressure of oxygen of 10^{-10} atm, active oxidation was observed. With the addition of 5% CH_4 to the mixture of N_2-2%SO_2, an initial (first hour) rapid weight loss was noted, presumably due to the formation of the volatile SiS.

Reaction of SiC in gas mixtures of 5% H_2/H_2O/Ar when at 1300°C was predicted by Jacobson et al. 1990 to fall within one of three regions: passive oxidation, active oxidation, or selective carbon removal depending upon the water content of the mixture. Gas phase diffusion (i.e., water transport to the SiC) was reported to be the rate-controlling step in the active oxidation region (oxygen partial pressures of 10^{-22} to 10^{-26} atm). In the carbon removal region (oxygen partial pressures less than 10^{-26} atm), iron impurities were found to react with the free silicon present to form iron silicides.

Maeda et al. 1989 investigated the oxidation of several different SiC materials in flowing humid air containing 1 to 40 vol% water vapor at a temperature of 1300°C for 100 hours. They found that water vapor greatly accelerated the oxidation of SiC and that a linear relationship existed between percent water vapor and weight gain. Lu et al. 1984 found that oxidation rates of SiC thin films were increased by 10 to 20 times in wet oxygen when compared to dry oxygen at temperatures of 950 to 1100°C. The active oxidation (i.e., weight loss) of SiC was reported to occur in one atmosphere hydrogen containing water vapor at pressures of 10^{-6} to 10^{-3} MPa between 1400 and 1527°C by Readey 1990. At high water vapor pressures, a reaction product of SiO_2 was formed; however, active oxidation continued, since this SiO_2 was reduced to SiO by the hydrogen present. The reactions that took place can be represented by the following equations:

$$SiC(s) + 3H_2O(g) \rightarrow SiO_2(s) + CO(g) + 3H_2(g) \qquad (15.39)$$

$$SiO_2(s) + H_2(g) \rightarrow SiO(g) + H_2O(g) \qquad (15.40)$$

In addition to water vapor, alkali vapors have been shown by Pareek and Shores 1991 to enhance oxidation rates. They studied the oxidation of α-SiC in flowing gas mixtures of dry CO_2-O_2 (9:1 ratio) containing small quantities of K_2CO_3 and K vapors at 1300 to 1400°C for times up to 42 hours. Water vapor was added in some tests; however, the vapor species in those cases was KOH. Pareek and Shores found, at low potassium levels, that the oxidation to SiO_2 followed a parabolic rate law; at higher potassium levels the growth followed a linear law; and when low levels of water vapor were also present (i.e., KOH vapors) the growth kinetics were intermediate between parabolic and linear, indicative of a possible transition from one rate law to another. At moderate to high levels of potassium in the presence of water vapor, the kinetics of oxidation again followed a linear rate law. The increased oxidation in atmospheres containing potassium vapors was suggested to be due to the enhanced mobility of the oxidant through the oxide layer containing dissolved potassium, although the reported activation energy of 225–463 kJ/mol was much higher than expected for oxygen diffusion through silica, which is about 115 kJ/mol. The scales were determined to be composed of cristobalite under most test conditions. At the higher potassium levels and higher temperatures, the scale was sufficiently fluid to flow from the samples.

Federer 1988b studied the effects of a vaporized solution of water containing 1 wt% NaCl in air upon sintered α-SiC under a mechanical load at 1200°C. He reported that a molten reaction layer of sodium silicate formed, causing premature failure under load within an average time of about 150 hours. The same material when exposed to a 1200°C in air and the same loading conditions could sustain the stress without failure for at least 1500 hours. In a similar test, Federer 1988a exposed several types of SiC to a flowing atmosphere containing sodium sulfate and water vapor in air at 1200°C. In these tests, Federer reported that the reaction layer contained tridymite embedded in a sodium silicate liquid. Enhanced oxygen diffusion through this liquid allowed continued corrosion to take place. No discussion was given for the effects of SO_3 gas upon the corrosion as Jacobson and coworkers did (Jacobson and Smialek 1985, Smialek and Jacobson 1986, Jacobson 1986, Fox and Jacobson 1988, Jacobson and Fox 1988, and Jacobson et al. 1986) other than to state that sodium sulfate vapor reacted with silica under low partial pressures of SO_3.

Park et al. 1990 investigated the corrosion of sintered α-SiC in a gas mixture containing 2 vol% chlorine and varying amounts of oxygen (0, 1, 2, and

4 vol%) in argon at temperatures of 900, 1000, and 1100°C. They concluded that small amounts of oxygen were necessary to facilitate active corrosion by removing carbon as CO, thus allowing access to the silicon for formation of $SiCl_4$ (or $SiCl_3$). Volatile SiO may also form. At 1000°C, the rate of active corrosion increased as the amount of oxygen increased. Some amorphous SiO_2 began to form at 1000°C and 2 vol% O_2, but it remained nonprotective even at 1100°C.

Lee et al. 2003 studied the effects of dry and moist KCl vapor on the corrosion of several SiC materials by bubbling premixed Ar, O_2, and Cl_2 through molten KCl. The premixed gas was bubbled through water to add either 2 or 3 vol% H_2O to the mixture. The predominant corrosion product was potassium silicate glass with many trapped bubbles. At 900°C the corrosion rate increased by about a factor of four when 3 vol% H_2O was present. Corrosion also increased with increased temperature and increased KCl vapor pressure.

Other Carbides

Probably the next most important carbide after silicon carbide is tungsten carbide. The principal application of this material is in cemented carbide cutting tools. The carbides of titanium, tantalum, and niobium are used as alloying additions to WC. Addition of TiC to WC cutting tools causes the formation of a titanium oxide surface layer that greatly increases the tool's wear resistance. When WC oxidizes, it forms the volatile WO_3 oxide that offers no protection to wear.

Most of the oxidation studies conducted on TiC are rather dated; however, reasonable agreement has been exhibited among the various studies. Stewart and Cutler 1967 found that the oxide layer that formed below 400°C was anatase and that above 600°C it was rutile. Single crystal studies indicated no difference in oxidation between the (100) and (110) faces at 1000°C. At low temperatures (752–800°C), the rate of oxidation exhibited a dependence upon the oxygen partial pressure to the 1/6 power, whereas at high temperatures, the dependence was to the 1/4 power. The actual mechanism of oxidation appeared to be mixed with a near-parabolic rate initially changing to a near-linear rate at longer times.

The use of ZrC at high temperatures (>450°C) has been limited due to excessive oxidation, although it possesses other excellent properties. The oxidation of powdered ZrC at low temperatures (between 380 and 550°C) was reported by Shimada and Ishii 1990. They reported that oxidation commenced at 300°C at all partial pressures of oxygen between 0.66 and 39.5 kPa and that complete oxidation occurred at different temperatures depending upon the oxygen pressure. Shimada and Ishii suggested that rapid initial oxidation occurred through the formation of an oxycarbide, $Zr(C_xO_{1-x})$. Following this initial oxidation, the mechanism changed around 470°C to one that formed cubic zirconia as the product of oxidation along with the generation of microcracks. Selected area electron diffraction around the edges of the ZrC grains oxidized below 470°C exhibited the presence of cubic ZrO_2 nuclei that were not observable by *XRD*. At low temperatures, when the reaction was about 75% complete, carbon was found to be present as hexagonal diamond. Hexagonal diamond was also produced initially (reaction 40% complete) at higher temperatures.

Arun et al. 1990 reported the following order of TiC > HfC > ZrC for the oxidation resistance for these three carbides at 1273 K. The oxidation of these materials is much greater when they are incorporated into hot-pressed compositions of $TiC-ZrO_2$, $ZrC-ZrO_2$, and $HfC-HfO_2$. Arun et al. also reported a greater oxidation of TiC when incorporated into ZrO_2 as opposed to Al_2O_3.

Early oxidation studies of Ti_3SiC_2 indicated that the products of oxidation in air were rutile, silica, and CO or CO_2. Some controversy existed as to whether the silica was amorphous or one of the crystalline phases (either tridymite or cristobalite). Nguyen et al. 2007 attributed this to the sample characteristics (i.e., impurities) and heat treatments among the various investigators. Nguyen et al. found an outer scale of TiO_2, which grew into pillar-like grains, with a layer of mixed SiO_2/TiO_2 underneath. After oxidation at 1200°C for 100 hrs, the silica crystallized into cristobalite. At the interface of the two layers were microscopic voids. The escape of CO/CO_2 disrupted the continuity of the outer oxide scale. Thus the resistance to oxidation of Ti_3SiC_2 was due to the continuity of the inner SiO_2/TiO_2 layer. Total scale thickness grew to 300 μm after 100 hrs at 1200°C. A previous study by Barsoum et al. 1997 found essentially the same results. They also mentioned that the oxidation resistance was a function of the TiC content of their samples (less being better). In addition to CO and/or CO_2 gases being formed, Sun et al. 2001 reported the formation of SiO gas that diffused outward, especially at high temperatures (>1100°C). The SiO was formed due to a low partial pressure of oxygen present at extended times when Ti_3SiC_2 reacted with the initial SiO_2 that formed. The outward diffusion of the SiO then oxidized as the partial pressure increased, leaving SiO_2 precipitates in the TiO_2 layer.

Wang and Zhou 2003 studied the oxidation of Ti_3AlC_2 at temperatures up to 1400°C. Below 1200°C the outer oxide layer was rutile, but at 1300°C this layer consisted of both TiO_2 and Al_2TiO_5. The inner layer at both temperatures was α-Al_2O_3. At 1400°C the outer layer was predominantly α-Al_2O_3 with some Al_2TiO_5.

Boron carbide is chemically very stable. It will dissociate in a vacuum above 2600°C into boron gas and solid carbon, Markarenko 1977. The oxidation of B_4C starts at about 600°C, forming a B_2O_3 film. Moisture in the air will lower this temperature to 250°C. Chlorine reacts with B_4C at 1000°C, forming BCl_3 and graphite.

Nanocrystalline (~30 nm) niobium carbide, NbC, starts to oxidize at about 350°C and is completed by about 550°C, Shi et al. 2005. The products of oxidation were assumed to be Nb_2O_5 and CO_2.

Oxides

Alumina

Among the ceramics community, alumina is considered to be one of the most inert materials toward a large number of environments. For this reason, alumina that is produced as a 95 to 100% Al_2O_3 material is used in many furnace applications. The one area where its reactivity is often overlooked is its application in laboratory furnaces. Most high-temperature laboratory furnaces use alumina as the standard lining. When the materials under investigation react to form gaseous species, and especially when the furnace atmosphere contains a low partial pressure of oxygen, one should be aware of the possible reactions that may occur with alumina.

The lining of a laboratory furnace can receive a much more severe usage than an industrial furnace. Generally this is due to repeated thermal cycling and to the investigation of a wide variety of materials that produce a wide range of corrosive environments. An example of the corrosion of sample crucible setter tiles is given below.

The test environment of a vertical tube furnace that used an alumina tube and horizontal alumina discs on which to place alumina sample crucibles included an input atmosphere of hydrogen and methane that gave an oxygen partial pressure of 10^{-14} Pa at the test temperature of 1300°C. The samples being tested were various silicon nitride samples containing small quantities of MgO, Y_2O_3, Fe_2O_3, ZrO_2, Al_2O_3, and CaO. A discussion of the furnace setup and operation can be found in de Jong et al. 1985.

The alumina discs in the upper portion of the furnace tube above the silicon nitride samples, along a temperature gradient that ranged from 1250 to 1185°C, exhibited a glazed surface layer of silicate glass containing crystals of cordierite and cristobalite (see Figure 15.14). The formation of cordierite was caused by the active oxidation of the silicon nitride and the vaporization of magnesia contained within the nitride samples that subsequently reacted with the alumina to form cordierite. The following equations describe the reaction:

$$2Si_3N_4 \text{ (s)} + 3O_2 \rightarrow 6SiO \text{ (g)} + 4N_2 \text{ (g) @1300°C} \qquad (15.41)$$

$$SiO \text{ (g)} + 1/2 \ O_2 \rightarrow SiO_2 \text{ (s) @1185–1200°C} \qquad (15.42)$$

$$MgO \text{ (s)} \rightarrow MgO \text{ (g) @1300°C} \qquad (15.43)$$

$$2MgO \text{ (g)} \rightarrow 2MgO \text{ (s)} + 2SiO_2 \text{ (s)} + Al_2O_3 \text{ (s)} \rightarrow 2MgO \cdot Al_2O_3 \cdot 2SiO_2 \text{ (s)}$$
$$\text{@ 1185–1250°C} \qquad (15.44)$$

Although Anderson 1979 reported that oxygen was necessary for the formation of potassium beta-alumina from sapphire used in vapor arc lamps, van Hoek et al. 1992 showed that potassium vapors (at 1 MPa) were able to reduce alumina at 1373 K in the absence of oxygen by the following reaction:

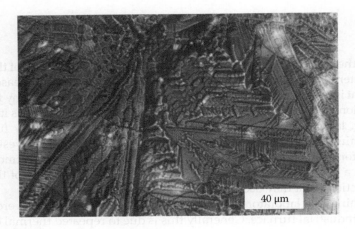

FIGURE 15.14
Cordierite and cristobalite formation on alumina. (Reflected light, differential interference contrast micrograph, magnification 500×.)

$$6K + 16Al_2O_3 \rightarrow 3(K_2O \cdot 5Al_2O_3) + 2Al \tag{15.45}$$

forming a potassium beta-alumina and metallic aluminum. Although approximately 3 wt% aluminum should form, they detected no metallic aluminum by XRD in their samples. They suggested that the presence of metallic aluminum was confirmed by the blackening of samples due to the formation of finely divided metallic aluminum. They found that the beta-alumina formed with the c crystallographic axis parallel to the substrate surface. This created an easy diffusion path perpendicular to the surface for diffusion of potassium and continued corrosion. They also suggested that this oriented growth was not the result of epitaxial growth, since the starting alumina was a polycrystalline material. It is unfortunate that many references can be found in the literature that refer to oriented growth on polycrystalline substrates as *epitaxy*. Although localized crystallographic matching may occur on a grain-to-grain basis forming an oriented polycrystalline layer on an oriented polycrystalline substrate, this was not the original meaning of epitaxy. It is enlightening that van Hoek et al. suggested another reason for oriented growth—the faster growth direction (easy diffusion path) being perpendicular to the reaction surface.

In a study of UF_6-fueled gas-core reactor systems, Wang et al. 1991 investigated the effects of UF_6 gas (at pressures of 20.0 to 22.7 kPa) upon alumina at temperatures of 973, 1073, 1273, and 1473 K for times up to 4 hours. At the three lower temperatures the following reaction was suggested to represent their findings:

$$6UF_6 + 2Al_2O_3 \rightarrow 4AlF_3 + 6UF_4 + 3O_2 \tag{15.46}$$

The AlF_3 formed on the alumina surface, whereas the UF_4 was found on the colder portions of the furnace chamber. At 1473 K several oxides of uranium were found in the surface scale and no AlF_3 was found due to its high vapor pressure (30.6 kPa) at that temperature. Weight gain was reported for the lower temperatures; however, at 1473 K, a large weight loss was exhibited due to vaporization of the AlF_3.

Wang et al. 1991 also performed a computer analysis of the expected reactions and found several differences between the experimental and calculated data. Although the differences that were found are not important, the cause of these differences is worth noting. The computer program used to predict chemical reactions at different combinations of temperature and pressure is dependent upon the database used and thus cannot predict products not contained in the database. Also the program used by Wang et al. could analyze only a closed system at equilibrium. Any reactions (e.g., formation of interfacial layer) that may retard further reaction would prevent equilibration. Insufficient time for reactions to proceed to completion would also contribute to the differences, since the computer program based upon minimization of total free energy of formation cannot predict the kinetics of the reactions. Thus one should remember that calculated reactions based upon thermodynamics is only a portion of any study and only an indication of what should be expected during actual experimentation.

Thermal barrier coatings are used to protect metals from the high temperatures, but then these coatings must also act as environmental barrier coatings when used in an environment where chemical attack can occur. Since water vapor can be a combustion product in turbine engines, materials resistant to water vapor attack at high temperature are required to protect the metal turbine blades and vanes. Alumina is a common protective material used alone or in combination with other materials. Thus the reaction of alumina with water vapor at temperatures between 1250 and 1500°C was investigated by Opila and Myers 2004. The partial pressure of the water vapor was between 0.15 and 0.68 atm in oxygen for a total pressure of 1 atm. Gas velocity was 4.4 cm/s. Under these conditions the corrosion product was $Al(OH)_3$ gas. Other vapor species are possible, but calculations indicated they were all of low probability under the conditions of the experiments. Opila and Myers reported another example where the furnace materials got involved in the reactions. In this case a fused quartz furnace tube was used that reacted with the water vapor forming $Si(OH)_4$ gas that subsequently reacted with the alumina samples at 1500°C, forming mullite on the surface of the samples. The alumina volatility was followed by weight loss measurements. The weight loss rate at 1250°C was 3.38×10^{-4} mg/cm$^2 \cdot$ hr and at 1450°C it was 3.09×10^{-3} mg/cm$^2 \cdot$ hr.

Alumino-Silicates

Professor Arnulf Muan has provided a considerable amount of experimental data concerning the atmospheric effects upon the phase equilibria of refractory materials. One such article was reprinted in the *Journal of the American*

Ceramic Society, Muan 1992, as a commemorative reprint. This article stressed the importance of the oxygen partial pressure in determining the phases present in the reaction of iron oxides with alumino-silicate refractories. Under oxidizing conditions, large amounts of ferric iron can substitute for aluminum in the various aluminum-containing phases; however, under reducing conditions this substitution is negligible. Large volume changes accompany some of the phase changes that occur with damaging results to the refractory. In addition, the temperature at which liquid phase develops decreases as the oxygen partial pressure decreases.

Reactions that have occurred between alumino-silicate refractories and the gaseous exhaust in glass furnace regenerators at about 1100 to 1200°C forming nepheline and noselite are shown below:

$$3Al_2O_3 \cdot 2SiO_2 + Na_2O \rightarrow Na_2O \cdot Al_2O_3 \cdot 2SiO_2 + 2Al_2O_3 \qquad (15.47)$$

The nepheline formed then reacts with SO_3 and more Na_2O vapor of the exhaust, forming noselite:

$$3(Na_2O \cdot Al_2O_3 \cdot 2SiO_2) + 2Na_2O + 2SO_3 \rightarrow$$
$$5Na_2O \cdot 3Al_2O_3 \cdot 6SiO_2 \cdot 2SO_3 \qquad (15.48)$$

Although the precise mechanisms that take place have not been determined, most likely the alumina and any silica available forms more nepheline. Large volume expansions (10–15%) accompany these reactions, resulting in spalling or shelling. Historically these reactions have been a serious problem to the glass manufacturer, since they cause plugging of the regenerator and a less-efficient combustion process. Various regenerator design and material changes have essentially eliminated this problem; however, it is a reaction that may still occur when the conditions are appropriate.

Other Silicates

$ASiO_4$ (A = Zr, Hf, and Ti)

Samples of $ASiO_4$ (A = Zr, Hf, and Ti) were tested at 1500°C for 50 hrs in an air:H_2O ratio of 70:30 wt% at a flow rate of 175 mL/min by Ueno et al. 2005. They found that the corrosion resistance decreased with increasing atomic number of the A cation. Although the Ti-containing sample exhibited essentially no weight change during the test, it did exhibit some surface corrosion (surface etch pits). $HfSiO_4$ and $ZrSiO_4$ exhibited weight losses of 7.08×10^{-6} and 1.33×10^{-6} g/cm^2 · hr, respectively. Corrosion occurred at grain surfaces that decomposed to AO_2. This reaction is represented by the following equation:

$$ASiO_2(s) + 2H_2O(g) \rightarrow AO_2(s) + Si(OH)_4(g) \qquad (15.49)$$

Chrome-Containing Materials

In a simulated industrial furnace atmosphere cycling between 8–10% combustibles (reducing) and 6–10% excess oxygen (oxidizing) at temperatures of 1260 and 1400°C, Mayberry et al. 1970 showed that refractories containing chrome ore developed a permanent expansion and a loss in strength. This was the result of spinel solution into periclase and then exsolution with the accompanying phase redistribution, recrystallization, and pore development during cycling. This example showed how a material may experience degradation due to atmosphere effects though not exhibiting any signs of classic corrosion (weight gain or loss, reaction product surface layer formation, etc.).

The double layer coating of Cr_2O_3 (inner layer) and $MnCr_2O_4$ (outer layer)[*] formed by preoxidation of various alloys was studied by Li et al. 2010 for the effect of CeO_2 on the high-temperature carburization behavior. In addition to CeO_2, Y_2O_3 and La_2O_3 have been reported to be beneficial to the stability of the Cr_2O_3 layer. Samples were tested at 1050°C in a gas mixture of 2% CH_4 + 98% H_2 at a flow rate of 55.4 mL/min for times up to about 250 hrs. Li et al. suggested the following equation for the carburization process (without CeO_2):

$$3MnCr_2O_4 + 13CH_4 \rightarrow 3MnO + 2Cr_3C_2 + 9CO + 26H_2 \qquad (15.50)$$

The presence of 5% CeO_2 (added to the $MnCr_2O_4$) accelerated the carburization of $MnCr_2O_3$ by forming $CeCrO_3$. Under the high-temperature reducing environment CeO_2 was reduced to Ce_2O_3 that subsequently reacted with Cr_2O_3 to form $CeCrO_3$. At longer carburization times $CeCrO_3$ decomposed to CeO_2 and Cr_3C_2. The addition of CeO_2 to Cr_2O_3 reduced the carburization rate compared to pure Cr_2O_3 for about the first 15 hrs. The weight loss for these samples was negligible at about 70 hrs; however, at extended times a weight gain was observed. The weight loss was due to the formation of Cr_3C_2. The weight gain was due to the formation of carbon.

Magnesia-Containing Materials

McCauley and coworkers, McCauley 1989, Parker and McCauley 1982a, Parker and McCauley 1982b, and McGarry and McCauley 1992 have studied the effects of vanadium upon the phase equilibria in magnesia-containing materials. This work was initiated in an effort to understand the effects of vanadium impurities in fuel oils upon basic refractories. It has been found that only small amounts of V_2O_5 are needed to alter the phase assemblages in high-magnesia materials. The reactions that occur generally form low-melting vanadates (i.e., tricalcium and trimagnesium vanadates with melting

---

[*] These double layer coatings are used for applications as interconnects for planar solid oxide fuel cells. The spinel layer blocks the intrusion of carbon from carburizing environments more effectively than Cr_2O_3 alone. CeO_2 is added to reduce the oxidation of the alloy and to enhance oxide scale adhesion, Alman and Jablonski 2007.

points of 1380 and 1145°C, respectively, and magnesium-calcium-vanadium garnet with its melting point of 1167°C) and depending upon the exact compositions can develop appreciable amounts of liquid at service temperatures. Although the initial reaction is a gaseous phase reaction, it quickly converts to a liquid attack.

Ready 1992 has shown that magnesia containing a small percentage of carbon black or graphite can form CO/CO_2 bubbles in oxidizing atmospheres at temperatures ranging from 1200 and 1500°C. These bubbles, which formed primarily at grain boundaries, were the source of intergranular fracture. For the unoxidized materials the fracture was 100% transgranular, since no grain boundary separation (caused by the presence of bubbles) was present.

Magnesia-carbon refractories have been studied for many years due to their excellent resistance to steelmaking slags. Several types of carbon sources have been tried, with graphite being the source of carbon in the latest versions of these refractories. The graphite can be oxidized directly by reaction with oxygen or indirectly through reaction with MgO. In the case of direct oxidation only CO gas is formed. In the case of indirect oxidation Mg gas is also formed. At temperatures below 1400°C the direct oxidation of graphite is important, whereas above 1400°C the reaction with MgO becomes important. Since very little is known about the oxidation kinetics at low temperatures, Sadrnezhaad et al. 2006 studied the reactions between 600 and 1250°C. Samples of magnesia containing 14.3 wt% graphite and 4.8 wt% phenolic resin as binder were tested at 600, 650, 700, 800, 950, 1100, and 1250°C in air. The weight loss with time was determined. Graphite grain size distribution and temperature distribution were determined to govern the rate of graphite oxidation. Below about 800°C soot[*] deposition was determined to be an important factor in blocking pores, thus minimizing the inter-diffusion of gases. A mathematical model developed to explain the experimental results indicated that a mixed mechanism governed the oxidation with three activation energies, one for gas adsorption onto the graphite, one for pore diffusion of gases within the decarburized layer, and one for diffusion of reactants toward active reaction sites.

SiAlON

Shimada et al. 2003 investigated the resistance to NaCl vapor attack of $\beta\text{-}Si_{6-z}Al_zO_zN_{8-z}$ (z = 0–3). Various concentrations of NaCl vapor carried by flowing argon gas at 1300°C for 3–24 hrs were studied. Corrosion was minimal for z = 0 (pure Si_3N_4), slight for z = 1 or 2, and severe for z = 3. The corrosion of the SiAlONs was accompanied by the formation of bubbles with z = 3 being the most severe. Shimada et al. attributed the bubbles to the escape of N_2 through a viscous melt. For z = 1 and 2 silica appeared and for z = 3

---

[*] Soot, a carbonaceous material generally in the 25 nm size range, is formed by the exothermic oxidation of CO to CO_2.

mullite also appeared. Scale thicknesses were about 10 μm for $z = 1$ and 2 and about 60 μm for $z = 3$. No distinct scale was formed on the $z = 0$ samples.

Titanates

The decomposition reactions of $CaCu_3Ti_4O_{12}$ in oxygen-free nitrogen for 6 hrs at 1000°C (sintered pellets) and in 5%H_2/95%Ar heated at 10°C/min (crushed pellet) were investigated by Adams et al. 2006. This material is of interest for internal barrier layer capacitor applications.[*] During the sintering process this material either loses oxygen (model 1), reducing the Ti^{4+} to Ti^{3+}, or in a more complex mechanism (model 2) Cu^{2+} is reduced to Cu^+ at temperature and reoxidizes on cooling, causing a reduction of Ti^{4+} to Ti^{3+}. The consequence of these reactions is that the material becomes an n-type semiconductor with insulating grain boundaries. The work of Adams et al. supported the cation nonstoichiometry model (#2) rather than the oxygen loss model (#1). Weight loss studies by hydrogen reduction on powered samples indicated commencement of weight loss at about 330°C and completion by 680°C. The original $CaCu_3Ti_4O_{12}$ had completely converted to $CaTiO_3$, TiO_2, and metallic Cu. In sintered pellets this reaction was incomplete when heated to 1000°C in N_2. The copper was present as Cu_2O at the interface of the bulk and the reaction layers. Closer to the surface the Cu_2O volatilized, forming a very porous outer layer of $CaTiO_3$ and TiO_2.

Zirconia

Lepistö et al. 1988 studied the effects of humid conditions at 150°C for up to 1000 hrs upon the stability of metastable tetragonal phase in TZP (i.e., tetragonal zirconia polycrystals). Several different materials containing a small amount of yttria were tested and all were found to contain increasing amounts of monoclinic zirconia as the exposure time increased. It was believed that surface finish along with the grain size and yttria content all had an effect upon the transformation of the tetragonal-to-monoclinic zirconia. In a later study Lepistö and Mäntylä 1989 concluded that the stability of yttria-containing TZP in humid atmospheres was through the dissolution of ZrO_2 at grain boundaries with the subsequent relief of localized stresses followed by the transformation of the tetragonal-to-monoclinic phase. The proposed mechanism of dissolution was through the dissociation of adsorbed water molecules on the zirconia surface. The oxygen ions formed are then proposed to anneal the oxygen vacancies present within the yttria-containing zirconia. As a result water is formed, which then allows continued dissolution. The water adsorption is greatest at defect sites and thus is dependent upon the yttria concentration. Lepistö and Mäntylä found that both yttria

---

[*] The details of the microstructure that causes the "giant" dielectric constant (>10^4) for this material are still being investigated.

and alumina increased the dissolution rate, whereas ceria did not since it did not change the vacancy concentration. Thus the tetragonal phase stability toward humid atmospheres can be increased by using ceria as the dopant.

Along with alumina, Wang et al. 1991 studied the effects of UF_6 gas upon partially stabilized zirconia containing 7 wt% CaO. Temperatures of 873, 973, and 1073 K were investigated at times up to 2 hours. The UF_6 gas pressure was maintained at 20 to 22.7 kPa. A weight increase was reported for only the lowest temperature. The surface scale was nonprotective for all temperatures. The compounds that formed on the sample surface after exposure at 873 K were ZrF_4, CaF_2, UO_3, and U_3O_8. Due to the hygroscopic nature of ZrF_4 this compound was identified as a hydrate. At 973 K the same compounds were found except for UO_3. At 1073 K additional compounds (UF_4, UO_2F_2, and $UO_2F_2 \cdot 1.5H_2O$) were found due to dissociation of the UF_6 and reaction with atmospheric moisture. In addition, some zirconium oxyfluorides were found. As in the analysis of the alumina reactions, the computer predictions were different from the experimental results.

Silicides

The oxidation of $MoSi_2$ has been reported to occur by several mechanisms by Fitzer 1990 depending upon the temperature and the oxygen partial pressure. Initially only MoO_3 formed, but volatilized, allowing the formation of SiO_2. The partial pressure of oxygen at the interface $MoSi_2/SiO_2$ then decreased, allowing oxidation of only silicon to continue. At very high temperatures (>1200°C) and low oxygen pressures (<10^{-6} atm), active oxidation occurred with the formation of volatile silicon monoxide, as long as the silicon content on the surface was sufficient. At low pressures of oxygen, selective oxidation of silicon occurred due to its greater affinity for oxygen than molybdenum. The selective oxidation of silicon led to the formation of a sublayer of Mo_5Si_3. At moderately high temperatures (around 1000°C) and high oxygen pressures ($\approx 10^{-2}$ atm) the evaporation of the molybdenum oxides formed led to a protective SiO_2 layer. During the volatilization of the molybdenum oxides, the SiO_2 layer was very porous, allowing rapid oxidation with temperature increase. At lower temperatures where the molybdenum oxides did not volatilize but remained as solid oxide reaction products, a continuous silica layer could not form. This occurs at temperatures below 600°C and is called *pesting*, which can lead to total destruction of the material. The actual amount of *pesting* that occurred, however, was dependent upon the microstructure, Berztiss et al. 1992. Lin et al. 1994 found that the oxidation of $MoSi_2$ in a combustion gas environment (i.e., natural gas) was very similar to that described above. Initially SiO_2 and volatile MoO_3 formed until the silica layer was of sufficient thickness to diminish the H_2O/CO_2 activity (equivalent to a low partial pressure of oxygen) to a level where Mo_5Si_3 formed. The change in mechanisms took only minutes at temperatures of 1370 and 1600°C. Compare the above data with that discussed in Chapter 17 on composites containing $MoSi_2$.

Superconductors

It is well known that $YBa_2Cu_3O_x$ (123) is unstable with respect to reaction with carbon dioxide or water vapor. This reaction is related to the large enthalpies of formation of barium carbonate and barium hydroxide, which are −64.4 and −35.4 kcal, respectively. Davison et al. 1987 reported the formation of barium carbonate when *123* was held over water for 48 hrs. Yan et al. 1987 reported the formation of $Cu(OH)_2$ when *123* was exposed to 85°C and 85% relative humidity for 90 minutes. After exposing *123* to 80°C and 100% relative humidity for times ranging from 15 minutes to 24 hours, Fitch and Burdick 1989 reported the formation of Y_2BaCuO_x, $BaCO_3$, CuO, $Cu(OH)_2$, and the possible formation of $Y(OH)_3$ and BaO. Fitch and Burdick, who noticed that corrosion was visibly present by a significant expansion of their samples, concluded that barium was leached first and then reacted with atmospheric CO_2 to form $BaCO_3$ on the surface.

Attack by Solids

The stability of various materials to graphite is a good example of a solid-solid reaction. In this case, however, at least one of the products is a gas. The stability of a few selected refractory oxides in contact with graphite increases in the order TiO_2, Al_2O_3, ThO_2, MgO, $MgAl_2O_4$, SiO_2, and BeO, as reported by Klinger et al. 1966.

Magnesia

Magnesia vaporization is important in basic refractories where it migrates to form a region rich in magnesia by vaporization and condensation and leaving behind a region of high porosity. The zone of high porosity causes a mechanically weak area that may crack or spall. Vaporization and condensation of magnesia can also occur in silicon nitride where it is used as a sintering aid (see equations 15.32 and 15.33).

Martin and Hunter 1974 found that the carbon in pitch-containing high-magnesia refractories can react with the magnesia to form magnesium gas according to the following equation:

$$C(s) + MgO(s) \rightarrow Mg(g) + CO(g) \tag{15.51}$$

This magnesium gas is then transported to the hot-face of the refractory where it can react with FeO in the slag forming iron metal liquid and a dense solid magnesia layer according to:

$$Mg(g) + FeO(l) \rightarrow Fe(l) + MgO\ (s) \qquad (15.52)$$

These same reactions take place when the source of carbon is graphite. The dense magnesia layer that forms provides a degree of protection against further corrosion.

Metals

Because of the prevalence of platinum metal in various research and manufacturing operations, the reactions of various refractory oxides with platinum is of considerable importance. Ott and Raub 1976 reported that platinum acts as a catalyst for the reduction of refractory oxides by hydrogen, carbon, CO, and organic vapors. These reactions can occur as low as 600°C and result due to the affinity of platinum for the metal of the oxide by forming intermetallic compounds and crystalline solutions.

Gutman and Gutmanas 1990 reported that titanium powder reacted with silicon nitride, forming a complex interfacial layer structure consistent with the Ti-N-Si and Ti-Si phase diagrams. A 1-hour heat treatment in evacuated stainless steel bag at 1273 K yielded two morphologically different main layers. A fine-grained layer (1.5 μm thick) next to the silicon nitride was composed of mostly Ti_5Si_3. A very thin layer adjacent to the silicon nitride contained TiN with some dissolved silicon. As the silicon increased away from the silicon nitride the phases changed into TiN plus Ti_5Si_3. The next layer was almost pure Ti_5Si_3 with some additional titanium adjacent to the next main layer. A coarse-grained layer came next containing a mixture of what was thought to be Ti_3Si particles embedded within Ti metal with a little dissolved silicon. Similar results were obtained when different times (20, 40, and 60 min) and temperatures (1173, 1223, and 1323 K) were used.

Silica

Miller et al. 1979 have shown that carbon reacted with SiO_2 to form the intermediate phase SiC, which then reacted with silica to form the gaseous phase SiO. The following equations were given to represent the reaction:

$$SiO_2 + 3C \rightarrow SiC + 2CO \qquad (15.53)$$

$$2SiO_2 + SiC \rightarrow 3SiO + CO \qquad (15.54)$$

They stated that these reactions were sufficiently rapid at 1000°C and in the presence of iron, which acts as a catalyst for the reduction of silica by SiC, to cause failure of silicate refractories in coal gasification atmospheres.

Probably one of the more severe reactions of the past that has taken place in commercial glass furnaces is that between silica and alumina or alumina-containing refractories. When these two materials are in direct physical contact at high temperature, an interface of mullite forms. This reaction is accompanied by a substantial volume increase that tends to push the two original materials apart. Separation of silica and alumina by the more neutral material zircon has prevented this deleterious reaction in modern furnaces.

Superconductors

The $YBa_2Cu_3O_x$ *(123)* superconductors, where $x = 6.5$ to 7.0, have been reported to exhibit reaction and/or decomposition when in contact with various materials. This has presented researchers with the problem of sample holders for the production of *123* materials. Williams and Chandhury 1988 have conducted a thermodynamic study of the various materials that might react with *123*. Based upon the heat of formation of CuO of –18.6 kcal/gram atom and the following equations:

$$M + 2CuO \rightarrow MO_2 + 2Cu \tag{15.55}$$

or

$$M + 3CuO \rightarrow M_2O_3 + 3Cu \tag{15.56}$$

they reported that the nine elements Ru, Rh, Pd, Ag, Os, Ir, Pt, Au, and Hg should not react with CuO, and most likely would not react with *123*. Murphy et al. 1988 reported that *123* was nonreactive toward silver and, to a lesser extent, gold.

One of the potential applications of superconductors is that of thin films on a semiconductor substrate; however, the most widely used semiconductor substrate material, silicon, reacts with *123*. An examination of the various phase equilibria indicated that $BaSi_2O_5$ does not react with *123*, since these two materials form a stable tie-line in the BaO-Y_2O_3-CuO-SiO_2 quaternary system. Thus this barium silicate could be used as a buffer layer for production purposes or during manufacture of thin films on semiconducting substrates.

Mikalsen et al. 1988 reported that no reaction occurred between thin film superconductors in the Bi-Sr-Ca-Cu-O system and MgO substrates even after annealing at 850°C for 30 min. Thin films on Al_2O_3, however, reacted and became insulating and transparent. Abe et al. 1988 and Ibara et al. 1989 have reported that melts of $BiSrCaCu_2O_x$ or $BiPb_ySrCaCu_2O_x$ reacted with alumina crucibles contaminating their samples.

Additional Recommended Reading

Amoroso, G.G. and Fassina, V. 1983. Stone Decay and Conservation, *Materials Science Monograph 11*, Elsevier, Amsterdam, 453 pp.

Cao, X.Q., Vassen, R and Stoever, D. 2004. Ceramic Materials for Thermal Barrier Coatings, *J. European Ceram. Soc.*, 24, 1–10.

Guthrie Jr., G.D. and Mossman, B.T. (eds.), 1993. Health Effects of Mineral Dusts, *Reviews in Mineralogy*, Min. Soc. Am., Washington, DC, Vol. 28, 584 pp.

Lea, F.M. 1970. *The Chemistry of Cement and Concrete*, Edward Arnold Publishers, London, UK.

Taylor, H.F.W. 1979. Mineralogy, Microstructure, and Mechanical Properties of Cements, *Proc. Br. Ceram. Soc.*, 29, 147–63.

Exercises, Questions, and Problems

1. Determine whether molten aluminum will react with cordierite by calculating the free energy of reaction. If so, what are the reaction products?

2. List the following oxides in the order of most thermodynamically stable to least stable: Al_2O_3, CaO, Fe_2O_3, MgO, and ZrO_2.

3. Discuss several cases where corrosion is beneficial.

4. Using the Al_2O_3-MgO-SiO_2 phase diagram, determine the interface, if any, that should form on an alumina crucible containing a silica melt at 1000, 1200, and 1400°C. What changes would one expect if a mullite or spinel crucible were substituted for the alumina? Which of these materials is best at each temperature?

5. Discuss how forced convection (erosion) affects the reaction interface layer thickness.

6. Calculate the critical slag layer thickness for the passive-to-active oxidation of SiC in air at 1400°C using the following equation:

$$pO_2 = RTC^*/(1 + 2X_1/A)$$

Assume $C^* = 0.0.80$ mol/m^3 and $A = 0.31$ μm.

7. Does the packing density of the tetrahedra of the different silica polymorphs affect their dissolution? If so, how?

8. Explain why the reported oxidation of silicon nitride and/or carbide by various investigators varies.

9. Describe the difference between active and passive oxidation of SiC.

10. Calculate the Gibbs free energy of reaction for the following reactions at 800°C:

$$1/2\ SiC + Mg \rightarrow 1/2\ Mg_2Si + 1/2\ C$$

$$1/6 \; Si_3N_4 + Mg \rightarrow 1/3 \; Mg_3N_2 + 1/2 \; Si$$

Which reaction is more likely to occur? Which material would be the best crucible to contain molten magnesium at 800°C: SiC, Si_3N_4, or a Si_3N_4-bonded SiC?

References

Abe, Y., Hosono, H., Hosoe, M., Iwase, J. and Kubo, Y. 1988. Superconducting Glass-Ceramic Rods in $BiCaSrCu_2O_x$ Prepared by Crystallization under a Temperature Gradient, *Appl. Phys. Lett.*, *53*(14), 1341–12.

Abid, A., Bensalem, R. and Sealy, B.J. 1986. The Thermal Stability of AlN, *J. Mater. Sci.*, *21* 1301–4.

Adams, T.B., Sinclair, D.C. and West, A.R. 2006. Decomposition Reactions in $CaCu_3Ti_4O_{12}$ Ceramics, *J. Am. Ceram. Soc.*, *89*(9), 2833–38.

Ado, G., Bernache, D., Billy, M., Hahn, K.S. and Lefort, P. 1985. Mécanisme de frittage sous charge du nitrure d'aluminium et de l'oxynitrure (AlON), *Rev. Chim. Mineral.*, *22*, 473–83.

Allaire, C. 1992. Refractory Lining for Alumina Electrolytic Cells, *J. Am. Ceram. Soc.*, *75*(8), 2308–11.

Allaire, C. and Desclaux, P. 1991. Effect of Alkalies and of a Reducing Atmosphere on the Corrosion of Refractories by Molten Aluminum, *J. Am. Ceram. Soc.*, *74*(11), 2781–85.

Alman, D.E. and Jablonski, P.D. 2007. Effect of Minor Elements and a Ce Surface Treatment on the Oxidation Behavior of an Fe–22Cr–0.5Mn (Crofer 22 APU) Ferritic Stainless Steel, *Int. J. Hydrogen Energy*, *32*(16), 3743–53.

Amadeh, A., Labbe, J.C., Laïmeche, A. and Quintard, P. 1996. Influence of Boron Nitride and Carbon Additives on the Behavior of Sintered AlN in a Steel-Making Environment, *J. European Ceram. Soc.*, *16*, 403–8.

Amadeh, A., Labbe, J.C. and Quintard, P.E. 2004. Wettability and Corrosion of AlN-BN-C by a SiC_a Treated Liquid Steel, *Corrosion Sci.*, *46*, 183–91.

Amoroso, G.G. and Fassina, V. 1983. *Stone Decay and Conservation*, Materials Science Monographs 11, Elsevier, Amsterdam, p. 453.

Anderson, N.C. 1979. Basal Plane Cleavage Cracking of Synthetic Sapphire Arc Lamp Envelopes, *J. Am. Ceram. Soc.*, *62*(1–2), 108–9.

Arun, R., Subramanian, M. and Mehrotra, G.M. 1990. Oxidation Behavior of TiC, ZrC, and HfC Dispersed in Oxide Matrices, in *Ceramic Transactions Vol. 10: Corrosion and Corrosive Degradation of Ceramics*, Tressler, R. E. and McNallan, M. (eds.), Am. Ceram. Soc., Westerville, OH, pp. 211–23.

Balat, M.J.H., Flamant, G., Male, G. and Pichelin, G. 1992. Active to Passive Transition in the Oxidation of Silicon Carbide at High Temperature and Low Pressure in Molecular and Atomic Oxygen, *J. Mater. Sci.*, *27*, 697–703.

Ballman, A.A. and Laudise, R.A. 1965. Crystallization and Solubility of Zircon and Phenacite in Certain Molten Salts, *J. Am. Ceram. Soc.*, *48*(3), 130–33.

Barsoum, M.W., El-Raghy, and Ogbuji, L.U.J.T. 1997. Oxidation of Ti_3SiC_2 in Air, *J. Electrochem. Soc., 144*(7), 2508–16.

Bates, J. L. 1987. Heterogeneous Dissolution of Refractory Oxides in Molten Calcium-Aluminum Silicate, *J. Am. Ceram. Soc., 70*(3), C55–57.

Baxter, D.J., Graziani, T., Wang, H.-M. and McCauley, R.A. 1998. Corrosion of a Dense, Low-Additive Si_3N_4 in High Temperature Combustion Gases, *J. European Ceram. Soc., 18*, 2323–30.

Baumgartner, C.E. 1984. Metal Oxide Solubility in Eutectic Li/K Carbonate Melts, *J. Am. Ceram. Soc., 67*(7), 460–62.

Bergstrom, L. and Pugh, R.J. 1989. Interfacial Characterization of Silicon Nitride Powders, *J. Am. Ceram. Soc., 72*(1), 103–9.

Berztiss, D.A., Cerchiara, R.R., Gulbransen, E.A., Pettit, F.S. and Meier, G.H. 1992. Oxidation of $MoSi_2$ and Comparison with Other Silicide Materials, *Mater. Sci. Eng., A155*, 165–81.

Bonar, J.A., Kennedy, C.R. and Swaroop, R.B. 1980. Coal-Ash Slag Attack and Corrosion of Refractories, *Ceram. Bull., 59*(4), 473–78.

Bonetti, G., Toninato, T., Bianchini, A. and Martini, P.L. 1969. Resistance of Refractories to Corrosion by Lead-Containing Glasses, *Proc. Brit. Ceram. Soc., 14*, 29–40.

Borom, M.P., Arendt, R.H. and Cook, N.C. 1981. Dissolution of Oxides of Y, Al, Mg, and La by Molten Fluorides, *Ceram. Bull., 60*(11), 1168–74.

Borom, M.P., Brun, M.K. and Szala, L.E. 1988. Kinetics of Oxidation of Carbide and Silicide Dispersed Phases in Oxide Matrices, *Adv. Ceram. Mat., 3*(5), 491–97.

Bowen, P., Highfield, J.G., Mocellin, A. and Ring, T.A. 1990. Degradation of Aluminum Nitride Powder in an Aqueous Environment, *J. Am. Ceram. Soc., 73*(3), 724–28.

Brady, P.V. and House, W.A. 1996. Surface-Controlled Dissolution and Growth of Minerals, Chp. 4, pp. 225–305 in *Physics and Chemistry of Mineral Surfaces*, P.V. Brady (ed.), CRC Press, New York, NY.

Bright, E. and Readey, D.W. 1987. Dissolution Kinetics of TiO2 in HF-HCl Solutions, *J. Am. Ceram. Soc., 70*(12), 900–6.

Broczkowski, M.E., Keech, P.G., Noël, J.J. and Shoesmith, D.W. 2010. Corrosion of Uranium Dioxide Containing Simulated Fission Products in Dilute Hydrogen Peroxide and Dissolved Hydrogen, *J. Electrochem. Soc., 157*(8), C275–81.

Brondyke, K.J. 1953. Effect of Molten Aluminum on Alumina-Silica Refractories, *J. Am. Ceram. Soc., 36*(5), 171–74.

Butterman, W.C. and Foster, W.R. 1967. Zircon Stability and the ZrO_2 – SiO_2 Phase Diagram, *Am. Mineral., 52*, 880–5.

Buykx, W.J., Hawkins, K., Levins, D.M., Mitamura, H., Smart, R.St.C., Stevens, G.T., Watson, K.G., Weedon, D. and White, T.J. 1988. Titanate Ceramics for the Immobilization of Sodium-Bearing High-Level Nuclear Waste, *J. Am. Ceram. Soc., 71*(8), 678–88.

Carney, C.M., Mogilvesky, P. and Parthasarathy, T.A. 2009. Oxidation Behavior of Zirconium Diboride Silicon Carbide Produced by the Spark Plasma Sintering Method, *J. Am. Ceram. Soc., 92*(9), 2046–52.

Charola, A.E. and Lazzarini, L. 1986. Deterioration of Brick Masonry Caused by Acid Rain, pp. 250–58 in *Materials Degradation Caused by Acid Rain*, ACS Symposium Series 318, Baboian, R. (ed.), Am. Chem. Soc., Washington, DC.

Chartier, T., Besson, J.L. and Goursat, P. 1986. Microstructure, Oxidation and Creep Behavior of a β'-Sialon Ceramic, *Int. J. High Tech. Ceram., 2*(1), 33–45.

Chen, H.C., Liu, Z.Y. and Chuang, Y.C. 1992. Degradation of Plasma-Sprayed Alumina and Zirconia Coatings on Stainless Steel during Thermal Cycling and Hot Corrosion, *Thin Solid Films, 223,* 56–64.

Chen, H., Liu, J. and Huang, W. 2006. Corrosion Behavior of Silicon Nitride Bonding Silicon Carbide in Molten Magnesium and AZ91 Magnesium Alloy, *Materials Sci. and Engr. A, 415,* 291–96.

Choi, D.J., Fischbach, D.B. and Scott, W.D. 1989. Oxidation of Chemically-Vapor-Deposited Silicon Nitride and Single-Crystal Silicon, *J. Am. Ceram. Soc., 72*(7), 1118–23.

Chung, Y-D. and Schlesinger, M.E. 1994 Interaction of CaO-FeO-SiO_2 Slags with Partially Stabilized Zirconia, *J. Am. Ceram. Soc., 77*(3), 611–16.

Cinibulk, M.K., Thomas, G. and Johnson, S.M. 1992. Oxidation Behavior of Rare-Earth Disilicate-Silicon Nitride Ceramics, *J. Am. Ceram. Soc., 75*(8), 2044–49.

Clauss, H. and Salge, H. 1974. Electron Micro-Probe Analysis of the Dissolution Behavior of Fusion Cast Tank Blocks, *Glastech. Ber., 47*(7–8), 159–81.

Clark, D.E., White, W.B. and Machiels, A.J. (eds.) *Advances in Ceramics Vol. 20: Nuclear Waste Management II,* 1986. Am. Ceram. Soc., Westerville, OH, 773 pp.

Clayton, J.C. 1988. In-Pile and Out-of-Pile Corrosion Behavior of Thoria- and Urania-Based Nuclear Fuels, *Ceram. Eng. Sci. Proc., 9* (9–10).

Cornie, J.A., Chiang, Y-M., Uhlmann, D.R., Mortensen, A. and Collins, J.M. 1986. Processing of Metal and Ceramic Matrix Composites, *Ceram. Bull., 65*(2), 293–304.

Costello, J.A. and Tressler, R.E. 1981. Oxidation Kinetics of Hot-Pressed and Sintered α-SiC, *J. Am. Ceram. Soc., 64*(6), 327–31.

Cree, J.W. and Amateau, M.F. 1987. Mechanical Behavior of SiC Exposed to Molten Lithium and Lithium Salts, *Ceram. Eng. Sci. Proc., 8*(7–8), 812–14.

Cubicciotti, D. and Lau, K.H. 1978. Kinetics of Oxidation of Hot-Pressed Silicon Nitride Containing Magnesia, *J. Am. Ceram. Soc., 61*(11–12), 512–17.

Davies, M.W. and Phennah, P.J. 1959. Reactions of Boron Carbide and Other Boron Compounds with Carbon Dioxide, *J. Appl. Chem., 9*(4), 213.

Davison, S., Smith, K., Zhang, Y-C., Liu, J-H., Kershow, R., Dwight, K., Rieger, P.H. and Wold, A. 1987. Chemical Problems Associated with the Preparation and Characterization of Superconducting Oxides Containing Copper, Chp. 7 in *Chemistry of High-Temperature Superconductors,* Nelson, D.L., Whittingham, M.S. and George, T.F. (eds.), Amer. Chem. Soc., Washington, DC, pp. 65–78.

Deal, B.E. and Grove, A.S. 1965. General Relationship for the Thermal Oxidation of Silicon, *J. Appl. Phys., 36*(12), 3770–78.

de Jong, R. 1991. Incorporation of Additives into Silicon Nitride by Colloidal Processing of Metal Organics in an Aqueous Medium, Univ. Microfilms Int. (Ann Arbor, MI), Order No. DA9123266, *Diss. Abstr. Int., B52*(3), 1660.

de Jong, R., McCauley, R. A., Fordham, R. J. and Riley, F.L. 1985. High Temperature Corrosion of Some Silicon Nitrides, *Proceedings of the European Materials Research Society Conference,* Nov. 26–29, Strasbourg, France.

de Jong, R., McCauley, R.A. and Tambuyser, P. 1987. Growth of Twinned β-Silicon Carbide Whiskers by the Vapor-Liquid-Solid Process, *J. Am. Ceram. Soc., 70*(11), C338–41.

Derobert, M. 1975. Microscope and X-Ray Diffraction Identification of Crystalline Phases in Refractories and Their Corrosion Products in Glass Tanks, *Bull. Soc. Fr. Ceram., 109,* 31–36.

Du, H., Tressler, R.E. and Spear, K.E. 1989. Thermodynamics of the Si-N-O System and Kinetic Modeling of Oxidation of Si_3N_4, *J. Electrochem. Soc.*, *136*(11), 3210–15.

Duchesne, D.J., Hipps, K.W., Grasher, B.A. and Norton, M.G. 1999. The Formation of Transition Aluminas during Oxidation of AlN, *J. Mater. Sci. Lett.*, *18*(11), 877–79.

Dupré, N., Martin, J-F., Oliveri, J., Soudan, P., Guyomard, D., Yamada, A. and Kanno, R. 2009. Aging of the $LiNi_{1/2}Mn_{1/2}O_2$ Positive Electrode Interface in Electrolyte, *J. Electrochem. Soc.*, *156*(5), C180–85.

Dutta, I., Mitra, S. and Rabenberg, L. 1992. Oxidation of Sintered Aluminum Nitride at Near-Ambient Temperatures, *J. Am. Ceram. Soc.*, *75*(11), 3149–53.

Ervin, G. 1958. Oxidation Behavior of Silicon Carbide, *J. Am. Ceram. Soc.*, *41*(9), 347–52.

Evans, T. and Phaal, C. 1962. *Proc. 5th Biennial Conf. on Carbon, 1*, Pergamon Press, New York, NY, p. 147.

Fahrenholtz, W.G. 2005. The ZrB_2 Volatility Diagram, *J. Am. Ceram. Soc.*, *88*(12), 3509–12.

Fahrenholtz, W.G. 2007. Thermodynamic Analysis of ZrB_2-SiC Oxidation: Formation of a SiC-Depleted Region, *J. Am. Ceram. Soc.*, *90*(1), 143–48.

Federer, J.I. 1988a. Corrosion of SiC Ceramics by Na_2SO_4, *Adv. Ceram. Mat.*, *3*(1), 56–61.

Federer, J.I. 1988b. Stress-Corrosion of SiC in an Oxidizing Atmosphere Containing NaCl, *Adv. Ceram. Mat.*, *3*(3), 293–95.

Ferber, M.K., Ogle, J., Tennery, V.J. and Henson, T. 1985. Characterization of Corrosion Mechanisms Occurring in a Sintered SiC Exposed to Basic Coal Slags, *J. Am. Ceram. Soc.*, *68*(4), 191–97.

Fergus, J.W. and Worrell, W.L. 1990. The Oxidation of Chemically Vapor Deposited Silicon Carbide, in *Ceramic Transactions, Vol. 10: Corrosion and Corrosive Degradation of Ceramics*, Tressler, R.E. and McNallan, M. (eds.), Am. Ceram. Soc., Westerville, OH, pp. 43–51.

Fitch, L.D. and Burdick, V.L. 1989. Water Corrosion of $YBa_2Cu_3O_{7-x}$ Superconductors, *J. Am. Ceram. Soc.*, *72*(10), 2020–23.

Fitzer, E. 1990. Oxidation of Molybdenum Disilicide, in *Ceramic Transactions Vol. 10: Corrosion and Corrosive Degradation of Ceramics*, Tressler, R. E. and McNallan, M. (eds.), Am. Ceram. Soc., Westerville, OH, pp. 19–41.

Fox, D.S. 1992. Oxidation Kinetics of CVD Silicon Carbide and Silicon Nitride, *Ceram. Eng. & Sci. Proc.*, *13*(7–8), 836–43.

Fox, D.S. and Jacobson, N.S. 1988. Molten-Salt Corrosion of Silicon Nitride: I, Sodium Carbonate, *J. Am. Ceram. Soc.*, *71*(2), 128–38.

Franz, I. and Langheinrich, W. 1972. Formation of Silicon Dioxide from Silicon Nitride, in *Reactivity of Solids: Proc. 7th International Symp. on Reactivity of Solids*, Anderson, J.S., Roberts, M.W. and Stone, F.S. (eds.), Hall and Chapman, London, UK, pp. 303–14.

Gajić-Krstajić, Lj.M., Trišović, T.Lj. and Krstajić, N.V. 2004. Spectrophotometric Study of the Anodic Corrosion of Ti/RuO_2 Electrode in Acid Sulfuric Solution, *Corrosion Sci.*, *46*, 65–74.

Graustein, W.C., Cromack Jr., K. and Sollins, P. 1977. Calcium Oxalate: Occurrence in Soils and Effect on Nutrient and Geochemical Cycles, *Science, New Series*, *198*(4323), 1252–54.

Gregory, O.J. and Richman, M.H. 1984. Thermal Oxidation of Sputter-Coated Reaction-Bonded Silicon Nitride, *J. Am. Ceram. Soc.*, *67*(5), 335–40.

Grjotheim, K., Holm, J.L., Krohn, C. and Thonstad, J. 1966. Recent Progress in the Theory of Aluminium Electrolysis, in *Selected Topics in High Temperature Chemistry*, Forland, T., Grjotheim, K., Motzfeldt, K. and Urnes, S. (eds.), Universitetsforlaget, Oslo, pp. 151–78.

Gu, Z., Edgar, J.H., Wang, C. and Coffey, D.W. 2006. Thermal Oxidation of Aluminum Nitride Powder, *J. Am. Ceram. Soc.*, 89(7), 2167–71.

Guthrie Jr., G.D. 1992. Biological Effects of Inhaled Minerals, *Amer. Mineral.*, 77(3/4), 225–43.

Gutman, I. and Gutmanas, E.Y. 1990. Interaction of Si_3N_4 with Titanium Powder, *J. Materials Sci. Letters*, 9, 813–15.

Harris, R.C. 1975. Oxidation of 6H-α Silicon Carbide Platelets, *J. Am. Ceram. Soc.*, 58(1–2), 7–9.

Harris, L.A., Cross, D.R. and Gerstner, M.E. 1977. Corrosion Suppression on Rutile Anodes by High Energy Redox Reactions, *J. Electrochem. Soc.*, 124(6), 839–44.

He, H., Zhu, R.K., Qin, Z., Keech, P., Ding, Z. and Shoesmith, D.W. 2009. Determination of Local Corrosion Kinetics on Hyper-Stoichiometric UO_{2+x} by Scanning Electrochemical Microscopy, *J. Electrochem. Soc.*, 156(3), C87–94.

Hilger, J.P., Babel, D., Prioul, N. and Fissolo, A. 1984. Corrosion of AZS and Fireclay Refractories in Contact with Lead Glass, *J. Am. Ceram. Soc.*, 64(4), 213–20.

Hinze, J.W., Tripp, W.C. and Graham, H.C. 1975. The High Temperature Oxidation of Hot-Pressed Silicon Carbide, in *Mass Transport Phenomena in Ceramics*, Cooper, A.R. and Heuer, A.H. (eds.), Plenum Press, New York, NY.

Hirayama, H., Kawakubo, T., Goto, A. and Kaneko, T. 1989. Corrosion Behavior of Silicon Carbide in 290°C Water, *J. Am. Ceram. Soc.*, 72(11), 2049–53.

Horkans, J. and Shafer, M.W. 1977a. Effect of Orientation, Composition, and Electronic Factors in the Reduction of O_2 on Single Crystal Electrodes of Conducting Oxides of Molybdenum and Tungsten, *J. Electrochem. Soc.*, 124(8), 1196–202.

Horkans, J. and Shafer, M.W. 1977b. An Investigation of the Electrochemistry of a Series of Metal Dioxides with Rutile-Type Structure: MoO_2, WO_2, ReO_2, RuO_2, OsO_2, and IrO_2, *J. Electrochem. Soc.*, 124(8), 1202–7.

Horton, R.M. 1969. Oxidation Kinetics of Powdered Silicon Nitride, *J. Am. Ceram. Soc.*, 52(3), 121–24.

Hume, L.A. and Rimstidt, J.D. 1992. The Biodurability of Chrysotile Asbestos, *Amer. Mineral.*, 77(9/10), 1125–28.

Huseby, I.C. and Klug, F.J. 1979. Chemical Compatibility of Ceramics for Directionally Solidifying Ni-Base Eutectic Alloys, *Ceram. Bull.*, 58(5), 527–35.

Ibara, Y., Nasu, H., Imura, T. and Osaka, Y. 1989. Preparation and Crystallization Process of the High-Tc Superconducting Phase in Bi, Pb-Sr-Ca-Cu-O Glass-Ceramics, *Japanese J. Appl. Phys.*, 28(1), L37–40.

Ibrahim, M.A.M., Korablov, S.F. and Yoshimura, M. 2002. Corrosion of Stainless Steel Coated with TiN, (TiAl)N and CrN in Aqueous Environments, *Corrosion Sci.*, 44, 815–28.

Irving, R.J. and Worsley, I.G. 1968. Oxidation of Titanium Diboride and Zirconium Diboride at High Temperatures, *J. Less-Common Metals*, 16(2), 102–12.

Jacobson, N.S. 1986. Kinetics and Mechanism of Corrosion of SiC by Molten Salts, *J. Am. Ceram. Soc.*, 69(1), 74–82.

Jacobson, N.S. and Fox, D.S. 1988. Molten-Salt Corrosion of Silicon Nitride: II, Sodium Sulfate, *J. Am. Ceram. Soc.*, 71(2), 139–48.

Jacobson, N.S. and Smialek, J.L. 1985. Hot Corrosion of Sintered α-SiC at 1000°C, *J. Am. Ceram. Soc., 68*(8), 432–39.

Jacobson, N.S., Stearns, C.A. and Smialek, J.L. 1986. Burner Rig Corrosion of SiC at 1000°C, *Adv. Ceram. Mat., 1*(2), 154–61.

Jacobson, N.S., Eckel, A.J., Misra, A.K. and Humphrey, D.L. 1990. Reactions of SiC with $H_2/H_2O/Ar$ Mixtures at 1300°C, *J. Am. Ceram. Soc., 73*(8), 2330–32.

Jaskowiak, M.H. and DiCarlo, J.A. 1989. Pressure Effects on the Thermal Stability of Silicon Carbide Fibers, *J. Am. Ceram. Soc., 72*(2), 192–97.

John, P., Polwart, N., Troupe, C.E. and Wilson, J.I.B. 2002. The Oxidation of (100) Textured Diamond, *Diamond and Related Materials, 11*, 861–66.

Jorgensen, P.J., Wadsworth, M.E. and Cutler, I.B. 1959. Oxidation of Silicon Carbide, *J. Am. Ceram. Soc., 42*(12), 613–16.

Jorgensen, P.J., Wadsworth, M.E. and Cutler, I.B. 1961. Effects of Water Vapor on Oxidation of Silicon Carbide, *J. Am. Ceram. Soc., 44*(6), 258–60.

Jose, P.D., Gupta, D.K. and Rapp, R.A. 1985. Solubility of $α-Al_2O_3$ in fused Na_2SO_4 at 1200 K, *J. Electrochem. Soc., 132*(3), 735–37.

Jennings, H.M. 1986. Aqueous Solubility Relationships for Two Types of Calcium Silicate Hydrate, *J. Am. Ceram. Soc., 69*(8), 614–18.

Joshi, P.N. 1992. Metal-Organic Surfactants as Sintering Aids for Silicon Nitride in an Aqueous Medium, MS thesis, Rutgers University.

Kastrissios, T., Stephenson, M. and Turner, P.S. 1987. Hydrothermal Dissolution of Perovskite: Implications for Synroc Formulation, *J. Am. Ceram. Soc., 70*(7), C144–46.

Katnani, A.D. and Papathomas, K.I. 1987. Kinetics and Initial Stages of Oxidation of Aluminum Nitride: Thermogravimetric Analysis and X-ray Photoelectron Spectroscopy Study, *J. Vac. Sci. Technol., A5*, 1335.

Klemm, H., Taut, C. and Wötting, G. 2003. Long-Term Stability of Nonoxide Ceramics in an Oxidative Environment at 1500°C, *J. European Ceram. Soc., 23*, 619–27.

Kiehle, A.J., Heung, L.K., Gielisse, P.J. and Rockett, T.J. 1975. Oxidation Behavior of Hot-Pressed Si_3N_4, *J. Am. Ceram. Soc., 58*(1–2), 17–20.

Kim, H-E. and Moorehead, A.J. 1990. High-Temperature Gaseous Corrosion of Si_3N_4 in H_2-H_2O and $Ar-O_2$ Environments, *J. Am. Ceram. Soc., 73*(10), 3007–14.

Kim, S.M., Lu, W-K., Nicholson, P.S. and Hamielec, A.E. 1974. Corrosion of Aluminosilicate Refractories in Iron-Manganese Alloys, *Ceram. Bull., 53*(7), 543–47.

Kitano, Y. 1962a. Polymorphic Formation of Calcium Carbonite in Thermal Springs with an Emphasis on the Effect of Temperature, *Bull. Chem. Soc. Japan, 35*, 1980.

Kitano, Y. 1962b. Behavior of Various Inorganic Ions in the Separation of Calcium Carbonate from a Bicarbonate Solution, *Bull. Chem. Soc. Japan, 35*, 1973.

Klinger, N., Strauss, E.L. and Komarek, K.L. 1966. Reaction between Silica and Graphite, *J. Am. Ceram. Soc., 49*(7), 369–75.

Kuriakose, A.K. and Margrave, J.L. 1964. The Oxidation Kinetics of Zirconium Diboride and Zirconium Carbide at High Temperatures, *J. Electrochem. Soc., 111*(7), 827–31.

Labbe, J.C. and Laïmeche, A. 1996. Study of the Behavior of Aluminium Nitride in the Iron and Steel Industry, *J. European Ceram. Soc., 16*, 893–98.

Lakatos, T. and Simmingskold, B. 1967. The Influence of Constituents on the Corrosion of Pot Clays by Molten Glass, *Glass Technology, 8*(2), 43–47.

Lakatos, T. and Simmingskold, B. 1971. Influence of Viscosity and Chemical Composition of Glass on Its Corrosion of Sintered Alumina and Silica-Glass, *Glastek. Tidskr.*, 26(4), 58–68.

Lange, F.F., Singhal, S.C. and Kuznicki, R.C. 1977. Phase Relations and Stability Studies in the Si_3N_4-SiO_2-Y_2O_3 Pseudoternary System, *J. Am. Ceram. Soc.*, 60(5–6), 249–52.

Larsen, D.C., Adams, J.W., Johnson, L.R., Teotia, A.P.S. and Hill, L.G. (eds.) 1985. *Ceramic Materials Advanced Heat Engines*, Noyes Publications, Park Ridge, NJ, p. 221.

Lavrenko, V.A. and Alexeev, A.F. 1986. High-Temperature Oxidation of Boron Nitride, *Ceramics Inter.*, 12(1), 25–31.

Lawson, M.G., Kim, H.R., Pettit, F.S. and Blachere, J.R. 1990. Hot Corrosion of Silica, *J. Am. Ceram. Soc.*, 73(4), 989–95.

Lawson, M.G., Pettit, F.S. and Blachere, J.R. 1993. Hot Corrosion of Al_2O_3, *J. Mater. Res.*, 8, 1964–71.

Lea, F.M. 1970. *The Chemistry of Cement and Concrete*, Edward Arnold Publishers, London, UK.

Lee, B.J. and Lee, D.N. 1989. Calculation of Phase Diagrams for the $YO_{1.5}$-BaO-CuO_x System, *J. Am. Ceram. Soc.*, 72(2), 314–19.

Lee, S.Y., Park, Y.S., Hsu, P.P. and McNallan, M.J. 2003. Synergistic Effects of Water Vapor and Alkali Chloride Vapors on the High-Temperature Corrosion of SiC-Based Ceramics, *J. Am. Ceram. Soc.*, 86(8), 1292–98.

Lepistö, T.T. and Mäntylä, T.A. 1989. A Model for Structural Degradation of Y-TZP Ceramics in Humid Atmosphere, *Ceram. Eng. Sci. Proc.*, 10(7–8), 658–67.

Lepistö, T.T., Lintula, P.V. and Mäntylä, T.A. 1988. TZP-Ceramics in Humid Conditions at 150°C, *Ceram. Eng. Sci. Proc.*, 9 (9–10), 1517–23.

Li, H., Cui, X. and Chen, W. 2010. Effect of CeO_2 on High Temperature Carburization Behavior of Mn-Cr-O Spinel and Chromium Oxide, *J. Electrochem. Soc.*, 157(10), C321–27.

Liang, D-T. and Readey, D.W. 1987. Dissolution Kinetics of Crystalline and Amorphous Silica in Hydrofluoric-Hydrochloric Acid Mixtures, *J. Am. Ceram. Soc.*, 70(8), 570–77.

Lin, W., Hsu, J-Y. and Speyer, R.F. 1994. Stability of Molybdenum Disilicide in Combustion Gas Environments, *J. Am. Ceram. Soc.*, 77(5), 1162–68.

Lin, Z., Li, M., Wang, J. and Zhou, Y. 2007. Microstructure and High-Temperature Corrosion Behavior of a Cr-Al-C Composite, *J. Am. Ceram. Soc.*, 90(12), 3930–37.

Lindsay, J.G., Bakker, W.T. and Dewing, E.W. 1964. Chemical Resistance of Refractories to Al and Al-Mg Alloys, *J. Am. Ceram. Soc.*, 47(2), 90–94.

Liu, G.M., Li, M.S. and Zhou, Y.C. 2002a. Hot Corrosion of Ti_3SiC_2-Based Ceramic Superficially Deposited with Na_2SO_4 at 900°C and 1000°C in Air, *Corrs. Sci.*, 45(6), 1217–26.

Liu, G.M., Li, M.S. and Zhou, Y.C. 2002b. Corrosion Behavior of Ti_3SiC_2 and Siliconized Ti_3SiC_2 in the Mixture of K_2CO_3 and Li_2CO_3 Melts at 750°C, *J. Mater. Sci. Lett.*, 21(22), 1755–57.

Liu, G.M., Li, M.S. and Zhou, Y.C. 2003. Corrosion Behavior and Strength Degradation of Ti_3SiC_2 in the Eutectic K_2CO_3 and Li_2CO_3 Mixture, *J. European Ceram. Soc.*, 23(11), 1957–62.

Liu, G.M., Li, M.S., Zhou, Y.C. and Zhang, Y. 2005. Hot Corrosion Behavior of Ti_3SiC_2 in the Mixture of Na_2SO_4-NaCl Melts, *J. European Ceram. Soc.*, 25, 1033–39.

Lu, W-J., Steckl, A.J., Choe, T.P. and Katz, W. 1984. Thermal Oxidation of Sputtered Silicon Carbide Thin Films, *J. Electrochem. Soc.*, 131(8), 1907–14.

Luthra, K.L. 1991(a). Some New Perspectives on Oxidation of Silicon Carbide and Silicon Nitride, *J. Am. Ceram. Soc.*, 74(5), 1095–103.

Luthra, K.L. 1991(b). A Mixed-Interface Reaction/Diffusion-Controlled Model for Oxidation of Si_3N_4, *J. Electrochem. Soc.*, 138(10), 3001–7.

Maeda, M., Nakamura, K. and Yamada, M. 1989. Oxidation Resistance Evaluation of Silicon Carbide Ceramics with Various Additives, *J. Am. Ceram. Soc.*, 72(3), 512–14.

Markarenko, G.N. 1977. Borides of the IVb Group, Chp. VII in *Boron and Refractory Borides*, Matkovich, V.I. (ed.), Springer-Verlag, New York, NY, pp. 310–30.

Martin, D.M. and Hunter Jr., O. 1974. Polymer Bonded Refractories, in *The Bond in Refractories*, Staley Jr., W. and Givan, G. (eds.), Tenth Annual Symposium on Refractories, Am. Ceram. Soc., 5 Apr.

Materials Research Society Symposia Proceedings, Scientific Basis for Nuclear Waste Management IV–XXXIII, 1982–2009, MRS, Pittsburgh, PA.

Matsuda, T. 1989. Stability to Moisture for Chemically Vapour-Deposited Boron Nitride, *J. Mater. Sci.*, 24, 2353–58.

Matsushima, M., Yadoomaru, S., Mori, K. and Kawai, Y. 1977. A Fundamental Study on the Dissolution Rate of Solid Lime into Liquid Slag, Trans. *Iron Steel Inst. Jpn.*, 17, 442–49.

Mattia, D., Desmaison-Brut, M., Dimovski, S., Gogotsi, Y. and Desmaison, J. 2005. Oxidation behavior of aluminum nitride – hafnium diboride ceramic composite, *J. Euro. Cer. Soc.*, 25, 1789–96.

Mayberry, M.L., Boyer, W.H., Martinek, C.A. and Neely, J.E. 1970. Effect of Alternating Oxidizing-Reducing Atmospheres on Basic Refractories, Presented at the Pacific Coast Regional Meeting of the Am. Ceram. Soc., Oct, San Francisco, CA.

McCauley, J.W. and Corbin, N.D. 1979. Phase Relations and Reaction Sintering of Transparent Cubic Aluminum Oxynitride Spinel (ALON), *J. Am. Ceram. Soc.*, 62(9–10), 476–79.

McCauley, R.A. 1975. Unpublished data.

McCauley, R.A. 1989. The Effects of Vanadium upon Basic Refractories, *UNITECR '89 Proceedings*, Trostel Jr., L.J. (ed.), Am. Ceram. Soc., Westerville, OH, pp. 858–63.

McConnell, J.D.C. 1955. The Hydration of Larnite (β-Ca_2SiO_4) and Bredigite (β'-Ca_2SiO_4) and the Properties of the Resulting Gelatinous Mineral Plombierite, *Mineral. Mag.*, 30, 672–80.

McGarry, M.J. and McCauley, R.A. 1992. Subsolidus Phase Equilibria of the MgO-V_2O_5-SiO_2 System, *J. Am. Ceram. Soc.*, 75(10), 2874–76.

McKee, D.W. and Chatterji, D. 1976. Corrosion of Silicon Carbide in Gases and Alkaline Melts, *J. Am. Ceram. Soc.*, 59(9–10), 441–44.

Mellinger, G.B. (ed.) *Ceramic Transactions Vol. 9: Nuclear Waste Management III*, 1990. Am. Ceram. Soc., Westerville, OH, 595 pp.

Mieskowski, D.M. and Sanders, W.A. 1985. Oxidation of Silicon Nitride Sintered with Rare-Earth Oxide Additions, *J. Am. Ceram. Soc.*, 68(7), C160–63.

Mikalsen, D.J., Roy, R.A., Yee, D.S., Shivashankar, S.A. and Cuomo, J.J. 1988. Superconducting Thin Films of the Bi-Sr-Ca-Cu-O System Prepared by Multilayer Metal Deposition, *J. Mater. Res.*, 3(4), 613–18.

Miller, P.D., Lee, J.G. and Cutler, I.B. 1979. The Reduction of Silica with Carbon and Silicon Carbide, *J. Am. Ceram. Soc.*, 62(3–4), 147–49.

Mohan, P., Yuan, B., Patterson, T., Desai, V.H. and Sohn, Y.H. 2007. Degradation of Yttria-Stabilized Zirconia Thermal Barrier Coatings by Vanadium Pentoxide, Phosphorous Pentoxide, and Sodium Sulfate, *J. Am. Ceram. Soc.*, 90(11), 3601–7.

Montintin, J. and Desmaison-Brut, M. 1990. Oxidation Behavior of Hot-Isostatic-Pressed Tantalum Nitride, in *High Temperature Corrosion of Technical Ceramics*, Fordham, R.J. (ed.), Elsevier Applied Science, London, UK, pp. 121–30.

Muan, A. 1984. Thermodynamics Aspects of the Application of Ceramics/Refractories in Advanced Energy Technologies, in *Ceramics in Advanced Energy Technologies*, Krockel, H., Merz, M., Van der Biest, O. (eds.), D. Reidel Pub. Co., Dorrecht, pp. 348–66.

Muan, A. 1992. Reactions between Iron Oxides and Alumina-Silica Refractories, *J. Am. Ceram. Soc.*, 75(6), 1319–30.

Mukerji, J. and Biswas, S.K. 1990. Synthesis, Properties, and Oxidation of Alumina-Titanium Nitride Composites, *J. Am. Ceram. Soc.*, 73(1), 142–45.

Murphy, D.W., Johnson Jr., D.W., Jin, S. and Howard, R.E. 1988. Processing Techniques for the 93°K Superconductor $Ba_2YCu_3O_7$, *Science*, 241, 19 August, 922–30.

Myhra, S., Savage, D., Atkinson, A. and Riviere, J.C. 1984. Surface Modification of Some Titanate Minerals Subjected to Hydrothermal Chemical Attack, *Am. Mineral.*, 69(9/10), 902–9.

Narushima, T., Goto, T., Iguchi, Y. and Hirai, T. 1991. High-Temperature Active Oxidation of Chemically Vapor-Deposited Silicon Carbide in an $Ar-O_2$ Atmosphere, *J. Am. Ceram. Soc.*, 74(10), 2583–86.

Nelson, H.G. 1993. A Challenge to Materials: Advanced Hypersonic Flight Hydrogen and High Temperature Materials, *Proc. 1993 Conf. on Processing, Fabrication and Applications of Advanced Composites*, Long Beach, CA, 9–11 Aug, Upadhya, K. (ed.), ASM, OH, pp. 11–20.

Nguyen, T.D., Choi, J-H., Park, S-W. and Lee, D-B. 2007. Comments on the High Temperature Oxidation Characteristics of Ti_3SiC_2 in Air, *J. Ceram. Processing Res.*, 8(6), 397–401.

Oda, K. and Yoshio, T. 1993. Oxidation Kinetics of Hexagonal Boron Nitride Powder, *J. Mater. Sci.*, 28, 6562–66.

Oishi, Y., Cooper Jr., A.R. and Kingery, W.D. 1965. Dissolution in Ceramic Systems: III, Boundary Layer Concentration Gradients, *J. Am. Ceram. Soc.*, 48(2), 88–95.

Ogbuji, L.U.J.T. 1992. Role of Si_2N_2O in the Passive Oxidation of Chemically-Vapor-Deposited Si_3N_4, *J. Am. Ceram. Soc.*, 75(11), 2995–3000.

Oliveira, F.C., Edwards, R.A.H., Fordham, R.J. and Riley, F.L. 1990. High Temperature Corrosion of $Si_3N_4-Y_2O_3-Al_2O_3$ Ceramics in $H_2S/H_2O/H_2$ Reducing Environments, in *High Temperature Corrosion of Technical Ceramics*, Fordham, R.J. (ed.), Elsevier Applied Science, London, UK, pp. 53–68.

Opila, E. and Myers, D.L. 2004. Alumina Volatility in Water Vapor at Elevated Temperatures, *J. Amer. Cer. Soc.*, 87 (9) 1701–5.

Orlova, R.G., Moroz, I. Kh. and Naidenova, G.A. 1979. Kinetics of Mullite Dissolution in Alkali and Alkaline-Earth Melts (Obtained during the Firing of Mullite Porcelain), *Neorg. Stekla, Pokrytiya Mater.*, No. 4, 103–9 (Russ).

Ott, D. and Raub, C.J. 1976. The Affinity of the Platinum Metals for Refractory Oxides, *Platinum Met. Rev.*, 20(3), 79–85.

Pareek, V. and Shores, D.A. 1991. Oxidation of Silicon Carbide in Environments Containing Potassium Salt Vapor, *J. Am. Ceram. Soc.*, 74(3), 556–63.

Park, D.S., McNallan, M.J., Park, C. and Liang, W.W. 1990. Active Corrosion of Sintered α-Silicon Carbide in Oxygen-Chlorine Gases at Elevated Temperatures, *J. Am. Ceram. Soc.*, 73(5), 1323–29.

Parker, F.J. and McCauley, R.A. 1982a. Investigation of the System $CaO-MgO-V_2O_5$: I, Phase Equilibria, *J. Am. Ceram. Soc.*, 65(7), 349–51.

Parker, F.J. and McCauley, R.A. 1982b. Investigation of the System $CaO-MgO-V_2O_5$: II, Crystalline Solutions and Crystal Chemistry, *J. Am. Ceram. Soc.*, 65(9), 454–56.

Paul, A., Jayaseelan, D.D., Venugopal, S., Zapata-Solvas, E., Binner, J., Vaidhyanathan, B., Heaton, A., Brown, P. and Lee, W.E. 2012. UHTC Composites for Hypersonic Applications, *Ceram. Bull.*, 91(1), 22–28.

Plucknett, K.P. and Lewis, M.H. 1992. Microstructure and Oxidation Behavior of HIPed Silicon Nitride, *Ceram. Eng. & Sci. Proc.*, 13(9–10), 991–99.

Podobeda, L.T., Tsapuk, A.K. and Buravov, A.D. 1976. Oxidation of Boron Nitride under Nonisothermal Conditions, *Sov. Powder Metall.*, 9(165), 44–47.

Raeder, C.H. and Knorr, D.B. 1990. Stability of $YBa_2Cu_3O_{7-x}$ in Molten Chloride Salts, *J. Am. Ceram. Soc.*, 73(8), 2407–11.

Ramaswamy, P., Seetharamu, S., Varma, K.B.R. and Rao, K.J. 1997. $Al_2O_3-ZrO_2$ Composite Coatings for Thermal Barrier Applications, *Comp. Sci. Technol.*, 57, 81–89.

Readey, D.W. 1990. Gaseous Corrosion of Ceramics, in *Ceramic Transactions, Vol. 10: Corrosion and Corrosive Degradation of Ceramics*, Tressler, R.E. and McNallan, M. (eds.), Am. Ceram. Soc., Westerville, OH, pp. 53–80.

Ready, D.W. 1992. High Temperature Gas Corrosion of Ceramic Composites, *Ceram. Eng. Sci. Proc.*, 13(7–8), 301–18.

Rosner, D.E. and Allendorf, H.D. 1970. High Temperature Kinetics of the Oxidation and Nitridation of Pyrolytic Silicon Carbide in Dissociated Gases, *J. Phys. Chem.*, 74(9), 1829–39.

Sadrnezhaad, S.K., Mahshid, S., Hashemi, B. and Nemati, Z.A. 2006. Oxidation Mechanism of C in MgO-C Refractory Bricks, *J. Am. Ceram. Soc.*, 89(4), 1308–16.

Sandhage, K.H. and Yurek, G.J. 1988. Indirect Dissolution of Sapphire into Silicate Melts, *J. Am. Ceram. Soc.*, 71(6), 478–89.

Sandhage, K.H. and Yurek, G.J. 1990. Indirect Dissolution of Sapphire into Calcia-Magnesia-Alumina-Silica Melts: Electron Microprobe Analysis of the Dissolution Process, *J. Am. Ceram. Soc.*, 73(12), 3643–49.

Sandhage, K.H. and Yurek, G.J. 1991. Indirect Dissolution of $(Al,Cr)_2O_3$ in CaO-MgO-Al_2O_3-SiO_2 (CMAS) Melts, *J. Am. Ceram. Soc.*, 74(8), 1941–54.

Sato, T., Tokunaga, Y., Endo, T., Shimada, M., Komeya, K., Komatsu, M. and Kameda, T. 1988a. Corrosion of Silicon Nitride Ceramics in Aqueous Hydrogen Chloride Solutions, *J. Am. Ceram. Soc.*, 71(12), 1074–79.

Sato, T., Tokunaga, Y., Endo, T., Shimada, M., Komeya, K., Komatsu, M. and Kameda, T. 1988b. Corrosion of Silicon Nitride Ceramics in Aqueous HF Solutions, *J. Mater. Sci.*, 23(10), 3440–46.

Sato, T., Kubato, K. and Shimada, M. 1991a. Corrosion Kinetics and Strength Degradation of Sintered α-Silicon Carbide in Potassium Sulfate Melts, *J. Am. Ceram. Soc.*, 74(9), 2152–55.

Sato, T., Sato, S. and Okuwaki, A. 1991b. Corrosion Behavior of Alumina Ceramics in Caustic Alkaline Solutions at High Temperatures, *J. Am. Ceram. Soc.*, 74(12), 3081–84.

Schnitzer, M. and Kodama, H. 1977. Reactions of Minerals with Soil Humic Substances, Chp. 21 in *Minerals in Soil Environments*, Dinauer, R.C. (ed.), Soil Sci. Soc. Am., Madison, WI, pp. 741–70.

Seshadri, S.G. and Srinivasan, M. 1988. Liquid Corrosion and High-Temperature Oxidation Effects on Silicon Carbide/Titanium Diboride Composites, *J. Am. Ceram. Soc.*, 71(2), C72–74.

Shao, N., Dai, J.W., Li, G.Y., Nakae, H. and Hane, T. 2004. Effect of La on the Wettability of Al_2O_3 by Molten Aluminum, *Materials Letters*, 58, 2041–44.

Shi, L., Gu, Y., Chen, L., Yang, Z., Ma, J. and Qian, Y. 2005. Synthesis and Oxidation Behavior of Nanocrystalline Niobium Carbide, *Solid State Ionics*, 176, 841–43.

Shimada, S. and Ishii, T. 1990. Oxidation Kinetics of Zirconium Carbide at Relatively Low Temperatures, *J. Am. Ceram. Soc.*, 73(10), 2804–8.

Shimada, S., Okuyama, T., Kiyono, H. and Takahasi, J. 2003. Corrosion of HIPed of $\beta\text{-}Si_{6-z}Al_zO_zN_{8-z}$ (z = 0, 1, 2, 3) Ceramics by NaCl Vapor, *J. European Ceram. Soc.*, 23, 1735–41.

Siljan, O-J. and Seltveit, A. 1991. Chemical Reactions in Refractory Linings of Alumina Reduction Cells, *UNITECR '91 CONGRESS*, 2nd edition, edited by the German Refractories Association, Bonn, Verlag Stahleisen mbH, Düsseldorf, pp. 59–65.

Siljan, O-J., Rian, G., Pettersen, D.T., Solheim, A. and Schøning, C. 2002. Refractories for Molten Aluminum Contact Part I: Thermodynamics and Kinetics, *Refractories Applications and News*, 7(6), 17–25.

Singhal, S.C. 1976. Thermodynamic Analysis of the High-Temperature Stability of Silicon Nitride and Silicon Carbide, *Ceramurgia Int.*, 2(3), 123–30.

Singhal, S.C. 1977a. Oxidation of Silicon Nitride and Related Materials, in *Nitrogen Ceramics*, Riley, R.L. (ed.), NATO Adv. Study Inst. Ser.:E, Appl. Sci., No. 23, Noordhoff, Leyden, pp. 607–26.

Singhal, S.C. 1977b. Oxidation of Silicon-Based Structural Ceramics, in *Properties of High Temperature Alloys with Emphasis on Environmental Effects*, Foroulis, A.Z. and Pettit, F.S. (eds.), Electrochem. Soc. Inc., Princeton, NJ, pp. 697–712.

Singhal, S.C. and Lange, F.F. 1977. Oxidation Behavior of Sialons, *J. Am. Ceram. Soc.*, 60(3–4), 190–91.

Smialek, J.L. and Jacobson, N.S. 1986. Mechanism of Strength Degradation for Hot Corrosion of $\alpha\text{-}SiC$, *J. Am. Ceram. Soc.*, 69(10), 741–52.

Spear, K.E., Tressler, R.E., Zheng, Z. and Du, H. 1990. Oxidation of Silicon Carbide Single Crystals and CVD Silicon Nitride, in *Ceramic Transactions, Vol. 10: Corrosion and Corrosive Degradation of Ceramics*, Tressler, R.E. and McNallan, M. (eds.), Am. Ceram. Soc., Westerville, OH, pp. 1–18.

Stewart, R.W. and Cutler, I.B. 1967. Effect of Temperature and Oxygen Partial Pressure on the Oxidation of Titanium Carbide, *J. Am. Ceram. Soc.*, 50(4), 176–81.

Sun, Z., Zhou, Y. and Li, M. 2001. Oxidation Behavior of Ti_3SiC_2-Based Ceramic at 900-1300°C in Air, *Corrs. Sci*, 43, 1095–1109.

Suryanarayana, D. 1990. Oxidation Kinetics of Aluminum Nitride, *J. Am. Ceram. Soc.*, 73(4), 1108–10.

Tang, Z., Wang, F. and Wu, W. 2000. Effect of Al_2O_3 and Enamel Coatings on 900°C Oxidation and Hot Corrosion Behaviors of Gamma-TiAl, *Mater. Sci Eng. A, 276*, 70–75.

Taylor, H.F.W. 1979. Mineralogy, Microstructure, and Mechanical Properties of Cements, *Proc. Br. Ceram. Soc.*, 29, 147–63.

Thomas, E.A. and Brock, W.W. 1974. A Post-Mortem Examination of Zircon and Bonded Alumina-Zirconia-Silica Paving, *Proc. 10th International Congress on Glass, Ceram. Soc. Japan, No. 2 Refractories & Furnaces*, 2-9 to 2-19, 9 Jul.

Tressler, R.E., Meiser, M.D. and Yonushonis, T. 1976. Molten Salt Corrosion of SiC and Si_3N_4 Ceramics, *J. Am. Ceram. Soc., 59*(5–6), 278–79.

Tripp, W.C. and Graham, H.C. 1968. Thermogravimetric Study of the Oxidation of ZrB_2 in the Temperature Range of 800–1500°C, *J. Electrochem. Soc., 118*(7), 1195–99.

Tripp, W.C. and Graham, H.C. 1976. Oxidation of Si_3N_4 in the Range 1300 to 1500°C, *J. Am. Ceram. Soc., 59*(9–10), 399–403.

Tripp, W.C., Davis, H.H. and Graham, H.C. 1973. Effect of an SiC Addition on the Oxidation of ZrB_2, *Ceram. Bull., 52*, 612–16.

Tsai, R.L. and Raj, R. 1982. Dissolution Kinetics of β-Si_3N_4 in an Mg-Si-O-N Glass, *J. Am. Ceram. Soc., 65*(5), 270–74.

Ueno, S., Jayaseelan, D.D., Ohji, T. and Lin, H.-T. 2005. Corrosion and Oxidation Behavior of $ASiO_4$ (A = Ti, Zr, and Hf) and Silicon Nitride with an $HfSiO_4$ Environmental Barrier Coating, *J. Ceram. Processing Res., 6*(1), 81–84.

Ueno, S., Ohji, T. and Lin, H.-T. 2006. Comparison of Hot-Corrosion Behavior of Al_2O_3, Lu_2O_3 and Their Silicates, *J. Ceram. Processing Res., 7*(3), 201–5.

Umakoshi, M., Mori, K. and Kawai, Y. 1980. Corrosion Kinetics of Refractory Materials in Molten CaO-FeO-SiO_2 Slags, *Kyushu Daigaku Kogaku Ih., 53*, 191–97.

van Hoek, J.A.M., van Loo, F.J.J. and Metselaar, R. 1992. Corrosion of Alumina by Potassium Vapor, *J. Am. Ceram. Soc., 75*(1), 109–11.

Vaughn, W.L. and Maahs, H.G. 1990. Active-to-Passive Transition in the Oxidation of Silicon Carbide and Silicon Nitride in Air, *J. Am. Ceram. Soc., 73*(6), 1540–43.

Vedula, K., Abada, A. and Williams, W.S. 1988. Potential for Diboride Reinforcement of Oxide Matrix Composites, in *Materials Research Society Symposium Proceedings, Vol. 125: Materials Stability and Environmental Degradation*, Barkatt, A., Verink Jr., E.D. and Smith, L.R. (eds.), Mat. Res. Soc., Pittsburgh, PA, pp. 61–69.

Walker, J.K. and Saha, C.K. 1988. Formation of a Surface Carbide Layer during Sintering of Titanium Diboride, *J. Am. Ceram. Soc., 71*(4), C207–9.

Wang, J., Zhang, L. and Zeng, Q. 2008. Theoretical Investigation for the Active-to-Passive Transition in the Oxidation of Silicon Carbide, *J. Am. Ceram. Soc., 91*(5), 1665–73.

Wang, X.H. and Zhou Y.C. 2003. Oxidation Behavior of Ti_3AlC_2 at 1000–1400°C in Air, *Corrosion Sci., 45*, 891–907.

Wang, X.H., Lejus, A-M. and Vivien, D. 1990. Oxidation Behavior of Lanthanide Aluminum Oxynitrides with Magnetoplumbite-Like Structure, *J. Am. Ceram. Soc., 73*(3), 770–74.

Wang, P.S., Hsu, S.M. and Wittberg, T.N. 1991a. Oxidation Kinetics of Silicon Carbide Whiskers Studied by X-Ray Photoelectron Spectroscopy, *J. Mater. Sci., 26*, 1655–58.

Wang, P.S., Hsu, S.M., Malghan, S.G. and Wittberg, T.N. 1991b. Surface Oxidation Kinetics of Si_3N_4-4%Y_2O_3 Powders Studied by Bremsstrahlung-Excited Auger Spectroscopy, *J. Mat. Sci., 26*, 3249–52.

Wang, S.C.P., Anghaie, S. and Collins, C. 1991c. Reaction of Uranium Hexafluoride Gas with Alumina and Zirconia at Elevated Temperatures, *J. Am. Ceram. Soc., 74*(9), 2250–57.

Wang, Y., An, L., Fan, Y., Zhang, L., Burton, S. and Gan, Z. 2005. Oxidation of Polymer-Derived SiAlCN Ceramics, *J. Am. Ceram. Soc.*, *88*(11), 3075–80.

Wang, Z. Skybakmoen, E. and Grande, T. 2009. Chemical Degradation of Si_3N_4-Bonded SiC Sidelining Materials in Aluminum Electrolysis Cells, *J. Am. Ceram. Soc.*, *92*(6), 1296–1302.

Webster, R.P. and Kukacka, L.E. 1986. Effects of Acid Deposition on Portland Cement Concrete, pp. 239–49 in *Materials Degradation Caused by Acid Rain*, ACS Symposium Series 318, Baboian, R. (ed.), Am. Chem. Soc., Washington, DC.

Wicks, G.G. and Ross, W.A. (eds.) *Advances in Ceramics Vol. 8: Nuclear Waste Management*, Am. Ceram. Soc., Columbus, OH, 1984, 746 pp.

Wicks, G.G., Bickford, D.F. and Bunnell, L.R. (eds.) *Ceramic Transactions Vol. 23: Nuclear Waste Management IV*, 1991. Am. Ceram. Soc., Westerville, OH, 799 pp.

Wilding, L.P., Smeck, N.E. and Drees, L.R. 1977. Silica in Soils: Quartz, Cristobalite, Tridymite, and Opal, Chp. 14 in *Minerals in Soil Environments*, Dinauer, R.C. (mngr. ed.), Soil Sci. Soc. Am., Madison, WI, pp. 471–552.

Williams, R.S. and Chandhury, S. 1988. Chemical Compatibility of High-Temperature Superconductors with Other Materials, Chp. 22 in *Chemistry of High-Temperature Superconductors II*, Nelson, D.L. and George, T.F. (eds.), Amer. Chem. Soc., Washington, DC, pp. 291–302.

Wills, R.R., Sekercioglu, I. and Niesz, D.E. 1980. The Interaction of Molten Silicon with Silicon Aluminum Oxynitrides, *J. Am. Ceram. Soc.*, *63*(7–8), 401–3.

Wu, N., Chen, Z. and Mao, S.X. 2005. Hot Corrosion Mechanism of Composite Alumina/Yttria-Stabilized Zirconia Coating in Molten Sulfate-Vanadate Salt, *J. Am. Ceram. Soc.*, *88*(3), 675–82.

Yamaguchi, A. 1974. Reactions between Alkaline Vapors and Refractories for Glass Tank Furnace, in *10th International Congress on Glass*, No. 2 *Refractories and Furnaces*, Ceram. Soc. Japan, 9 Jul., pp. 2-1 to 2-8.

Yan, M.F., Barns, R.L., O'Bryan Jr., H.M., Gallagher, P.K., Sherwood, R.C. and Jin, S. 1987. Water Interaction with the Superconducting YBa_2Cu_3O7 Phase, *Appl. Phys. Lett.*, *51*(7), 532–34.

Yao, Z., Jiang, Z. and Zhang, X. 2006. Effect of Na_2SO_4 on Structure and Corrosion Resistance of Ceramics Coatings Containing Zirconium Oxide on Ti-6Al-4V Alloy, *J. Am. Ceram. Soc.*, *89*(9), 2929–32.

Yoshimura, M., Hiuga, T. and Somiya, S. 1986. Dissolution and Reaction of Yttria-Stabilized Zirconia Single Crystals in Hydrothermal Solutions, *J. Am. Ceram. Soc.*, *69*(7), 583–4.

Yu, H.S., Min, G.H., Liu, Y.X. and Chen, X.C. 1999. *J. Nonferrous Met. 9*, 785 (in Chinese). (as cited in Chen et al. 2006)

Xu, Z., He, L., Mu, R., He, S., Huang, G. and Cao, X. 2010. Hot Corrosion Behavior of Rare Earth Zirconates and Yttria Partially Stabilized Zirconia Thermal Barrier Coatings, *Surface & Coatings Technology*, *204*, 3652–61.

16

Specific Glassy Materials

Perhaps the preceding millennia have not had a Glass Age because it is still to come.

Hubert Schroeder

Introduction

The corrosion of glassy materials is predominantly through the action of aqueous media. The attack by gases quite often is that of water vapor or some solution after various species condense and dissolve in the water. Therefore this chapter is devoted mostly to aqueous attack.

In general, very high silica (>96% SiO_2), aluminosilicate, and borosilicate compositions have excellent corrosion resistance to a variety of environments. Silicate glasses in general are less resistant to alkali solution than they are to acid solution. A list of about 30 glass compositions with their resistance to weathering, water, and acid has been published by Hutchins and Harrington 1966 and is shown in Tables 16.1 and 16.2. The dissolution rate versus pH for several composition types is depicted in Figure 16.1.

Borosilicate Glasses

The durability of borosilicate glasses has been extensively investigated by the nuclear waste glass community. No attempt will be made here to review all the literature related to nuclear waste glasses; however, an article by Jantzen 1992 described quite well the use of Pourbaix diagrams in predicting the dissolution of nuclear waste glasses. Jantzen has done a very thorough job in explaining the interrelationship of pH, Eh, activity, free energy of hydration, and glass dissolution. It was shown that solution Eh had an effect upon network dissolution that was 20 times less than that of pH. But when redox-sensitive elements were leached from the glass, the solution Eh had a much larger effect. Jantzen also concluded that less durable glasses had a more negative free energy of hydration and thus released more silicon and boron

TABLE 16.1

Properties of Commercial Glasses

Glass Code[a]	Type	Forms Usually Available[b]	Weathering	Water	Acid
0010	Potash-soda-lead	T	2	2	2
0080	Soda-lime	BMT	3	2	2
0120	Potash-soda-lead	TM	2	2	2
1720	Aluminosilicate	BT	1	1	3
1723	Aluminosilicate	BT	1	1	3
1990	Potash-soda-lead		3	3	4
2475	Soda-zinc	T	3	2	2
3320	Borosilicate		1	1	2
6720	Soda-zinc	P	1	2	
6750	Soda-barium	BPR	2	2	
6810	Soda-zinc	BPR		1	2
7040	Borosilicate	BT	3	3	4
7050	Borosilicate	T	3	3	4
7052	Borosilicate	BMPT	2	2	4
7056	Borosilicate	BTP	2	2	4
7070	Borosilicate	BMPT	2	1	2
7250	Borosilicate	P	1	2	2
7570	High lead		1	1	4
7720	Borosilicate	BPT	2	2	2
7740	Borosilicate	BPSTU	1	1	1
7760	Borosilicate	BP	2	2	2
7900	96% silica	BPTUMF	1	1	1
7913	96% silica	BPRSTF	1	1	1
7940	Fused silica	UF	1	1	1
8160	Potash-soda-lead	PT	2	2	3
8161	Potash-lead	PT	2	1	4
8363	High lead	LC	3	1	4
8871	Potash-lead		2	1	4
9010	Potash-soda-barium	P	2	2	2
9700	Borosilicate	TU	1	1	2
9741	Borosilicate	BTU	3	3	4

[a] The 4-digit glass codes (e.g., 0010) refer to Corning Glass Works glasses.

[b] B = blown glass; P = pressed ware; S = plate glass; M = sintered slip cast ware (multiform); R = rolled sheet; T = tubing and cane; LC = large castings; F = fibers; U = panels.

Note: The ratings listed are as follows: 1 = high resistance, 2 = occasionally troublesome; and 3 = careful consideration for use necessary.

Source: Hutchins and Harrington, 1966, Copyright © 1966 by John Wiley & Sons, Inc., reprinted by permission of John Wiley & Sons, Inc.

TABLE 16.2

Corrosion of Glass, Applies Only to Durable Compositions

Type of Reagent	Temp	Degree of Attack	Remarks
Water	Boil	Negligible	No absorption or swelling
	100–260°C	0.001 to 0.01 mg/cm^2, in 6 hr	Depends on glass
Seawater, 5% sea salt	Boil	0.03–0.08 mg/cm^2 24 hr	1 yr in ocean, no visible effect
Acids			
HF	All	Severe	Not recommended
21% H$_3$PO$_4$	100°C	0.005 mg/cm^2, 24 hr	Glass satisfactory except at high concentrations or raw acid with fluorides
85% H$_3$PO	100°C	0.014 mg/cm^2, 24 hr	
Other inorganic	Boil	Negligible	
Organic	Boil	Negligible	
Bases			
Strong, 5% NaOH	80°C	0.3 mg/cm^2, 6 hr	
6.9% KOH	80°C	0.2 mg/cm^2, 6 hr	
Weak, 3% NH$_4$OH	80°C	0.33 mg/cm^2, 100 hr	
Halogens	to 150°C	Negligible	Dry fluorine questionable
Metal salts			
Acid	to 150°C	Negligible	
Neutral	to 150°C	Negligible	
Basic 0.5N Na$_2$CO$_3$	100°C	0.18 mg/cm^2, 6 hr	
5% Na$_2$CO$_3$	150°C	10 mg/cm^2, 6 hr	
Inorganic nonmetallic halides	to 150°C	Negligible	Fluorides excepted
Sulfur dioxide	to 150°C	Negligible	Slight bloom may appear
Ammonia (dry)	to 150°C	Negligible	See bases for NH4OH
Oxidizing chemicals	to 150°C	Negligible	
Reducing chemicals	to 150°C	Negligible	
Hydrocarbons	to 150°C	Negligible	Includes chlorinated compounds
Amines	to 150°C	Negligible	Those with pronounced basic reaction questioned
Polyhydroxyl aliphatics	to 150°C	Negligible	
Mercaptans	to 150°C	Negligible	
Oils and fats	to 150°C	Negligible	

Note: A weight loss of 1 mg/cm^2 is equivalent to a depth loss of 0.01 mm/(specific gravity of glass) for those cases where the attack is not selective.

Source: Hutchins III, J.R. and Harrington, R.V. 1966. Glass, in *Encyclopedia of Chemical Technology*, 2nd ed., Vol. 10, Wiley, New York, NY, p. 572. Copyright © 1966 John Wiley & Sons, reprinted with permission.

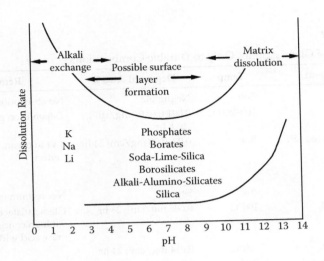

FIGURE 16.1
Dissolution rate vs. pH.

into solution. Higher boron release over that of silicon was attributed to the greater solution activity of vitreous boria compared to that of vitreous silica at any given pH. A good source of information for the reader interested in the aqueous attack upon borosilicate glasses and nuclear waste materials in general are the references listed in Chapter 15 under Materials Research Society: *Scientific Basis for Nuclear Waste Management* 1982a and b, 1983–1988, 1989a and b, and 1991; Advances in Ceramics: *Nuclear Waste Management*, 1984 and 1986; and Ceramic Transactions: *Nuclear Waste Management*, 1990 and 1991.

In borosilicate glasses that require a heat treatment step after initial melting and cooling to produce phase separation, a surface layer were formed by selective evaporation of Na_2O and B_2O_3. These surface layers have been observed by several workers. This silica-rich surface layer can influence the subsequent leaching process that would be needed to produce Vycor™-type glass, Hood and Nordberg 1940. If the hydrated surface layer were removed before heat treatment, the silica-rich layer would be almost entirely eliminated.

The leaching rate in 3 N HCl solution for borosilicate glasses with an interconnected microstructure was shown by Takamori and Tomozawa 1978 to be dependent upon the composition of the soluble phase. The composition and size of this interconnected microstructure was also dependent upon the temperature and time of the phase separation heat treatment process. Taylor et al. 1987 have shown that phase-separated low-soda borosilicate glasses form a less durable Na_2O plus B_2O_3-rich phase dispersed within a more durable silica-rich phase. The overall durability in distilled deionized water was strongly dependent upon the soda content and was best for a composition containing about 3 mol% Na_2O. The durability was also dependent upon the SiO_2/B_2O_3 ratio, with the higher silica content glasses being more durable. In a study of soda borosilicate glasses, Kinoshita et al. 1991 related the effects of the Si/B ratio to the dissolution rates. At low Si/B ratios, the glasses dissolved congruently at rapid constant rates at a $pH = 2$ in HCl/glycine solutions. Higher Si/B ratios caused the selective leaching of sodium and boron, leaving behind a silica-rich layer that caused the dissolution rate to decrease with time.

In a study closely related to borosilicate glasses, El-Hadi et al. 1985 investigated the addition of soda to B_2O_3 and the effect upon durability, which is generally very poor for borate glasses. Increased durability toward both acids and bases was related to the change in coordination of the boron from three to four as the alkali level was increased. Alkali borate glasses also increase in density as the alkali content is increased, suggesting that the change in coordination causes a more compact, more difficult to leach structure. Addition of various divalent metal oxides to a lithium borate glass also increased the durability in the following order: $CdO > ZnO > PbO > SrO > BaO$. Tait and Jensen 1982 found an order of magnitude increase in durability (in deionized water) of a sodium borosilicate glass containing 8.5 mol% ZnO. CaO and Al_2O_3 also increased the durability.

The attack by various acids was studied by Katayama et al. 1978 who determined that the corrosion of a barium borosilicate glass decreased in the order acetic, citric, nitric, tartaric, and oxalic acid, all at a pH of 4 at 50°C. The mechanism of attack by orthophosphoric acid was shown to vary with temperature by Walters 1983. The considerable degradation above 175°C was attributed to acid dehydration. At the higher temperatures, the acid condensed and reacted with the glass, forming a protective layer of SiP_2O_7. The formation of this barrier layer formed sufficient stresses to produce strength loss and caused mechanical failure.

Metcalfe and Schmitz 1972 studied the stress corrosion of E-glass (borosilicate) fibers in moist ambient atmospheres and proposed that ion exchange of alkali by hydrogen ions led to the development of surface tensile stresses that could be sufficient to cause failure.

The effect of dissolved water content upon the resistance of borosilicate glasses to acid vapor attack (over boiling 20% HCl) was investigated by Priest and Levy 1960. Increasing water contents correlated with increasing corrosion resistance.

The use of borosilicate foamed glass blocks to line the outlet ducts of coal-burning power plants was reported by Koch and Syrett 1985 to perform better than silicate cement gunite as well as nickel-based or titanium alloys in an 18-month test. This was attributed to the high concentration of aluminum in the outlet flue gas that formed soluble complexes with fluorine that are not detrimental to borosilicate glass.

Fast ion conduction glasses, such as lithium-borate and lithium-chloroborate glasses, were studied by Velez et al. 1982 to determine their resistance to molten lithium at temperatures between 180 and 250°C. They found that those compositions with a minimum B_2O_3 content resulted in the best resistance to attack.

Recently Conzone et al. 2002 have reported the development of borate glasses for use in treatment of rheumatoid arthritis, since these glasses are potentially more reactive with physiological liquids. Borate glasses containing only alkali ions dissolve uniformly (i.e., congruently) in simulated physiological liquids at temperatures ranging from 22 to 75°C. When the borate glasses contain other cations (such as Ca, Mg, Fe, Dy, Ho, Sm, and Y) in amounts ranging from 2 to 30 wt% dissolution is nonuniform (i.e., incongruent) with the formation of new compounds. Day 2002/03 gave an example of Dy_2O_3-containing borate solid glass microspheres that reacted to form hollow spheres, shells of concentric layers, or microspheres filled with homogeneous gel-like material depending upon the Dy_2O_3 content. The dissolution mechanism involved the selective leaching of lithium and boron, allowing the rare earth (i.e., Dy) to react and form an insoluble phosphate.[*] When calcium-containing borate glasses were reacted a semicrystalline or gel calcium phosphate formed that had a composition very similar to hydroxyapatite. Although early work by Hench and colleagues indicated the need for the formation of a silica gel surface layer for silicate glasses to be bioactive, the work of Day and colleagues indicated that a silica gel was not always necessary for bioactivity.

Sealing glasses for molten carbonate fuel cells was studied by Pascual et al. 2003. A Li_2O-K_2O-B_2O_3-SiO_2 and a Na_2O-B_2O_3-SiO_2 glass were immersed into molten lithium and potassium carbonate eutectic at temperatures between 500 and 700°C for times ranging from 5 min to 1 day. Commercial fused silica and Pyrex® were tested for comparison. Corrosion of all these glasses was considerable at 650°C. Three stages of corrosion were present: first, a diffusion-controlled process, then network dissolution, and finally layer formation. The surface layer that formed consisted mostly of Li_2SiO_3. Alkali borates that also formed as corrosion products were dissolved into the molten carbonates. The network dissolution stage appeared to be the reaction of $CO_3^=$ with Si-O-Si and B-O-B bonds.

Chalcogenide-Halide Glasses

Lin and Ho 1963 studied the chemical durability of As-S-I glasses exposed to neutral, acidic, and basic solutions. These glasses exhibited excellent resistance to neutral and acidic (pH 2–8) solutions; however, in basic solutions they formed thioarsenites or thioarsenates:

---

[*] Phosphorus is from a phosphate-buffered saline simulated physiological liquid.

$$As_2S_3 + 3NaOH \rightarrow Na_3AsS_3 + As(OH)_3 \qquad (16.1)$$

$$2As(OH)_3 \rightarrow As_2O_3 + 3H_2O \qquad (16.2)$$

or

$$8As_2S_5 + 30NaOH \rightarrow 10Na_3AsS_4 + 3As_2O_5 + 15H_2O \qquad (16.3)$$

As pH increased from 10 to 14, the rate of attack increased about 400 times. Higher iodine contents lowered the durability. For a given iodine content, increased arsenic contents also lowered durability. Plots of weight loss versus the square root of time were linear, indicative of a diffusion-controlled process. The rate of attack on alkaline solutions increased linearly with temperature. Lin and Ho concluded that the low solubility of these glasses was consistent with the fact that the As-S bond is highly covalent in nature.

Fluoride Glasses

The corrosion of fluoride glasses has become rather important recently due to their potential application as optical components because of their excellent IR transmission properties, Murch 1985, and their application as membranes in fluoride-ion-selective electrodes, Ravaine and Perera 1986. The corrosion of these glasses is generally characterized by a double interfacial layer, an inner portion of hydrated species and an outer nonprotective layer of crystalline precipitates, generally ZrF_4, Doremus et al. 1984, except when highly soluble compounds are present, Ravaine and Perera 1986 and Simmons and Simmons 1986. The reaction:

$$F^-_{(glass)} + OH^-_{(aq)} \leftrightarrow F^-_{(aq)} + OH^-_{(glass)} \qquad (16.4)$$

reported by Ravaine and Perera 1986 depicts the exchange reaction that forms this interfacial hydrated layer.

Simmons and Simmons 1986 studied the corrosion of fluorozirconate glasses in water (pH = 5.6). A direct correlation was found between the solubility of the modifier additive and the glass durability. Those additives with the greatest water solubility (AlF_3, NaF, LiF, and PbF_2) were determined to cause the greatest solubility of the glasses. ZrF_4, BaF_2, and LaF_3 exhibited lower solubilities. The corrosion behavior of all the glasses was controlled by the Zr and Ba contents and the pH drift of the solution. The other modifier additives had only a limited effect upon corrosion. The order of leach rate for ZBL glass was Zr > Ba >> La. The order when Al was added changed to Al > Zr >> Ba > La and when Li was added changed to Li > Al > Zr > Ba >> La. When Na replaced Li, the

Al leach rate was lower than the Na, and the others remained the same. The addition of Pb had the greatest effect by not exhibiting the marked decrease in the leach rate with time for the various components.

The major difference between fluorozirconate and silicate glasses is the drift in pH during the corrosion process. The fluorozirconate exhibits a solution pH drift toward acidic values. The equilibrium solution pH for a ZrBaLaAlLi-fluoride glass was found to be 2.6. Additional studies upon crystalline forms of the various additives indicated that the main cause of the drop in pH was the hydrolysis of ZrF_4 forming the complex species $[ZrF_x(OH)_y]^{+4-x-y}[H_2O]_n$. It is interesting that these glasses exhibited minimal corrosion from atmospheric moisture, even when exposed to 100% RH at 80°C for up to 1 week. Gbogi et al. 1981 reported similar results for a ZBL glass exposed to ambient conditions for 30 days, and Robinson and Drexhage 1983 reported no corrosion for ThF_4-containing fluoride glasses up to 200°C.

The time dependency of leaching rates varied with the composition of the heavy metal fluoride additive, Ravaine and Perera 1986. Compositions containing Zr, Ba, and Th; U, Ba, and Mn; and Sc, Ba, and Y displayed a continuous decrease in corrosion rate with time. Those containing Th, Ba, Mn, and Yb or Th, Ba, Zn, and Yb displayed a minimum. Those containing Pb, K, Ga, Cd, Y, and Al displayed a plateau. Ravaine and Perera also reported a direct relationship between fluoride ion conductivity and corrosion rate. Only the Sc, Ba, and Y composition did not form the outer layer of crystalline precipitates.

Thorium-based glasses containing Zn-Ba-Y-Th, Zn-Ba-Yb-Th, or Zn-Ba-Yb-Th-Na have been reported to be 50 to 100 times more resistant to dissolution than the corresponding zirconium-based glasses, Simmons et al. 1983.

Lead-Containing Glasses

Yoon 1971 found that lead release was a linear function of pH when testing lead-containing glasses in contact with various beverages. Low pH beverages such as orange juice or colas leached lead more slowly than did neutral pH beverages such as milk. This dependence upon pH was also reported by Das and Douglas 1967 and by Pohlman 1974. Later, Yoon 1973 reported that if the ratio of moles of lead plus moles of alkali per moles of silica were kept below 0.7, release in 1 hr was minimized. If this ratio was exceeded, lead release increased linearly with increasing PbO content. Lehman et al. 1978 reported a slightly higher threshold for more complex compositions containing cations of Ca^{2+} and Al^{3+} or B^{3+} in addition to the base Na_2O-PbO-SiO_2 composition. The lead release in these complex compositions was not linear but increased upward with increased moles of modifiers. Lehman et al. related the mechanism of release or corrosion to the concentration of nonbridging oxygens. A threshold concentration was necessary for easy diffusion of the

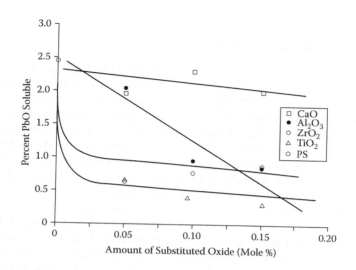

FIGURE 16.2
Lead solubility in 0.25% HCl from a lead bisilicate glass vs. additions of ZrO_2, TiO_2, CaO, and Al_2O_3. (After Haghjoo and McCauley 1983.)

modifier cations. This threshold was reported to be where the number of nonbridging oxygens per mole of glass-forming cations equaled 1.4.

Krajewski and Ravaglioli 1982 correlated the release of Pb^{2+} by acid attack to the site coordination of the network modifiers. The presence of cations with cubic coordination produced increased Pb^{2+} release, whereas cations with antiprismatic coordination produced a decreased Pb^{2+} release.

In general, it has been determined that mixed alkalies lower the release of lead by attack from acetic acid below that of a single alkali-PbO-silicate glass; lead release increased with increasing ionic radius of the alkaline earths, but combinations of two or more alkaline earths exhibited lower lead release; Al_2O_3 and ZrO_2 both lowered the lead release; and B_2O_3 increased the lead release. Thinner glaze coatings on clay-based ceramic bodies decreased lead release due to interaction of the glaze and the body, providing higher concentration of Al_2O_3 and SiO_2 at the glaze surface, *Lead Glazes for Dinnerware* 1974.

Haghjoo and McCauley 1983 found that small substitutions (0.05–0.15 mol%) of ZrO_2 and TiO_2 to a lead bisilicate glass lowered the solubility of lead ion in 0.25% HCl by an order of magnitude. Additions of Al_2O_3 had a lesser effect, while additions of CaO had essentially no effect (see Figure 16.2).

The mechanism of release or corrosion for these glasses containing lead is similar to those proposed by Charles 1958 for alkali-silicate glasses. The rate of this reaction depends upon the concentration gradient between the bulk glass and the acid solution and the diffusion coefficient through the reacted layer. In general, maximum durability can be related to compact, strongly bonded glass structures, which in turn exhibit low thermal expansion coefficients and high softening points, Moore 1956.

Nitrogen-Containing Glasses

Nitrogen-containing glasses in the M-Si-Al-O-N (M = Ca, Mg, Ba, La, Ce, Nd, and Y) system have long been known to be present as residual glassy grain boundary phases in silicon nitride-based ceramics. In some cases various rare earth elements have been added. These glassy grain boundary phases are a consequence of the additives used to produce a dense ceramic for these covalently bonded materials. When these silicon nitride materials are exposed to various environments it is invariably the grain boundary phase that corrodes first and leads to ultimate failure. Thus it is of interest to study the effects of various environments upon these glasses.

The oxidation behavior of M-Si-Al-O-N glasses with M = Nd or Y was studied by Sebai et al. 1995. Various ratios of Y/Al and Si/Al along with substitutions of Nd for Y and various nitrogen contents were evaluated at 1075°C in air under various partial pressures of oxygen. Sebai et al. found that the oxidation resistance of the Y-containing glasses was better than the Nd-containing glasses. Oxidation resistance also increased with increasing nitrogen or aluminum content. All these glasses oxidized rapidly above the glass transition temperature forming a porous oxide scale composed of yttrium silicates and mullite. These results confirmed the earlier work of O'Meara et al. 1991.

Tulliani et al. 2002 studied the oxidation behavior of LnSiAlON (Ln = Y or La) glasses at temperatures between 990 and 1150°C and water vapor pressures between 360 and 2690 Pa. Three different gaseous atmospheres were used—oxygen, N_2/H_2O, and O_2/H_2O, all at a total partial pressure of 100 kPa. Oxidation proceeded by the replacement on the surface of $2N^{3-}$ by $3O^=$. This resulted in a net weight gain. Complete replacement of all nitrogen by oxygen resulted in a weight gain of 1.66 wt% for the La-containing glass and 3.48 wt% for the Y-containing glass. In the N_2/H_2O mixture a porous oxide scale formed on the surface of the La-containing glass. Up to 1050°C no crystallization of the bulk glass occurred; however, the scale was composed of $La_4Si_8Al_6O_{31}$ and $La_2Si_2O_7$. The porous scale on the Y-containing glass was composed of $Y_2Si_2O_7$ and mullite.

Phosphorus-Containing Glasses

The study of phosphate glass corrosion has shown that the glass structure plays a very important role in the rate of dissolution. Phosphate glasses are characterized by chains of PO_4 tetrahedra. As the modifier (alkalies or alkaline earths) content of these glasses is increased, there is increased cross-linking between the chains. An indication of increased cross-linking is higher T_g^* values. *When very little cross linkage exists, corrosion is high. When the amount of cross linkage is high, corrosion is low.* Similar phenomena should exist for other glass-forming cations that form chain structures (B^{3+} and V^{5+}). During the study of aqueous attack of soda-lime-silica glasses containing P_2O_5, Clark et al. 1976 found that a double reaction layer was formed, consisting of a silica-rich region next to the glass and a Ca-P-rich reaction next to the water solution. This Ca-P film eventually crystallized into an apatite structure and provided a good mechanism to bond the glass to bone in implant applications. In order for these compositions to be highly active toward aqueous media, the bioactive glass composition must contain less than 60 mol% SiO_2, a high content of Na_2O and CaO, and a high CaO/P_2O_5 ratio, Hench 1992. When the SiO_2 content was greater than 60 mol%, the hydroxyapatite reaction layer did not form within 2–4 weeks. For a glass to be beneficial as an implant, the reactions leading to the formation of the $CaO-P_2O_5$-rich surface film must occur within minutes of implantation. The dependency of bioactivity upon the structure of the glass was thus a very important concern in the development of these materials. When the silica content exceeded 60 mol%, the glass structure changed from one of two-dimensional sheets containing chains of polyhedra to a three-dimensional network common to the high-silica glasses. The two-dimensional structure, being a more open structure, allowed more rapid ion exchange and thus faster hydroxyapatite film formation.

Potassium phosphate glasses containing various oxide additions were tested for water solubility by Minami and Mackenzie 1977, with Al_2O_3 and WO_3 additions yielding the greatest improvement. In alkali phosphate glasses containing Al_2O_3 or WO_3, the durability increased as the ionic radius of the alkali cation decreased, a trend that was common in most glasses.

Reis et al. 2001 investigated the durability of zinc-iron phosphate glasses in distilled water at 90°C for up to 32 days. They found the durability to be 100 times better than window glass and the dissolution rate to decrease with increasing iron content. Excellent durability of glasses containing more than 30 mol% Fe_2O_3 was related to the presence of the Fe-O-P bond.

According to Hench 2002/03 in his discussion of bioactive glasses, the dissolution kinetics are a function of the following variables:

1. composition,
2. particle size,
3. pore size distribution, average size, and volume%,
4. surface area,
5. thermal stabilization temperature, and
6. chemical stabilization temperature.

---

[*] T_g is the term used to represent the transformation temperature where the glass changes from an elastic solid to a visco-elastic one.

The alumina content of bioactive glasses is very important in controlling the durability of the glass surface. The bioactivity, although dependent upon the bulk composition of the glass, decreases beyond acceptable levels once the alumina content rises above 1.0 to 1.5 wt%, Hench and Wilson 1993. This same phenomenon was present for glass compositions containing cations such as Ta_2O_5 except higher levels were tolerable (1.5–3.0 wt%).

The incorporation of nitrogen into phosphate glasses and its improvement of corrosion resistance has been described quite well by Prof. Delbert Day and his colleagues, Wilder et al. 1983, Reidmeyer et al. 1986, Bunker et al. 1987, Rajaram and Day 1987, and Rajaram and Day 1988. The incorporation of nitrogen yields P–N bonds that exhibit a greater covalent character than the P–O bonds, yielding glasses with greater toughness than oxide glasses. These glasses are of interest in a variety of applications: as sealing glasses, in nuclear waste disposal, and as laser hosts. Although data is available in the literature on the oxidation of oxynitride glasses, very little has been reported on oxynitride phosphate glasses. Munoz et al. 2006 evaluated the oxidation below 500°C in air of Li-Na-Pb-P-O-N glasses of different particle sizes. The surface oxidized first, followed by bulk oxidation controlled by diffusion. Since there was no evidence of metallic lead in the original glasses, it was believed that the weight gain experienced by these glasses was due only to the replacement of nitrogen by oxygen. As the nitrogen content of these glasses is increased, the viscosity increases as a consequence of increased network cross-linking. The incorporation of lead into these glasses reduced oxygen diffusion and thus the weight gain by oxidation.

Silicate Glasses

Generally, silicate glass corrosion processes are typified by diffusion-controlled alkali ion exchange for H^+ or H_3O^+, followed by matrix dissolution as the solution pH drifts toward higher values. This concept was perhaps first reported in 1958 by Wang and Tooley 1958. The initial exchange reaction produces a transformed gel-like surface layer. This surface layer may contain various crystalline phases depending upon the overall glass composition and solution pH. Diffusion through this layer becomes the rate-controlling step. This layer is formed through the process of network hydrolysis and condensation of network bonds releasing alkali, a process that is very similar to the second, essentially simultaneous, step of network dissolution. Thus the dissolution of silicate glasses is dependent upon the test conditions of time, temperature, pH, and the sample composition (i.e., structure). Although many references are made to the effects of glass composition upon dissolution, the actual correlation is with glass structure, not composition. This is due to the fact that composition determines structure. An example of this was indicated by Brady and House 1996. They determined that glasses that were silica-rich and highly polymerized dissolved more slowly than those containing large amounts of other cations. The key structural factor was that highly polymerized glasses dissolved more slowly.

The deterioration of a glass surface by atmospheric conditions, commonly called *weathering*, is very similar to that described above. If droplets of water remain on the glass surface ion exchange can take place with a subsequent increase in the *p*H. Since the volume of the droplets is normally small compared to the surface area they contact, drastic increases in *p*H can occur, causing severe etching of the surface. The rough surface formed can then collect additional solution, causing further deterioration. In some cases, the alkali-rich droplets can react with atmospheric gases, forming deposits of sodium and calcium carbonate, Simpson 1958. Chen and Park 1981 have identified crystals on the surface of a Na-disilicate glass exposed to room temperature air as $Na_3H(CO_3)_2 \cdot 2H_2O$. These deposits can act as a barrier to further weathering; however, they detract from the visual aesthetics of the glass. Whether dealkalization or matrix dissolution is the predominant mode of attack depends upon the volume and flow of water in contact with the surface.

The manufacturers of flat glass have for many years known of the beneficial effects of SO_2 gas treatment in increasing the weatherability of their products. This treatment, prior to the glass being annealed, allowed the sodium in the surface layers to react with the SO_2, forming sodium sulfate. The sulfate deposit was then washed off prior to inspection and packing. The first step in weathering was then diminished due to the low alkali content of the surface.

According to Charles 1958, the corrosion of an alkali-silicate glass by water proceeds through three steps. These are as follows:

1. H^+ from the water penetrates the glass structure. This H^+ replaces an alkali ion, which goes into solution. A nonbridging oxygen is attached to the H^+.

2. The OH^- produced in the water destroys the Si-O-Si bonds, forming nonbridging oxygens.

3. The nonbridging oxygens react with an H_2O molecule, forming another nonbridging oxygen-H^+ bond and another OH^- ion. This OH^- repeats step 2. The silicic acid thus formed is soluble in water under the correct conditions of *p*H, temperature, ion concentration, and time.

It is questionable as to whether the first step described above involves the penetration of a proton or a hydronium, H_3O^+ ion. There is evidence that supports the exchange of hydronium for alkalies, Scholze et al. 1975. In addition, the dissolution of silicate minerals, which is very similar to silicate glasses, has been reported to take place by exchange of hydronium ions for alkalies, Lasaga 1990. The development of films on the glass surface has been described by Sanders and Hench 1973. They showed that a 33 mol% Li_2O glass corroded more slowly than a 31 mol% Na_2O glass by two orders of magnitude. This difference was caused by the formation of a film on the Li_2O glass with a high silica content. Scratching the glass surface produced an unusually high release of silica. The nonbridging oxygen-H^+ groups may form surface films or go into solution. The thickness of this film and its adherence greatly affected the corrosion rate. In Na_2O-SiO_2 glasses, Schmidt 1958 found that films formed only on glasses containing more than 80 mol% SiO_2 at 100°C for 1 hour.

Several workers have investigated the concentration profiles of glass surfaces after leaching by water and attempted to explain the variations observed. Boksay et al. 1967 postulated a theory that fit the profiles observed in K_2O-SiO_2 glass but did not explain the profiles in Na_2O-SiO_2 glass, presumably due to a concentration-dependent diffusion coefficient. Doremus 1975 developed a theory that included a concentration-dependent diffusion coefficient to explain the profiles in Li_2O-SiO_2 glass; however, his theory still did not fit the observations for sodium determined by Boksay et al. 1968. Das 1979 attributed the differences in the profiles between the sodium and potassium glasses as being due to a difference in the structure of the leached layer caused by the relative difference in size between the H_3O^+ and the Na^+ ions and the similarity in size between H_3O^+ and K^+ ions. In general, the dissolution rate (i.e., dealkalization) decreased as the ion radius of the alkali decreased.

Douglas and coworkers, Rana and Douglas 1961a and b, Das and Douglas 1967, and Douglas and El-Shamy 1967 found that alkali removal was a linear function of the square root of time in alkali-silicate glass attacked by water. At longer times, the alkali removal was linear with time. Silica leached from alkali-silicate glasses decreased as the amount of silica in the glass increased, unlike that of the alkalies. Wood and Blachere 1978 investigated a 65SiO_2-10K_2O-25PbO (mol%) glass and did not find a square root of time dependence for removal of K or Pb but found a dependence that was linear with time. This behavior has also been reported by Eppler and Schweikert 1976 and by Douglas and coworkers. Wood and Blachere proposed that an initial square root of time dependence occurred but that the corrosion rate was so great that it was missed experimentally.

The pH of the extracting solution is also very important, as found by Douglas and El-Shamy 1967. They found that above $pH = 9$ the leaching rate of alkalies decreased with increasing pH, whereas below $pH = 9$ the leaching rate was independent of pH. A somewhat different relationship was found for the leaching rate of silica; above $pH = 9$ the rate increased with increasing pH, whereas below $pH = 9$ the amount of silica extracted was close to the detection limits of the apparatus. Two reactions were identified: one where alkalies passed into solution as a result of ion exchange with protons from the solution and one where silica passed into solution as a consequence of the breaking of siloxane bonds by attack from hydroxyl groups from the solution. Thus removal of silica was favored by an increase in hydroxyl ion activity (i.e., increased pH), which was accompanied by a reduction in proton activity and thus a reduction in alkali extraction.

The dependence of dissolution upon pH can be seen by an examination of equation 1.4 (Chapter 1) for the dissolution of minerals. Similarly glasses in contact with aqueous solutions can be represented by the following ion exchange reaction:

$$\text{MSiO}_2 \text{ (glass)} + n\text{H}^+ \text{ (aq)} \leftrightarrow \text{H}_2\text{SiO}_3 + \text{M}^{n+} \text{ (aq)} \tag{16.5}$$

which has as the equilibrium constant

$$k = \frac{a_{\text{H}_2\text{SiO}_3} a_{\text{M}^{n+}}}{a_{\text{MSiO}_3} a_{\text{H}^+}} \tag{16.6}$$

Expressing this in logarithm form then gives

$$\log a_{\text{H}_2\text{SiO}_3} = \log k - \log a_{\text{M}^{n+}} - n \ pH \tag{16.7}$$

Thus it should be obvious that the exchange reaction of a proton for the leachable ionic species in the glass is dependent upon the pH of the solution and also the leached ion activity in the solution.

Das 1981 has shown that substitutions of Al_2O_3 or ZrO_2 for SiO_2 in sodium silicate glasses shifted the pH at which increased dissolution occurred to higher values, creating glasses that were more durable and less sensitive to pH changes. Paul 1982 has also reported the beneficial effects of alumina and zirconia upon durability.

The manufacturers of soda-lime-silicate glasses have known for a long time that the addition of lime to sodium silicate glass increased its durability. Paul 1982 reported that substitutions of up to 10 mol% CaO for Na_2O rapidly decreased the leaching of Na_2O. Above about 10 mol% substitution, the leaching of Na_2O remained constant. With the larger amounts of CaO devitrification problems during manufacture also occur, requiring the substitution of MgO for some of the CaO. According to Paul 1982 calcium-containing glasses should exhibit good durability up to about $pH = 10.9$. He also indicated that replacement of ZnO for CaO could extend this durability limit to about $pH = 13$, although these compositions are attacked in acid solutions at $pH < 5.5$.

The effects of MgO, CaO, SrO, and BaO upon the leaching of Na_2O at 60 and 98°C in distilled water were reported by Paul 1982. At the higher temperature, the durability decreased with increasing ionic size, whereas at the lower temperature, the durability was relatively the same for all four alkaline earths. This was attributed to the restricted movement at the lower temperature for the larger ions.

Expanding upon the ideas originally proposed by Paul and coworkers, Paul 1977, Paul and Youssefi 1978, and Newton and Paul 1980, Jantzen and coworkers, Jantzen 1992, Plodinec et al. 1984, and Jantzen and Plodinec 1984, have shown that network or matrix dissolution was proportional to the summation of the free energy of hydration of all the glass components as given by the equation:

$$\Delta G° = A\,RT \log L \qquad (16.8)$$

where A is the proportionality constant and L is a normalized loss by leaching in mass per unit area. Jantzen 1992 has shown that high-silica glasses exhibited weak corrosion in acidic-to-neutral solutions and that low-silica glasses exhibited active corrosion at pH from <2 to 3. Between pH 2 and 10 in an oxidizing solution, hydrolysis occurred through nucleophilic attack with the formation of surface layers by reprecipitation or chemisorption of metal hydroxides from solution. In reducing solutions, surface layers tended to be silicates that exhibited weak corrosion or were even immune. In alkaline solutions at pH greater than about 10, both low- and high-silica glasses exhibited active corrosion with low-silica glasses having a potential for surface layer formation.

Ernsberger 1960 has described the attack of silica or silicate glasses by aqueous hydrofluoric acid in detail and related it to the structure of silica glasses. The silicon-oxygen tetrahedra are exposed at the surface in a random arrangement of four possible orientations. Protons from the water solution will bond with the exposed oxygens, forming a surface layer of hydroxyl groups. The hydroxyl groups can be replaced by fluoride ions in aqueous hydrofluoric solutions. Thus the silicon atoms may be bonded to an OH$^-$ or and F$^-$ ion. The replacement of the exposed oxygens of the tetrahedron by 2F$^-$ causes a deficiency in the silicon atom coordination, which is six with respect to fluorine. This causes the additional bonding of fluoride ions, with a particular preference for bifluoride. Thus the four fluoride ions near the surface provide an additional four-coordinated site for the silicon. A shift of the silicon to form SiF$_4$ can take place by a small amount of thermal energy. The ready availability of additional fluoride ions will then cause the (SiF$_6$)$^{2-}$ ion to form. This mechanism is supported by data that show a maximum in corrosion rate with bifluoride ion concentration. Although giving a slightly different description of the possible reactions, Liang and Readey 1987 reported that the dissolution of fused silica varied with HF concentration and was controlled by a surface reaction rather than diffusion through the liquid.

The solubility in nitric acid has been reported by Elmer and Nordberg 1958 with increasing concentration (from 0.8 to 7.0 N), just the opposite as that found in HF. In concentrations greater than 3 N, saturation was reached in about 24 hours. At 0.1 N, the rate was considerably lower than the other concentrations, not reaching saturation even after 96 hours.

White et al. 1986 found that for Na_2O-SiO_2 (33/67% composition) and Li_2O-SiO_2 glass compositions, environments that caused surface corrosion also caused enhanced crack growth. The environments studied were distilled water, hydrazine, formamide, acetonitrite, and methyl alcohol. White et al. found that acetonitrite was noncorrosive and that water was the most effective in leaching alkali, while hydrazine was the most effective in leaching silica. Formamide was only mildly effective in leaching alkali. The mechanism of corrosion for water, formamide, and hydrazine was reported to be alkali ion exchange with H^+ or H_3O^+.

The durability of gel-derived 20 mol% Na_2O–80 mol% SiO_2 glass subjected to various temperatures in deionized water was studied by Hench et al. 1983. They concluded that both lower soda contents (compared to a 33 mol% Na_2O glass tested in a previous study, Hench et al. 1982) and higher densities improved the durability.

The effect of dissolved water in soda-lime glass upon the rate of dissolution in water was related to the influence of absolute humidity at the time of forming and annealing by Bacon and Calcamuggio 1967. Very high resistance was obtained by use of very dry air. Similar results were obtained by Wu 1980 on a soda-silica glass containing K_2O, Al_2O_3, and ZnO with dissolved water contents between 4 and 8 wt%. Wu, however, reported leach rates independent of water contents at concentrations less than 4 wt%. Tomozawa et al. 1982 concluded that many Si-O bonds in the glass are possibly hydrolyzed by the dissolved water content, thus eliminating some steps during the dissolution of the glass in water and increasing the rate of attack.

Little information seems to have been published in the area of molten salt attack on glasses. The dissolution of several glass compositions was reported by Bartholomew and Kozlowski 1967 to be extensive and nonuniform in molten hydroxides. Samples attacked by sodium hydroxide exhibited an opaque and frosted surface, whereas those attacked by potassium hydroxide were transparent. Bartholomew and Kozlowski used the mechanism proposed by Budd 1961 to interpret the attack shown in their studies. Considering the hydroxide ion as basic, a vigorous reaction should take place with an acidic glass. This was confirmed experimentally by testing glasses of different chemistries.

Loehman 1979 reported no trends in leaching with nitrogen content for several Y-Al-Si-O-N glasses, although two of his compositions exhibited lower weight losses by at least a factor of two than fused silica when tested in distilled water at 95°C for 350 hours. In their study of soda-lime-silicate glasses, Frischat and Sebastian 1985 reported that a 1.1 wt% addition of nitrogen considerably increased the leach resistance to 60°C water for 49 hours. The release of sodium was 55% less and calcium 46% less for the nitrogen-containing glass. An additional indication of the greater resistance of the nitrogen-containing glass was the change in pH of the leaching solution with time. Starting with a solution pH of 6, the solution pH drifted to 9 for the nitrogen-free glass after 7 hours but reached 9 for the nitrogen-containing glass after only 25 hours. The improved leach resistance of this glass was attributed to a greater packing density for the nitrogen-containing glass.

White and Day 1994 reported no detectable weight loss of a 1 × 1 × 0.2 cm rare earth aluminosilicate (REAS) glass sample before 6 weeks in 100 mL of distilled water ($pH = 7$) or saline ($pH = 7.4$) at 37, 50, or 70°C. Dissolution rates of $\leq 3 \times 10^{-9}$ g/cm^2 · min were determined after 6 weeks. In a comparison study of fused silica, a Corning glass (CGW-1723™),[*] and yttria aluminosilicate (YAS), Oda and Yoshino 1991 showed that YAS was significantly more durable than fused silica in saturated steam at 300°C and 8.6 MPa. The dissolution mechanism is very important for applications in the human body; however, it is very difficult to determine whether these glasses exhibit congruent or incongruent dissolution. Surface analyses of microspheres and bulk glasses indicated that the mechanism was congruent, White and Day 1994. Using inductively coupled plasma and atomic adsorption spectroscopy it has been determined that the yttrium release from YAS microspheres in distilled water or saline at 37 or 50°C was below detectable limits, Erbe and Day 1991.

In the manufacture of flat glass by the float process[†] a cooperative diffusion process takes place where tin diffuses into the glass and the constituents of the glass diffuse into the tin. The reaction zone in the glass is about 25 μm thick. Many investigators have studied the tin oxide gradient of float glass and have reported a rather complex behavior, Sieger and Parsons 1974, Sieger 1975, Williams et al. 1997, Baitinger et al. 1980, Colombin et al. 1980, Swift 1984, Wang 1997, Kondrashov et al. 1998, Meisel 1998, and Paulson et al. 1998. Stannous tin is dominant at the near surface. A typical hump occurs in the tin profile at between 5 and 10 μm where stannic (or oxidized tin) is predominant (see Figure 16.3). This hump has been attributed to the additional tin from the ion exchange with calcium by Franz 1995. Investigation of extremely thin layers of glass has indicated tin oxide contents as high as 36% at the surface, Swift 1984. The amount of the tin contained within the glass surface and the depth to which it penetrates was dependent upon the exposure time and temperature (which relates to glass production tonnage and thickness) and the amount and type of impurities (especially sulfur) contained in the tin. Thin glass travels through the bath faster than thick glass and therefore has less time for the various reactions to take place.

At the hot end of the bath iron oxide in the glass will migrate toward the bottom surface where it is reduced (by reaction with either stannous oxide or hydrogen) to iron metal and dissolves into the tin. At the cooler end of the bath this tin/iron alloy will oxidize (oxygen coming from air ingress), forming both iron oxide and tin oxide. Iron has a greater potential to oxidize than tin and therefore acts as a scavenger for oxygen. Being

---

[*] CGW-1723™ is a clear aluminosilicate glass.
[†] The float process for the manufacture of flat glass involves floating molten glass onto molten tin in a chamber, called the float bath, containing a reducing atmosphere.

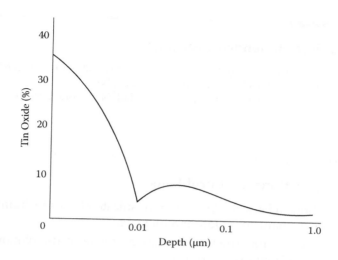

FIGURE 16.3
Tin oxide penetration into bottom surface of float glass. (After Swift 1984.)

essentially insoluble in tin, these two oxides will enter into the glass either by diffusion or by exchange for calcium oxide, Franz 1995. Calcium oxide is also insoluble in molten tin and will therefore form a deposit on the bottom surface of the glass. The deposit can be washed from the bottom surface of the glass by a vinegar solution.* Thus iron that enters the tin at the hot end of the bath will reenter the glass at the cold end, setting up an equilibrium concentration of iron in the tin. This equilibrium can be altered if the glass composition is changed from one of high iron content to one of lower iron content (or vice versa).

Although the interaction layer thickness is quite small, the presence of tin in the surface of the glass ribbon causes some secondary fabrication problems. Many fabrication methods require that the flat glass piece be bent. This is done by reheating the glass on a metal frame and allowing the glass to sag to the desired shape. This reheating process can provide additional oxidation of the tin (from stannous to stannic oxide) in the bottom surface.

This oxidation is accompanied by an expansion of the tin-rich layer causing a micro-wrinkled surface. This wrinkled surface becomes visible as a faint iridescent haze, known as the defect *bloom*, Mishra et al. 1998. This phenomenon can also occur when glass is reheated for tempering.

---

* The deposit of calcium oxide reacts with atmospheric carbon dioxide, forming calcium carbonate on the glass surface that is insoluble in water and must be washed off with a vinegar solution.

Additional Recommended Reading

Clark, D. E. and Zoitos, B. K. (eds.). 1992. *Corrosion of Glass, Ceramics and Ceramic Superconductors*, Noyes Publications, Park Ridge, NJ.

Paul, A. 1982. *Chemistry of Glasses*, Chapman and Hall, New York, 293 pp.

Exercises, Questions, and Problems

1. Discuss how pH affects dissolution of silicate glasses including the different mechanisms at low and high pH.

2. Discuss how glass structural variations relate to dissolution and how this is related to composition.

3. What structural factor and what pH relates to the minimum dissolution rate?

4. Describe the surface area/volume ratio of the attacking fluid effects upon dissolution rate.

5. How does a surface treatment of SO_2 gas diminish dissolution rates?

6. Why do Al_2O_3 and/or ZrO_2 substitutions for SiO_2 increase durability?

7. How does the Si/B ratio affect dissolution in borosilicate glasses?

8. Why is the number of nonbridging oxygens important to dissolution?

9. Explain how softening points and/or thermal expansion coefficients may relate to dissolution.

10. Discuss the penetration of tin into the bottom surface of float glass and how it may affect subsequent processing into various commercial products.

References

Bacon, F.R. and Calcamuggio, G.L. 1967. Effect of Heat Treatment in Moist and Dry Atmospheres on Chemical Durability of Soda-Lime Glass Bottles, *Am. Ceram. Soc. Bull.*, 46(9), 850–55.

Baitinger, W.E., French, P.W. and Swarts, E.L. 1980. Characterization of Tin in the Bottom Surface of Float Glass by Ellipsometry and EPS, *J. Non-Cryst. Solids*, 38–39, 749–54.

Bartholomew, R.F. and Kozlowski, T.R. 1967. Alkali Attack of Glass Surfaces by Molten Salts, *J. Am. Ceram. Soc.*, 50(2), 108–11.

Boksay, Z., Bouquet, G. and Dobos, S. 1967. Diffusion Processes in the Surface Layer of Glass, *Phys. Chem. Glasses*, 8(4), 140–44.

Boksay, Z., Bouquet, G. and Dobos, S. 1968. The Kinetics of the Formation of Leached Layers on Glass Surfaces, *Phys. Chem. Glasses*, 9(2), 69–71.

Brady, P.V. and House, W.A. 1996. Surface-Controlled Dissolution and Growth of Minerals, Ch. 4, pp. 225–305 in *Physics and Chemistry of Mineral Surfaces*, Brady, P.V. (ed.), CRC Press, New York, NY.

Budd, S.M. 1961. Mechanism of Chemical Reaction between Silicate Glasses and Attacking Agents: I, Electrophilic and Nucleophilic Mechanism of Attack, *Phys. Chem. Glasses*, 2(4), 111–14.

Bunker, B.C., Arnold, G.W., Rajaram, M. and Day, D.E. 1987. Corrosion of Phosphorus Oxynitride Glasses in Water and Humid Air, *J. Am. Ceram. Soc.*, 70(6), 425–30.

Chen, H. and Park, J.W. 1981. Atmospheric Reaction at the Surface of Sodium Disilicate Glass, *Phys. Chem. Glasses*, 22(2), 39–42.

Charles, R.J. 1958. Static Fatigue of Glass: I, *J. Appl. Phys.*, 29(11), 1549–53.

Clark, A.E., Pantano, C.G. and Hench, L.L. 1976. Spectroscopic Analysis of Bioglass Corrosion Films, *J. Am. Ceram. Soc.*, 59(1–2), 37–39.

Colombin, L., Charlier, H., Jelli, A. et al. 1980. Penetration of Tin in the Bottom Surface of Float Glass—A Synthesis, *J. Non-Cryst. Solids*, 38–39(Pt. II), 551–56.

Conzone, C.D., Brown, R.F., Day, D.E. and Ehrhardt, G.J. 2002. *In Vitro* and *In Vivo* Dissolution Behavior of a Dysprosium Lithium Borate Glass Designed for the Radiation Synovectomy Treatment of Rheumatoid Arthritis, *J. Biomedical Materials Research*, 60(2), 260–68.

Das, C. R. 1979. Reaction of Dehydrated Surface of Partially Leached Glass with Water, *J. Am. Ceram. Soc.*, 62(7–8), 398–402.

Das, C.R. 1981. Chemical Durability of Sodium Silicate Glasses Containing Al_2O_3 and ZrO_2, *J. Am. Ceram. Soc.*, 64(4), 188–93.

Das, C.R. and Douglas, R.W. 1967. Studies on the Reaction between Water and Glass. Part 3, *Phys. Chem. Glasses*, 8(5), 178–84.

Day, D.E. 2002/03. Reactions of Bioactive Borate Glasses with Physiological Liquids, *Glass Researcher*, 12(1&2), 21–22.

Doremus, R.H. 1975. Interdiffusion of Hydrogen and Alkali Ions in a Glass Surface, *J. Non-Cryst. Solids*, 19, 137–44.

Doremus, R.H., Bansal, N.P., Bradner, T. and Murphy, D. 1984. Zirconium Fluoride Glass: Surface Crystals Formed by Reaction with Water, *J. Mater. Sci. Lett.*, 3(6), 484–88.

Douglas, R.W. and El-Shamy, T.M.M. 1967. Reaction of Glass with Aqueous Solutions, *J. Am. Ceram. Soc.*, 50(1), 1–8.

El-Hadi, Z.A., Gammal, M., Ezz-El-Din, F.M. and Moustaffa, F.A. 1985. Action of Aqueous Media on Some Alkali-Borate Glasses, *Cent. Glass Ceram. Res. Inst. Bull.*, 32(1–2), 15–19.

Elmer, T.H. and Nordberg, M.E. 1958. Solubility of Silica in Nitric Acid Solutions, *J. Am. Ceram. Soc.*, 41(12), 517–20.

Eppler, R.A. and Schweikert, W.F. 1976. Interaction of Dilute Acetic Acid with Lead-Containing Vitreous Surfaces, *Ceram. Bull.*, 55(3), 277–80.

Erbe, E.M. and Day, D.E. 1991. *Proceedings: Science and Tech of New Glasses*, Sakka, S. and Soga, N. (eds.), Ceram. Soc. Japan, Tokyo, p. 105.

Ernsberger, F.M. 1960. Structural Effects in the Chemical Reactivity of Silica and Silicates, *J. Phys. Chem. Solids*, 13(3–4), 347–51.

Franz, H. 1995. Surface Chemistry of Commercial Glasses, *Ceram. Eng. Sci. Proc.*, 16(2), 221–27.

Frischat, G.H. and Sebastian, K. 1985. Leach Resistance of Nitrogen-Containing Na_2O-CaO-SiO_2 Glasses, *J. Am. Ceram. Soc.*, 68(11), C305–7.

Gbogi, E.O., Chung, K.H., Moynihan, C.T. and Drexhage, M.G. 1981. Surface and Bulk -OH Infrared Absorption in ZrF_4– and HfF_4-Based Glasses, *J. Am. Ceram. Soc.*, 64(3), C51–53.

Haghjoo, M. and McCauley, R.A. 1983. Solubility of Lead from Ternary and Quaternary Silicate Frits, *Ceram. Bull.*, 62(11), 1256–58.

Hench, L.L. 1992. Surface Modification of Bioactive Glasses and Ceramics, Ch. 9 in *Corrosion of Glass, Ceramics and Ceramic Superconductors*, Clark, D.E. and Zoitos, B.K. (eds.), Noyes Publications, Park Ridge, NJ, pp. 298–314.

Hench, L.L. 2002/03. Bioactive Glasses Help Heal, Repair and Build Human Tissue, *Glass Researcher*, 12(1&2), 18.

Hench, L.L. and Wilson, J. 1993. Introduction, in *An Introduction to Bioceramics, Advanced Series in Ceramics, Vol. 1*, World Scientific, pp 1–24.

Hench, L.L., Prassas, M. and Phalippou, J. 1982. Surface Behavior of Gel-Derived Glasses, *Ceramic Engineering and Science Proceedings*, 3(9–10), 477–83.

Hench, L.L., Wallace, S., Wang, S. and Prassas, M. 1983. The Processing and Environmental Behavior of a 20-mol% Na_2O-80 mol% SiO_2 (20N) Gel-Glass, *Ceramic Engineering and Science Proceedings*, 4(9–10), 732–39.

Hood, H.P. and Nordberg, M.E. 1940. Method of Treating Borosilicate Glasses, U.S. Patent 2,215,039, September 17.

Hutchins III, J.R. and Harrington, R.V. 1966. Glass, in *Encyclopedia of Chemical Technology*, 2nd ed., Vol. 10, Wiley, New York, NY, p. 572.

Jantzen, C.M. 1992a. Nuclear Waste Glass Durability: I, Predicting Environmental Response from Thermodynamic (Pourbaix) Diagrams, *J. Am. Ceram. Soc.*, 75(9), 2433–48.

Jantzen, C.M. 1992b. Thermodynamic Approach to Glass Corrosion, Ch. 6 in *Corrosion of Glass, Ceramics and Ceramic Superconductors*, Clark, D.E. and Zoitos, B.K. (eds.), Noyes Publications, Park Ridge, NJ, pp. 153–215.

Jantzen, C.M. and Plodinec, M.J. 1984. Thermodynamic Model of Natural, Medieval and Nuclear Waste Glass Durability, *J. Non-Cryst. Solids*, 67, 207–23.

Katayama, J., Fukuzuka, M. and Kawamoto, Y. 1978. Corrosion of Heavy Crown Glass by Organic Acid Solutions, *Yogyo Kyokai Shi*, 86(5), 230–37.

Kinoshita, M., Harada, M., Sato, Y. and Hariguchi, Y. 1991. Percolation Phenomenon for Dissolution of Sodium Borosilicate Glasses in Aqueous Solutions, *J. Am. Ceram. Soc.*, 74(4), 783–87.

Koch, G.H. and Syrett, B.C. 1985. Progress in EPRI Research on Materials for Flue Gas Desulphurization Systems, in *Dewpoint Corrosion*, Holmes, D.R. (ed.), Ellis Horwood Ltd., Chichester, UK, pp. 105–24.

Kondrashov, V., Bezlyudnaya, V. and Zverev, Y. 1998. Effect of Regimes and Methods of Glass Forming on the Tin Content in Float Glass, *Glass Ceram.*, 55(5/6), 174–75.

Krajewski, A. and Ravaglioli, A. 1982. Lead-Ion Stability in Vitreous Systems, *J. Am. Ceram. Soc.*, 65(5), 265–69.

Lasaga, A.C. 1990. Atomic Treatment of Mineral-Water Surface Reactions, Ch. 2 in *Reviews in Mineralogy, Vol. 23: Mineral-Water Interface Geochemistry*, Hochella, M.F. Jr. and White, A.F. (eds.), Mineral. Soc. Am., Washington, DC, pp. 17–85.

Lead Glazes for Dinnerware, 1974. International Lead Zinc Research Organization Manual, Ceramics I, International Lead Zinc Research Organization and Lead Industries Association, New York, NY.

Lehman, R.L., Yoon, S.C., McLaren, M.G. and Smyth, H.T. 1978. Mechanism of Modifier Release from Lead-Containing Glasses in Acid Solution, *Ceram. Bull.*, 57(9), 802–5.

Liang, D-T. and Readey, D.W. 1987. Dissolution Kinetics of Crystalline and Amorphous Silica in Hydrofluoric-Hydrochloric Acid Mixtures, *J. Am. Ceram. Soc.*, 70(8), 570–77.

Lin, F.C. and Ho, S. -M. 1963. Chemical Durability of Arsenic-Sulfur-Iodine Glasses, *J. Am. Ceram. Soc.*, 46(1), 24–28.

Loehman, R.E. 1979. Preparation and Properties of Yttrium-Silicon-Aluminum Oxynitride Glasses, *J. Am. Ceram. Soc.*, 62(9–10), 491–94.

Meisel, W. 1998. Depth Profile of Tin in Float Glass—A CEMS Study, *Glastech. Ber. Glass Sci. Technol.*, 72(9), 291.

Metcalfe, A.G. and Schmitz, G.K. 1972. Mechanism of Stress Corrosion in E Glass Filaments, *Glass Technol.*, 13(1), 5–16.

Minami, T. and Mackenzie, J.D. 1977.Thermal Expansion and Chemical Durability of Phosphate Glasses, *J. Am. Ceram. Soc.*, 60(5–6), 232–35.

Mishra, A., Pecoraro, G.A., Paulson, T.E. and Pantano, C.G. 1998. Glass-Tin Interactions during the Float Glass Forming Process, *Cer. Trans.*, 82, 205–17.

Moore, H. 1956. The Structure of Glazes, *Trans. Brit. Ceram. Soc.*, 55, 589–600.

Muñoz, F., Pascual, L., Durán, A., Rocherullé, J. and Marhand, R. 2006. Oxidation Behavior of Li-Na-Pb-P-O-N Oxynitride Phosphate Glasses, *J. European Ceram. Soc.*, 26, 1455–61.

Murch, G.E. (ed.). 1985. *Materials Science Forum, Halide Glasses I & II*, Proceedings of the 3rd International Symposium on Halide Glasses, Rennes, France, Trans Tech Publications, Aedermannsdorf, Switzerland.

Newton, R.G. and Paul, A. 1980. A New Approach to Predicting the Durability of Glasses from Their Chemical Composition, *Glass Tech.*, 21(6), 307–9.

Oda, K. and Yoshino, T. 1991. Properties of Y_2O_3-Al_2O_3-SiO_2 Glasses as a Model System of Grain Boundary Phase of Si_3N_4 Ceramics: Part 2. Leaching Characteristics, *J. Ceram. Soc. Japan Int. Ed.*, 99, 1110.

O'Meara, C., Dunlop, G.L. and Pompe, R. 1991. Formation, Crystallization and Oxidation of Selected Glasses in the Y-Si-Al-O-N System, *J. European Ceram. Soc.*, 8(3), 161–70.

Pascual, M.J., Pascual, L., Valle, F.J. and Durán, A. 2003. Corrosion of Borosilicate Sealing Glasses for Molten Carbonate Fuel Cells, *J. Am. Ceram. Soc.*, 86(11), 1918–26.

Paul, A. 1977. Chemical Durability of Glasses, *J. Mater. Sci.*, 12(11), 2246–68.

Paul, A. 1982. Chemical Durability of Glass, Ch. 4 in *Chemistry of Glasses*, Chapman and Hall, New York, NY, pp. 108–47.

Paul, A. and Youssefi, A. 1978. Alkaline Durability of Some Silicate Glasses Containing CaO, FeO, and MnO, *J. Mater. Sci.*, 13(1), 97–107.

Paulson, T.E., Spear, K.E. and Pantano, C.G. 1998. Thermodynamic Analysis of the Tin Penetration Profile in High-Iron Float Glass, *Proc. ICG*, Am. Cer. Soc.

Plodinec, M.J., Jantzen, C.M. and Wicks, G.G. 1984. Thermodynamic Approach to Prediction of the Stability of Proposed Radwaste Glasses, in *Advances in Ceramics, Vol. 8: Nuclear Waste Management*, Wicks, G.G. and Ross, W.A. (eds.), Am. Ceram. Soc., Columbus, OH, pp. 491–95.

Pohlman, H.J. 1974. Corrosion of Lead-Containing Glazes by Water and Aqueous Solutions, *Glastech. Ber.*, 47(12), 271–76.

Priest, D.K. and Levy, A.S. 1960. Effect of Water Content on Corrosion of Borosilicate Glass, *J. Am. Ceram. Soc.*, 43(7), 356–58.

Rajaram, M. and Day, D.E. 1987. Nitrogen Dissolution in Sodium Alkaline-Earth Metaphosphate Melts, *J. Am. Ceram. Soc.*, 70(4), 203–7.

Rajaram, M. and Day, D.E. 1988. Preparation and Properties of Oxynitride Glasses Made from $27R_2O$-$20BaO$-$3Al_2O_3$-$50P_2O_5$ Glass, *J. Non-Cryst. Solids*, 102, 173–80.

Rana, M.A. and Douglas, R.W. 1961a. The Reaction between Glass and Water. Part 1. Experimental Methods and Observations, *Phys. Chem. Glasses*, 2(6), 179–95.

Rana, M.A. and Douglas, R.W. 1961b. The Reaction between Glass and Water. Part 2. Discussion of Results, *Phys. Chem. Glasses*, 2(6), 196–205.

Ravaine, D. and Perera, G. 1986. Corrosion Studies of Various Heavy-Metal Fluoride Glasses in Liquid Water: Application to Fluoride-Ion-Selective Electrode, *J. Am. Ceram. Soc.*, 69(12), 852–57.

Reidmeyer, M.R., Rajaram, M. and Day, D.E. 1986. Preparation of Phosphorus Oxynitride Glasses, *J. Non-Cryst. Solids*, 85, 186–203.

Reis, S.T., Karabulut, M. and Day, D.E. 2001. Chemical Durability and Structure of Zinc-Iron Phosphate Glasses, *J. Non-Cryst. Solids*, 292(1–3), 150–57.

Robinson, M. and Drexhage, M.G. 1983. A Phenomenological Comparison of Some Heavy Metal Fluoride Glasses in Water Environments, *Mater. Res. Bull.*, 18, 1101–12.

Sanders, D.M. and Hench, L.L. 1973. Mechanisms of Glass Corrosion, *J. Am. Ceram. Soc.*, 56(7), 373–77.

Schmidt, Yu. A. 1958. *Structure of Glass*, Vol. 1, translated from the Russian, Consultants Bureau, New York, NY.

Scholze, V.H., Helmreich, D. and Bakardjiev, I. 1975. Investigation of the Behavior of Soda-Lime-Silica Glasses in Dilute Acids (Gr), *Glass Tech. Ber.*, 48(12), 237–47.

Sebai, M., Sjöberg, J, Goursat, P., Nestor, E., Flynn, R., Ramesh, R. and Hampshire, S. 1995. Oxidation Behavior of Yttrium and Neodymium Oxynitride Glasses, *J. European Ceram. Soc.*, 15, 1015–24.

Sieger, J.S. 1975. Chemical Characteristics of Float Glass Surfaces, *J. Non-Cryst. Solids*, 19, 213–20.

Sieger, J.S. and Parsons, J.M. 1974. Tin Penetration Profiles in Float Glass, *Am. Cer. Soc. Bull.*, 53(4), 350; abstract of a presentation at the 76th Annual Mtg. of the Am. Cer. Soc.

Simmons, C.J. and Simmons, J.H. 1986. Chemical Durability of Fluoride Glasses: I, Reaction of Fluorozirconate Glasses with Water, *J. Am. Ceram. Soc.*, 69(9), 661–69.

Simmons, C.J., Azali, S. and Simmons, J.H. 1983. *Chemical Durability Studies of Heavy Metal Fluoride Glasses*, Extended Abstract #47, 2nd International Symp. on Halide Glasses, Troy, NY.

Simpson, H.E. 1958. Study of Surface Structure of Glass as Related to Its Durability, *J. Am. Ceram. Soc.*, 41(2), 43–49.

Swift, H.R. 1984. How Surface Chemistry Affects Float Glass Properties, *The Glass Industry*, 65(5), 27–30.

Tait, J.C. and Jensen, C.D. 1982. Effect of Zn(II) Ion Adsorption on the Durability of Sodium Borosilicate Glasses, *J. Non-Cryst. Solids*, 49(1–3), 363–77.

Takamori, T. and Tomozawa, M. 1978. HCl Leaching Rate and Microstructure of Phase-Separated Borosilicate Glasses, *J. Am. Ceram. Soc.*, 61(11–12), 509–12.

Taylor, P., Ashmore, S.D. and Owen, D.G. 1987. Chemical Durability of Some Sodium Borosilicate Glasses Improved by Phase Separation, *J. Am. Ceram. Soc., 70*(5), 333–38.

Tomozawa, M., Erwin, C.Y., Takata, M. and Watson, E.B. 1982. Effect of Water Content on the Chemical Durability of $Na_2O \cdot 3SiO_2$ Glass, *J. Am. Ceram. Soc., 65*(4), 182–83.

Tulliani, J.M., Audoin, L. and Goursat, P. 2002. The Role of Water Vapour on the Oxidation of Two Ln-Si-Al-O-N Glasses (Ln = Y, La), *J. Non-Cryst. Solids, 306*, 99–109.

Velez, M.H., Tuller, H.L. and Uhlmann, D.R. 1982. Chemical Durability of Lithium Borate Glasses, *J. Non-Cryst. Solids, 49*(1–3), 351–62.

Walters, H.V. 1983. Corrosion of a Borosilicate Glass by Orthophosphoric Acid, *J. Am. Ceram. Soc., 66*(8), 572–74.

Wang, F.F-Y. and Tooley, F.V. 1958. Influence of Reaction Products on Reaction between Water and Soda-Lime-Silica Glass, *J. Am. Ceram. Soc., 41*(12), 521–24.

Wang, T-J. 1997. Penetration of Tin in the Surface of Float Glass, *Glass Technol., 38*(3), 104–6.

White, J.E. and Day, D.E. 1994. Rare Earth Aluminosilicate Glasses for In Vivo Radiation Delivery, pp. 181–208 in *Rare Elements in Glasses*, Shelby, J.E. (ed.), Key Engineering Materials, Vols. 94–95, Trans Tech Pub., Switzerland.

White, G.S., Greenspan, D.C. and Freiman, S.W. 1986. Corrosion and Crack Growth in 33% Na_2O-67%SiO_2 and 33% Li_2O-67% SiO_2 Glasses, *J. Am. Ceram. Soc., 69*(1), 38–44.

Wilder Jr., J.A., Day, D.E. and Bunker, B.C. 1983. Preparation and Properties of Phosphorus Oxynitride Glasses, *Glastech. Ber., 56K*, 845–49.

Williams, K., Johnson, C., Greengrass, J., Tilley, B., Gelder, D. and Johnson, J. 1997. Tin Oxidation State, Depth Profiles of Sn^{2+} and Sn^{4+}, and Oxygen Diffusivity in Float Glass by Mossbauer Spectroscopy, *J. Non-Cryst. Solids, 211*(1–2), 164–72.

Wood, S. and Blachere, J. R. 1978. Corrosion of Lead Glasses in Acid Media: I, Leaching Kinetics, *J. Am. Ceram. Soc., 61*(7–8), 287–92.

Wu, C.K. 1980. Nature of Incorporated Water in Hydrated Silicate Glasses, *J. Am. Ceram. Soc., 63*(7–8), 453–57.

Yoon, S.C. 1971. Lead Release from Glasses in Contact with Beverages, MS thesis, Rutgers University, New Brunswick, NJ.

Yoon, S.C. 1973. Mechanism for Lead Release from Simple Glasses, Univ. Microfilms Int. (Ann Arbor, MI), Order No. 73-27,997, *Diss. Abstr. Int., B34*(6), 2599.

17

Composite Materials

The whole is most always better than the sum of the parts.

Anonymous

Introduction

Although the term *composite* historically meant any product made from a combination of two (or more) materials, the modern meaning is less broad in scope. In general a composite is manufactured in an attempt to obtain the best properties of two materials, or at least to capture a specific property of each material that is potentially better in the composite. It is also possible for the composite to have a particular property that neither component exhibited individually. According to Holmes and Just 1983 a true composite is where distinct materials are combined in a nonrandom manner to produce overall structural characteristics superior to those of the individual components. Although, in a very broad sense, products such as glazed ceramic tile, enameled metal, and ceramic-coated metal (e.g., thermal barrier coatings) could be considered composites, they will not be considered as such here. Only those materials where a substantial intermixing of the different materials exists on a microscopic scale will be considered composites.

The concept of composite materials is not a new idea and is definitely not limited to ceramics. Nature has provided us with several excellent examples of composite materials. Wood is a composite of cellulose fibers contained in a matrix of lignin. Bone, another example, is composed of the protein collagen and the mineral apatite. In all these materials the result is a product that is lighter and stronger than either of the components individually. Because of this, they can be used in more severe environments (e.g., space exploration). A list of the more desirable properties of a composite is given in Table 17.1. Many materials contain additions used as sintering aids. These materials in general are not called composites; however, some contain more than 10 wt or vol% of these additives. The early work on silicon nitride ceramics investigated materials containing large portions of sintering aids; however, they were never called composites. In contrast, materials with additions used for reinforcement of mechanical properties, although called composites, contained as little as 2 vol% of the additive. So the question arises, how much of a second phase

TABLE 17.1

Desirable Properties of Composites

1. Very strong in tension
2. Very light weight
3. Very stiff
4. High strength-to-weight ratio
5. High stiffness-to-weight ratio
6. Good corrosion resistance
7. Low thermal expansion
8. High abrasion resistance
9. High melting point

is required for a material to be called a composite?* The answer to this question is related more to why the second phase was added. If added to improve properties, especially mechanical, it should be called a composite. If added for processing reasons, it is not called a composite. Both, however, fall under the definition given above by Holmes and Just 1983. In a very broad sense, all engineering materials are composites of one kind or another.

The composite matrix and the reinforcement, quite often fibrous, provide two different functions. The reinforcement is most often a discontinuous phase whether it be a fibrous material or a particulate material. It is important that the reinforcement be discontinuous, especially if it is a ceramic, so that cracks will not be able to propagate through it. The matrix must not damage the reinforcement and it must transmit any stresses to the reinforcement. Thus the adhesion of the matrix to the reinforcement is of prime importance for mechanical integrity and is the region of greatest importance related to corrosion. Since it is necessary to have weak interfaces to maximize toughness (i.e., resistance to crack propagation), the development of optimum fiber/matrix interfaces is quite difficult. To obtain these optimum characteristics it is sometimes required to coat the reinforcement fibers with various materials to obtain the proper debonding, sliding, and/or reaction characteristics. Fibers that do not debond do not enhance toughening and lead only to increased brittle fracture of the composite, Aveston and Kelly 1973, Curtin 1991, Curtin 1994, Marshall and Evans 1985, Davidge and Briggs 1989, and Evans and Marshall 1990.

A recent development in composites is that of a nano-sized second phase or reinforcement material. The second phase particles are generally less than 300 nm and are present in amounts equal to 1–30 vol%. These new composites unfortunately have been called *nanocomposites*.

Before going into the specifics of corrosion of composite materials, a few words must be said about those materials that have been called *cermets*. Historically the term *cermet* was derived to cover those materials composed

---

* A similar problem exists with the chemistry of metals and alloys.

of cobalt-bonded tungsten carbide and used as cutting tools. Since cermets contain both ceramics and metals, some confusion has existed in the literature as to an exact meaning. The term, however, has been used to cover a broad list of materials. It appears that the ceramic community confines cermets to essentially cutting tool materials, whatever the matrix or reinforcement, whereas the metals community confines cermets to only those materials with a metal matrix. Since the broader concept of composites includes those materials called cermets, only the term *composite* will be used in the discussion below.

The actual corrosion of composite materials quite often begins with reaction of the reinforcement material and especially with any interface material (called the *interphase*) used to coat the reinforcement for debonding. One property that exacerbates this is a mismatch in thermal expansion coefficients between the reinforcement and the matrix, leading to microcracks. These microcracks allow the ingress of corrosive gases (e.g., oxygen). Courtright 1993 has given the value of 10^{-12} $gmO_2/cm \cdot s$ for the limit of oxygen ingress that causes nonoxide fiber deterioration. Microcracks are also quite often a product of sample preparation techniques and thus great care must be used in cutting and grinding/polishing samples for testing. If the composite is cut or machined, any exposed fiber reinforcement will be susceptible to attack by the environment. Because of this inherent problem, protective coatings are often applied to the exterior surfaces. Actually the whole corrosion process of composite materials is not unlike that of other polyphase ceramic materials where the grain boundary phase is the first to corrode. See Chapters 15 and 16 for corrosion of specific crystalline and glassy materials. A complete understanding of all the phases that make up the microstructure of the composite must also be known for an accurate interpretation of any corrosion. For example, Munson and Jenkins 1997 reported that their samples were actually attacked internally by molten metal from a small amount of free aluminum present as a residue during the manufacture of Dimox™[*] (a melt-infiltrated alumina). Actually a large amount of the literature on composites is concerned with an evaluation of the internal reactions that take place among the various reinforcement, interphase, and matrix materials. The time-dependent loss of strength due to the corrosive nature of moist environments at room temperature is a major concern for composites containing glass or glass-ceramics as either the matrix or the reinforcement, Wu et al. 1995. As temperatures are increased, the concern shifts toward oxidation problems associated with nonoxide materials. (See the discussion about nitrides and carbides in Chapter 15 and also Chapters 18 and 19 for more details of oxidation and its effects upon the properties of nonoxides.)

---

[*] DIMOX™ (directed metal oxidation) is the name given to composites manufactured by a process developed by Lanxide Corp., Newark, DE, in 1986.

TABLE 17.2

Acronyms Used in the Discussion of Composites

BMC—brittle matrix composite
C/C—carbon-carbon composite
CCC—carbon-carbon composite
CFCC—continuous fiber ceramic composite
CFCMC—continuous fiber ceramic matrix composite
CFRC—continuous fiber reinforced ceramic
CFRCMC—continuous fiber reinforced ceramic matrix composite
CFRP—carbon fiber reinforced plastic, polymer, or polyester
CMC—ceramic matrix composite
CPC—ceramic particulate composite
FRCMC—fiber reinforced ceramic matrix composite
FRP—fiber reinforced plastic, polymer, or polyester
GCMFC—glass-ceramic matrix fiber composite
GFRP—glass fiber reinforced plastic
GRC—glass fiber reinforced cement
GRP—glass reinforced plastic, polymer, or polyester
IMC—intermetallic matrix composite
MMC—metal matrix composite
PMC—polymer matrix composite
RCC—reinforced carbon-carbon composite
SFRC—short fiber reinforced ceramic
TMC—titanium metal matrix composite
WRA—whisker reinforced alumina

With the advancement of the development of composites there is an increasing number of acronyms with which one must contend. To aid the reader a list is given in Table 17.2 of the most common acronyms.

Reinforcement

Fibers

Various types of materials have been used as the fibrous reinforcement. These include various glasses, metals, oxides, nitrides, carbides, and carbon and either in the amorphous or crystalline state. The surface chemistry and morphology of fibers is very important in determining their adherence to the matrix. Fiber internal structure and morphology determines the mechanical strength. A tremendous amount of literature is available that discusses the degradation of mechanical properties as temperatures are increased in various atmospheres; however, there is very little interpretation of any corrosion mechanisms that may be involved. Although many composites are classified as continuous-fiber reinforced, some composites contain fibers that are actually not continuous but of a high aspect ratio (i.e., length-to-width). The actual matrix material will determine the aspect ratio required to obtain a certain set of properties. Thus the term *high aspect ratio* is a relative term.

Boron fibers can generally be heated in air to temperatures of about 500°C without major strength deterioration. Above 500°C the oxide that formed at lower temperatures becomes fluid, increasing the oxidation rate and drastically reducing the strength, Wu et al. 1995. Galasso 1989 discussed the benefits of coating boron fibers by either SiC or by nitriding the surface. The SiC coating was more protective than the nitride with strength retention even after 1000 hrs at 600°C in air. Boron carbide (B_4C) is stable to 1090°C in an oxidizing atmosphere, whereas boron nitride is stable to only 850°C.

Carbon or graphite fibers have been used since the early 1970s as reinforcement for composites. Strength loss due to oxidation occurs at temperatures above 500°C in air. An interesting structural feature of carbon fibers is that they have a relatively large negative axial thermal expansion coefficient.

Glass fibers generally are used as reinforcement for composites that are to be used at low temperatures (i.e., < 500°C) due to the softening of glasses at elevated temperatures. These composites are generally of the polymer matrix type and are used for marine or at least moist environments. It is well known that glass is attacked by moist environments with the specific mechanism dependent upon the pH (see Chapter 16). It has been shown by Metcalfe and Schmitz 1972 that borosilicate glass fibers when exposed to moist ambient environments developed surface tensile stresses caused by exchange of alkali for hydrogen sufficient to cause failure.

A large portion of the CMC today contains SiC fiber reinforcement. This is mainly due to the excellent properties of SiC—low reactivity to many matrix materials, its strength at elevated temperatures, and its oxidation resistance. It is this latter property (i.e., oxidation resistance) that generally causes deterioration in these materials. SiC will oxidize readily when heated to temperatures greater than 1000°C. As discussed in Chapter 2, at low partial pressures of oxygen active corrosion takes place with the formation of gaseous products of CO and SiO. At higher partial pressures passive oxidation occurs with the formation of CO and SiO_2 that may be protective if cracks do not form. The formation of cracks is dependent upon the heat treatment and whether the oxide layer is crystalline or amorphous. These reactions generally result in the decrease of fiber strength. Nicalon™ fiber,[*] being formed by the pyrolysis of organometallics, actually contains some remnant oxygen (~9%) and carbon (~11%) that will affect the subsequent oxidation of the fiber. Two different grades of Nicalon™ fiber have been examined by various investigators, Clark et al. 1985, Clark et al. 1986, and Sawyer et al. 1986. Clark et al. 1985 reported these fibers to exhibit weight losses of 13 and 33%

---

[*] Nicalon™ (Nippon Carbon Co., Tokyo, Japan) is mostly amorphous.

after being treated in argon at 1400°C. Both grades of fiber gained weight (on the order of 2–3%) when treated in flowing wet air at 1000, 1200, and 1400°C. As-received Nicalon™ fibers have protective sizing (i.e., polyvinyl acetate) on their surfaces. When heated in air this sizing will burn off at temperatures between 250 and 500°C. At temperatures above about 1250°C the SiC_xO_y amorphous phase contained in these fibers decomposed to SiO and CO, Filipuzzi et al. 1994.

Titanium nitride (TiN) resists attack from iron or nickel aluminides better than does SiC and thus is a better reinforcement for these metal matrix composites, Nolan et al. 1991.

Fiber Coatings or Interphases

Protective coatings such as graphite or BN (also called *interphases*), in addition to providing proper debonding and pullout, Caputo et al. 1985 and Fareed et al. 1993, are used to provide some degree of oxidation resistance, Studt 1991 and Bender et al. 1986, for fibers such as SiC. The first step in degradation is the oxidation of the carbon interface that subsequently allows oxygen access to the entire reinforcement fiber surface, Kerans et al. 2002. A self-sealing layer of silica forms at temperatures above 1000°C that diminishes considerably the diffusion of oxygen through the interface to the fiber.

Bender et al. 1986 concluded that the BN protects the SiC fiber from the matrix, since BN will not react with SiO_2, which is generally present on the surface of the fibers. Boron nitride-coated mullite, carbon, and SiC fibers were tested in a mullite matrix with varying degrees of success by Singh and Brun 1987. Boron nitride–coated SiC fibers have shown an improvement over carbon-coated fibers with an increase of about 100–200°C in composite embrittlement (for a discussion concerning embrittlement see the section below on oxide-matrix composites) temperatures, French 1995.

Since CVD-deposited BN coatings contain some oxygen impurity, a thin silica layer forms on the SiC fibers, followed by internal oxidation and formation of a borosilicate glass, Kerans et al. 2002. At temperatures between 600 and 900°C BN oxidation was rapid but the formation of the borosilicate glass was slow, presenting the problem of continued oxidation. At lower temperatures BN oxidized to the volatile boria. Ogbuji 2003 reported that the BN first oxidized to B_2O_3, which then dissolved some of the SiC fiber and matrix, forming a borosilicate liquid. If any moisture were present, the boria may be volatilized by hydrolysis, releasing $B(OH)_4$ gas. This reaction resulted in a silica residue that cemented the fibers together, embrittling the composite. BN has been reported to react directly with H_2O at moisture levels greater than 10% to form the volatile species HBO_2, H_3BO_3, and $H_3B_3O_6$, Kerans et al. 2002. At lower moisture contents (20 ppm H_2O), however, the water reacts with the borosilicate glass present to form volatile HBO_3. At elevated temperatures even small amounts of moisture ($pH_2O \approx 0.003$ atm) are detrimental, removing boron from the glass, Zawada et al. 2003.

Since some matrices are grown *in situ*, techniques to coat fibers becomes problematic. A combination coating of BN and SiC was developed by Fareed et al. 1993 to eliminate the undesirable reaction of molten aluminum in contact with Nicalon™ fibers forming alumina and aluminum carbide during the directed metal oxidation method (at 900 to 1000°C) of forming an alumina matrix. When used alone as a coating, BN oxidation inhibited complete oxidation of the aluminum. In combination with SiC, however, Fareed et al. believed that any oxidation of BN led to the formation of boria glass that acted as a sealant to any microcracks, thus minimizing oxygen ingress and protection of the composite. The SiC outer coating protected the BN inner coating during growth of the matrix.

In an effort to find an interphase or coating for alumina and mullite fibers, Cooper and Hall 1993 developed a synthetic fluorophlogopite,[*] based upon their geochemical approach, that when reacted with alumina formed an intermediate spinel phase that was stable after heating to 1200°C in air for 150 hrs. Thus by coating alumina fibers with spinel and then using the fluorophlogopite as an interphase, an alumina matrix composite proved successful. Above about 1280°C the alumina reaction with fluorophlogopite produced forsterite, leucite, and spinel along with the volatile fluorides SiF_4, AlF_3, and KF (see equation below), making the spinel coated alumina fiber/ fluorophlogopite laminate unstable at those high temperatures.

$$4KMg_3AlSi_3O_{10}F_2 + 2Al_2O_3 \rightarrow 5Mg_2SiO_4 + 3KAlSi_2O_6 + 2MgAl_2O_4 + AlF_3 + SiF_4 + KF \tag{17.1}$$

Reactions between mullite and fluorophlogopite formed cordierite in addition to the phases mentioned above. This was not successful as a mullite fiber composite since the cordierite allowed potassium diffusion from the fluorophlogopite, continually deteriorating the mullite. In the alumina fiber case, the spinel coating acted as a barrier to potassium diffusion.

Cooper and Hall reported that reaction 17.1 occurred at temperatures above 1230°C in flowing dry argon, although thermodynamic calculations indicated that the reaction proceeded only after the temperature reached 1279°C. This was attributed to the partial pressures of the gaseous phases not summing to 1 atm during the experiment in flowing argon.

Since carbon oxidizes readily above 800°C, carbon fibers must be coated to improve their oxidation resistance. Wang et al. 2005 discussed the silica sol infiltration of a 3-D braided carbon fiber that formed a SiC coating upon carbothermic reduction at temperatures of 1200–1500°C. Dynamic TGA results indicated that the SiC-coated carbon fiber retained 28% of its weight once the temperature reached 940°C. Xiao et al. 2004 described one technique to coat carbon fibers with TiC/SiC mixed crystals. A polytitanocarbosilane (PTC)

---

[*] Cooper and Hall use the term *fluorophlogopite* interchangeably with the terms *mica*, *fluoromica*, and *fluorophyllosilicate*, which may cause some confusion unless the reader is well versed in mineralogy.

was used as the precursor for solution coating of the carbon fibers. In a comparison weight loss test among PTC-, SiO_2-, and SiC-coated carbon fibers oxidized at 800°C in air, the PTC-coated fibers performed the best.

The oxidation of carbon fibers that are coated progresses through the following steps according to Xiao et al.:

1. oxygen diffusion through surface coating,
2. oxygen diffusion through fissures and cracks,
3. reaction at surface active site,
4. oxidation at coating/fiber interface,
5. diffusion of CO/CO_2 out through fissures and defects,
6. diffusion of CO/CO_2 through the coating, and
7. CO/CO_2 leaving the surface to finish process.

Many other materials have been investigated as fiber coatings; however, most of the literature is concerned with only their mechanical properties and not their corrosion or oxidation characteristics. One material that has received a lot of attention is monazite ($LaPO_4$). A monazite coating on several commercially available SiC fibers was evaluated by Hay et al. 2006. A good review article on the properties and design of composite interfaces is that by Faber 1997.

Particulates

For a discussion of the various mechanisms involved in toughening composites when particulates are used as the reinforcement, one should read the article by Rice 1990. Particulate reinforcement can be oxides, carbides, nitrides, borides, or even metals. Some of the common materials used are listed below:

$$Al_2O_3, \ ZrO_2, \ SiC, \ TiC, \ BN, \ TiB_2, \ NbB_2, \ Co$$

The corrosion of most of these materials has been discussed in Chapter 15 (Specific Crystalline Materials) and will not be repeated here.

Actually most of the ceramic products manufactured today could be considered ceramic particulate reinforced ceramic matrix composites. For example, one product not generally considered a composite by the ceramics community is MgO:carbon refractories. This product contains a substantial quantity of carbon particles in an effort to improve the corrosion resistance of the MgO in molten metal applications. In addition to the obvious oxidation of carbon to monoxide and/or dioxide, carbon will also react with MgO at temperatures above 1400°C, forming magnesium vapor and carbon monoxide.

Other Structures

A reinforcement structure that is neither fibrous nor particulate is one termed a *3D network*. A technique called *squeeze casting* is used to infiltrate the 3D network with the matrix metal. The interfaces of a SiC 3D network reinforced copper alloy matrix was investigated by Xing et al. 2005. They found an interface of Cu_3Si between the 3D SiC and the Cu matrix formed during squeeze casting at 1150°C. Their conclusion was that some remnant free silicon reacted with the copper to form the interphase, since a Si particulate/Cu composite formed Cu_3Si but a SiC particulate/Cu composite did not.

Ceramic Matrix Composites (CMC)

Ceramic matrix composites are characterized by a high modulus of elasticity, excellent high temperature and corrosion resistance, but generally poor crack propagation resistance. The composite systems that have probably received the most attention are those of SiC or carbon fiber–reinforced SiC. The major problem with these materials is one of oxidation of either the carbon or the SiC. Below about 600°C oxidation is generally not a problem. Above 1000°C the oxidations of both carbon and SiC are rapid. It is the temperature range of 600 to 1000°C that is the most difficult where the carbon oxidation can be rapid but the SiC is relatively inert, Cawley et al. 1993. Any microcracks or pores can allow ingress of oxygen for continued oxidation of the carbon.

Stress corrosion cracking (discussed in Chapter 18) of the matrix is a major problem in CMC. Another problem is that of *oxidation embrittlement* that can manifest itself through the oxidation of the interface between the fiber reinforcement and the matrix subsequently causing a strong bond between the two, leading to embrittlement. Embrittlement may not be noticed when samples are tested in flexure due to load redistribution, thus requiring that samples be tested in tension, Heredia et al. 1995. Zawada et al. 2003 compared the high temperature fatigue durability of several CMCs reinforced with SiC fibers with and without interface coatings to intermittent exposure to moisture and salt fog at high temperatures. The matrix materials included Al_2O_3, carbon, SiNC, and aluminosilicate. Salt fog exposure decreased the durability. They concluded that fiber coatings of BN/SiC or BN/Si_3N_4 were of no value in protecting the CMC from decreased durability. If matrix cracks are present, constant contact with moisture could increase the loss of the BN interphase, producing gradual degradation with time.

Oxide-Matrix Composites

Al_2O_3-Matrix Composites

A BN/SiC-coated Nicalon™ fiber/Al_2O_3 composite has been reported by Heredia et al. 1995 to become embrittled when heated to temperatures between 650 and 850°C. The tensile strength after heat treatment for 24 hrs was reduced by one-half. Only the outer regions of the ~3 mm thick samples were fully embrittled.

A 21 vol% SiC in alumina composite was reported by Borom et al. 1988a to form a reaction zone upon oxidation at 1530°C for 150 hours that contained mullite and an amorphous aluminosilicate phase containing bubbles from the formation of CO. The SiO_2 formed by the oxidation of the SiC reacted with the alumina matrix to form the mullite. It is important that the formation of silica in the outer layer is sufficient for complete conversion of the alumina to mullite, Borom et al. 1988b. Insufficient silica causes a rigid scale that delaminates. Too much silica forms a scale containing mullite and silica on an alumina substrate that may also delaminate due to expansion mismatch during thermal cycling. A matrix of mullite works much better than alumina, since the scale is more compatible with the substrate, both containing mullite, and thus forms a protective layer. Luthra 1987 reported that the products of reaction of SiC with alumina should be mullite and alumina when the SiC content is below 24.4 vol% and silica and mullite when it is greater than 24.4 vol%. In practice this limit will vary due to mullite forming over a range of compositions.

A TiN/Al_2O_3 composite was reported by Mukerji and Biswas 1990 to exhibit linear oxidation kinetics above 820°C after a short (<120 min) parabolic induction period. The change from parabolic to linear kinetics was reported to be due to the difference in specific volumes between TiN and TiO_2 that caused an expansion of the oxidized layer forming cracks, which allowed oxidation to continue. The rutile that formed above 820°C was reported to grow epitaxially with a preferential growth direction of [211] and [101]. At 820 and 710°C this oriented growth was not present. Tampieri and Bellosi 1992 reported this oriented growth to occur in the [221] and [101] directions and only above 900°C. Contrary to Mukerji and Biswas, Tampieri and Bellosi reported parabolic growth between 900 and 1100°C for times up to 1200 minutes. These differences must be attributed to differences in starting materials and experimental conditions, since the authors did not report any specific reasons that one may assign to the variation in results.

TiN decomposes to form titanium oxides and aluminum titanates at temperatures in excess of 1550°C, Revankar et al. 1995. TiN will also react with alumina to form titanium oxides and aluminum titanates at temperatures as low as 1450–1500°C. Therefore during processing by hot pressing the temperature must be kept below 1500°C and the pressure must be high.

The oxidation at 1500°C of TiC-containing (25 vol%) alumina matrix composite has been reported to form Al_2TiO_5 as the reaction product by Borom et al. 1988(b). Approximately a 30 vol% expansion accompanied this reaction that caused delamination of the oxide reaction product layer.

In a previous study, Borom et al. 1988a reported the oxidation at 1520°C of $MoSi_2$ (10 vol%) dispersed within a matrix of alumina to form a reaction layer of mullite and volatile MoO_3 that completely escaped. It was suggested that this reaction layer contained an interconnected network of porosity through which the MoO_3 escaped, although no evidence of such porosity was given. Linear growth kinetics was reported for the formation of this nonprotective layer of mullite. A unique-appearing periodic change in density (porosity) was developed at about 200 µm intervals within the mullite reaction layer along with a slight bulging of the layer, both of which were reported to be due to volume changes during reaction and thermal expansion mismatch among the phases present during cooling.

Several investigators have shown that by adding particulate nickel to alumina a composite with improved mechanical properties can be obtained. Volume percentages ranging from 5 to 35% Ni have been studied. Although the oxidation resistance of nickel is quite good, compared to other metals, its resistance is decreased when added to alumina. Wang et al. 2003 studied the oxidation of Ni-toughened alumina from 1000 to 1300°C for up to 1000 hrs with nickel contents ranging from 5 to 15 vol%. They found that oxygen diffused into the composite and nickel diffused out to the surface forming NiO first and then a dense $NiAl_2O_4$ spinel on the surface.

One of the most promising alumina matrix composites is a melt-grown composite from the $Al_2O_3/Y_3Al_5O_{12}$ eutectic composition, Otsuka et al. 2005. This technique allows for the crystallization of two phases simultaneously. Eutectic solidification often results in the formation of lamellar or interconnected microstructures with minimum free energy that possess interesting mechanical properties. The microstructure was approximately 52/48 volume ratio of $Al_2O_3/Y_3Al_5O_{12}$ (YAG[*]). Corrosion studies were carried out in moist oxygen at 1700°C for 200 hrs and in combustion gases at 1500°C/30 min. Tests results indicated no degradation in mechanical properties although some microstructural changes occurred including a slight weight gain due to incorporation of OH ions in the eutectic lattice. Exposure to the combustion gases at 1500°C resulted in only slight discoloration due most likely to the incorporation of some carbon.

Although the reinforcement is the problem area most of the time, the alumina matrix can also deteriorate under some conditions. One of those is in an environment of hydrogen. The upper limit of usage has been estimated to be between 1200 and 1300°C by Nelson 1993.

Other Oxide-Matrix Composites

Glass and Glass-Ceramics (Alkali and Alkaline-Earth Aluminosilicates)

Probably the first glass matrix composites were those reinforced with graphite fibers, Sambell et al. 1972. Due to the oxidation problems with graphite other fiber reinforcements were developed, predominantly Al_2O_3 and SiC. Due to the inherent softening of glasses, glass matrix composites are relegated to temperatures that are generally less than 1000°C. To improve upon this temperature limitation, researchers are now investigating glass-ceramics

---

[*] $Y_3Al_5O_{12}$ has the garnet structure, thus the YAG designation.

that are formed in the glassy state and then subsequently crystallized to obtain maximum temperature stability.

Nicalon™ SiC fibers (i.e., generally any polymer-derived fibers of the Si-C-O and Si-C-N-O types) have been reported by several investigators to react with various alkali and alkaline-earth aluminosilicate glass matrices (LAS,[*] Brennan 1986; CAS,[†] Cooper and Chyung 1987 and Bonney and Cooper 1990; BaMAS,[‡] Chaim and Heuer 1991) during fabrication to form carbon at the fiber/matrix interface. Even though the oxygen contained within the as-fabricated SiC fibers can react to form SiO, SiO_2, C, CO, or CO_2 depending upon which of the following equations:

$$SiC + 3/2\ O_2 \rightarrow SiO + CO_2 \tag{17.2}$$

$$SiC + 3/2\ O_2 \rightarrow SiO_2 + CO \tag{17.3}$$

$$SiC + O_2 \rightarrow SiO_2 + C \tag{17.4}$$

$$SiC + 2CO \rightarrow SiO_2 + 3C \tag{17.5}$$

or

$$C + 1/2\ O_2 \leftrightarrow CO \tag{17.6}$$

is operative, it is also possible for the reaction to be caused by oxygen and/or CO that is dissolved in the glass or glass-ceramic matrix material, Shin et al., 1993. Pantano et al. 1993 developed a model, based upon stoichiometric SiC, that explained many experimental observations when Nicalon™ fiber was used, showing that it was the effective pO_2 of the glass that was the driving force for the reaction. Longtime exposure in air eventually oxidized the carbon and the SiC fibers, resulting in deterioration of the composite.

Glass-ceramics, being materials composed of several phases, are considered composites by many investigators and by all definitions they should be. These are materials that are formed as a glass and then either heated to an intermediate temperature or held at some intermediate temperature upon cooling to crystallize the glass either completely or partially. These are composites where the reinforcement is generally particulate and crystalline and the matrix is the remaining glassy phase. Therefore any corrosion that takes place attacks the matrix glassy phase first and generally more severely, McCracken et al. 1980.

A lithia-aluminosilicate glass-ceramic containing a small percentage of Nb_2O_5 was reported by Prewo et al. 1986 to react with the SiC reinforcement

---

[*] LAS = lithia-aluminosilicate.
[†] CAS = calcia-aluminosilicate.
[‡] BaMAS = baria-magnesia-aluminosilicate.

fibers, forming particles of NbC on the surface of the fibers. A very thin carbon-rich layer formed between the SiC fibers and the NbC that contributed to the excellent toughness and crack deflection of these composites.

The reactions between Nicalon™ fiber and Solaramic™* glass (a baria and chromia silicate) were examined by Mendelson 1985. He was concerned that reaction between the two might cause sufficient fiber/matrix bonding to degrade the mechanical properties. Based upon the assumption that glass constituents of less than ~ 5 wt% would not degrade the fibers or the composite, Mendelson calculated the free energy of reaction at 1350°C and 1 atm pressure between the remaining constituents and SiC. His calculations indicated that the formations of Cr_3C_2, $CrSi_2$, SiO_2, and BaC_2 were probable and that SiO formation was not. Upon heat treatment of composite samples at 1150 and 1350°C Mendelson found that Cr_3C_2, $CrSi_2$, and SiO_2 did indeed form and that barium diffused into the fibers at the higher temperature, causing embrittlement and degraded strengths. Mendelson attributed this incompatibility to the excess carbon and oxygen in the fibers and to the Cr- and Ba-containing glass matrix. The diffusion of barium into Nicalon™ fibers was reported also by Herron and Risbud 1985. Their matrix glass was a BaSiAlON. Diffusion of the barium extended into the fibers to a depth of ~15 μm. Aluminum was also found to diffuse into the fibers but to a lesser depth (~2–3 μm). Herron and Risbud offered no explanation as to the effect of this diffusion of Ba and Al upon the mechanical properties of the composites.

Heredia et al. 1995 reported the embrittlement of a carbon/boron-coated Nicalon™ fiber/magnesium aluminosilicate composite when heat-treated to temperatures between 500 and 750°C. Similar results were reported by Wetherhold and Zawada 1991 for a Nicalon™ fiber/alkaline-earth aluminosilicate resulting from the oxidation at 650°C in air of the carbon interface and subsequent embrittlement of the composite. This embrittlement was less severe if heat treatment was performed at a higher temperature (800 or 850°C for 15 min). This was attributed to the higher temperature causing the formation of a glassy phase, presumably amorphous SiO_2 that flowed and sealed the matrix inhibiting oxygen ingress, as reported in an earlier study by Bischoff et al. 1989. Wetherhold and Zawada suggested that a short high-temperature heat treatment could protect these composites against embrittlement when exposed to lower temperatures.

The reaction of various small quantities of materials contained within a composite must not go unexamined, especially when reinforced with SiC. Pannhorst et al. 1990 found that a composite containing a small amount of TiO_2 formed TiC during fabrication. This reaction was sufficiently severe to degrade the SiC fibers, causing the composite to exhibit bend strengths of about 100 MPa.

---

* Solaramic™, Arcilla Research, Epen, the Netherlands.

Mullite, Spinel, Titania, and Zirconia

The oxidation at temperatures between 1310 and 1525°C of a 30 vol% SiC in a mullite composite was reported by Borom et al. 1988b to obey parabolic kinetics and form a reaction layer of mullite and an amorphous aluminosilicate phase containing bubbles from CO evolution. Similar results were obtained by Luthra and Park 1990 and Hermes and Kerans 1988. By changing the matrix from mullite to a strontium-aluminosilicate (SrO · Al$_2$O$_3$ · 2SiO$_2$) phase, Borom et al. also showed that the presence of alkaline earth cations increased the oxidation rates by one to two orders of magnitude, presumably due to the formation of nonbridging oxygens in the silicate glass that allowed much higher transport rates.

A 50 vol% SiC particulate mullite matrix was found to exhibit excellent oxidation resistance in flowing air at temperatures of 1515, 1620, 1650, 1675, and 1700°C after 100 hrs by Tian and Shobu 2003. The surface oxide was determined to be cristobalite below 1650°C. Above that temperature an aluminosilicate glass formed.

In a composite containing 18.5 vol% SiC in a matrix of mullite (40 vol%), alumina (26 vol%), zirconia (12 vol%), and spinel (3.5 vol%), Baudin and Moya 1990 reported passive oxidation at 1200, 1300, and 1400°C in air. No weight changes were reported for 800 and 900°C and minimal changes were noted at 1000 and 1100°C. The oxidized layer contained cordierite along with mullite, zirconia, and alumina at 1200 and 1300°C. At 1400°C, mullite and zircon were detected along with a viscous amorphous phase. The silica oxidation product apparently reacted with the free alumina and zirconia present to form additional mullite and zircon.

Hermes and Kerans 1988 found that the magnesium from a spinel matrix composite containing 30 vol% SiC heated to 1250°C in air diffused to the surface faster than the aluminum or silicon, forming an outer layer of MgO over a dense intermediate layer of spinel. This is an example of the demixing of a mixed oxide (i.e., spinel) along an oxygen chemical potential gradient. Since diffusion of magnesium and aluminum is much greater than that of oxygen, the metals diffuse from low partial pressures of oxygen to high partial pressures. A third innermost layer was composed of porous cordierite. None of the scale layers contained SiC. At a temperature of 1450°C, the nonprotective scale was essentially one porous layer composed of cordierite and small grains of spinel. Panda and Seydel 1986 found that a spinel (prepared from hydrated magnesium nitrate and aluminum hydroxide) matrix reacted with SiC fibers (Versite-grade single crystal fibers) to form a composition they assigned to sapphirine. They concluded that the SiC fibers oxidized during the calcining of the composite forming surface silica that reacted with the spinel to form the sapphirine. Although the identification of sapphirine was not conclusive (undetected by XRD), Panda and Seydel based their result upon the chemistry obtained from energy dispersive spectroscopy. If one were to examine the MgO - Al$_2$O$_3$ - SiO$_2$ phase diagram, it should be apparent that, if the spinel were stoichiometric, cordierite should have formed. A reasonable amount of spinel nonstoichiometry (i.e., Al_2O_3-rich) would be required for any sapphirine to form. Thus the reaction product that Panda and Seydel thought to be sapphirine was most likely cordierite.[*]

Molybdenum disilicide is the most oxidation resistant of all the silicides. Borom et al. 1988a reported the oxidation of $MoSi_2$ (8 vol%) dispersed within a matrix of mullite at 1500°C for 6 hours. At the low partial pressure of oxygen near the original surface, the silicon from the $MoSi_2$ was selectively oxidized, similar to that reported by Fitzer 1990, leaving behind a region of metallic Mo and silica dispersed within the mullite matrix. The addition of silica to this region increased the optical transparency that was very noticeable with examination by optical microscopy. As one proceeded toward the surface with increasing oxygen pressure, the molybdenum was oxidized first to MoO_2 and then to MoO_3. The additional silica that formed was incorporated into the matrix by dissolution and diffusion in the liquid state. Since the MoO_3 that formed was volatile, it mechanically forced this aluminosilicate liquid toward the surface. Mullite was present throughout all the various zones; however, the crystal size and quantity changed due to the other reactions taking place. Thus the oxidation of the $MoSi_2$-mullite composite initially exhibited a weight gain but then shifted to one of weight loss.

The flexural behavior of alkali-resistant high zirconia glass fiber reinforced cement composites was evaluated by Bentur et al. 1985 after exposure to water at 20 and 50°C for times up to 2 years. The principal mode of degradation was not the etching of the glass fiber surfaces (indicated by their smooth surfaces) but by the growth of hydration products between the glass filaments. Each glass fiber strand was composed of about 200 individual filaments. Initially, crack stresses and MOR of the unreinforced matrix changed very little with exposure. In general, the MOR of the composites degraded considerably approaching the MOR of the matrix. After 6 months, considerable differences existed among the several types of fibers investigated. One exhibited embrittlement, one degraded to 50% of its original toughness, and one degraded very little. These differences were attributed by Bentur et al. to the differences in growth of hydration products, predominantly $CaO \cdot H_2O$, between the glass filaments. Within the first year of exposure, chemical attack of the glass fibers did not appear to play a role in the degradation of the mechanical properties. Even after 2 years it was minimal. The degree of hydration product growth and its density was directly related to the degree of embrittlement. This embrittlement was attributed to an increase in pullout bond strength due to the growth of $CaO \cdot H_2O$. In a composite where no hydration products formed, ductile fracture occurred as evidenced by fiber pullout.

---

[*] Both these phases contain the same elements but in different ratios (sapphirine $M_4A_5S_2$ and cordierite $M_2A_2S_5$) and in small quantities could easily be confused with one another by EDS analysis.

Ready 1992 suggested that the water pressures developed during the reaction of hydrogen with particulate NiTiO₃ in a matrix of titania at temperatures between 700 and 1000°C were sufficient to cause microstructural degradation (i.e., grain boundary cracks). Thermodynamic data indicated that pressures as high as 6 MPa could be developed. The proposed mechanism involved the diffusion of hydrogen through TiO_2 grains, reduction of $NiTiO_3$ producing Ni metal and H_2O gas at the $TiO_2/NiTiO_3$ interface, and subsequent grain boundary separation. The separation of the grain boundaries allowed additional hydrogen ingress. According to Ready, the location of pores (i.e., water vapor bubbles) at only the matrix/particulate interface suggested that the reduction of the $NiTiO_3$ was controlled by oxygen diffusion out of the TiO_2 grains toward the interface.

Arun et al. 1990 reported the following order of TiC > HfC > ZrC for the oxidation resistance for these three carbides at 1273 K. The oxidation of these materials was much greater when they were incorporated into hot-pressed compositions of $TiC-ZrO_2$, $ZrC-ZrO_2$, and $HfC-HfO_2$. Arun et al. also reported a greater oxidation of TiC when incorporated into ZrO_2 as opposed to Al_2O_3.

Oxynitrides

Li et al. 2005 investigated the corrosion in acidic solutions of a SiAlON (containing some yttria) composite reinforced with TiN. Tests were performed in 8M HCl at 80°C for 100 hrs. Corrosion took place predominantly at the grain boundaries. TiN contents from 0 to 10 wt% were evaluated with the finding that increased TiN increased corrosion. The grain boundary phase was a combination of amorphous phases in the Al-Y-Si-O-N and Y-Ti-O systems. The yttrium of the grain boundary phase forms a soluble YCl_3 phase that leads to degradation of the grain boundary phase, Sato and Tokunaga 1988.

Nonoxide-Matrix Composites

Carbon-Carbon Composites

Carbon-carbon (i.e., carbon fiber reinforcement and carbon matrix) composites are probably the only materials that possess a combination of high strength/weight ratio, very low thermal expansion, excellent thermal shock resistance, and strength retention over a wide temperature range. This combination of properties makes them highly desirable in the aerospace industry. The major drawback for widespread use of C-C composites, however, is their poor oxidation resistance above 500°C. Only through the use of oxygen barrier coatings can C-C composites be useful at elevated temperatures in oxidative environments. Silicon nitride applied by CVD has proven to work well as an oxygen barrier for applications in rapid thermal cycles up to 1800°C, Strife 1991. For less rapid cycling to lower temperatures (<1500°C) and thermal soaking at temperatures between 600 and 1000°C multilayer coatings containing boron with CVD Si_3N_4 or SiC overlays have been tested. Although the boron can form a low-temperature liquid that will seal microcracks, its use at high temperatures may be limited due to volatility and the high fluidity of the glass formed. Labruquere et al. 2002a reported that silicon-rich coatings on the carbon fibers enhanced their oxidation resistance. Although the coating did not stop oxidation, it oxidized at a slower rate than the carbon fibers and was confined to the region between the coating and the fiber and between the coating and the matrix. After testing silicon-rich and boron-rich Si-B-C coatings, Labruquere et al. concluded that to protect the carbon/carbon composite efficiently the following parameters were required:

1. the coating must have a Pilling & Bedworth* coefficient greater than 1,
2. the coating must oxidize at a faster rate than the carbon,
3. the coating must have a minimum thickness (~ 100 nm),
4. the coating must not react with the carbon fibers, and
5. the coating must form a stable oxide.

In another study Labruquere et al. 2002b found that wet air (500 L/hr of 3 vol% H_2O) increased the oxidation rate by a factor of three.

Cawley et al. 1993 reported that the crystallographic orientation of well-crystallized carbon determines the rate of oxidation. Thomas 1965 showed that oxidation parallel to the c-axis was lower by 30 times compared to the basal plane and that the rate exhibited about 10% anisotropy in the basal plane at 800°C.

The reaction of carbon with hydrogen forms methane at low temperatures. Other products begin to form as the temperature is raised above 1400°C. The severity of the reaction is dependent upon the amount of atomic hydrogen present. Molecular hydrogen is relatively inert.

SiC-Matrix Composites

Oxidation

As discussed in the previous section on fiber coatings, an interphase material is either deposited onto the fibers before composite fabrication or formed *in situ*. If the interphase is carbon, the composite must receive an exterior surface protective coating. This is the case for SiC fiber/SiC matrix composites. Once the carbon interphase has been oxidized, leaving behind an annular cavity surrounding the fibers, continued oxidation fills the cavity with silica. The amount of silica present is dependent upon the proximity of the reaction site to the location of oxygen ingress. The time/temperature schedule required for complete filling of the cavities with silica is also dependent upon

---

* The Pilling and Bedworth coefficient is defined as the ratio of the volume of silica formed to the volume of the ceramic consumed, Pilling and Bedworth 1923.

the interphase layer thickness. Filipuzzi et al. 1994 reported that a time of 10 hrs in flowing oxygen at temperatures between 900 and 1300°C was required to consume completely a 1 μm thick carbon interphase in a 13 × 3 × 3 mm sample. Filipuzzi et al. reported that composites with thin interphase layers (on the order of 0.1 μm) resulted in microcracking due to the volume increase associated with the SiC to SiO₂ conversion. Microcracking was not observed at high temperatures (i.e., 1200°C) presumably due to stress release through the lower-viscosity silica glass, nor was it observed in composites with a thick interphase presumably due to the cavity not being filled with silica.

In a graphite-coated Nicalon™ fiber/SiC composite tested at 600 and 950°C in air, Lin and Becher 1997 found that lifetimes were more dependent upon open porosity (15–25%) than on parameters such as graphite coating thickness or fiber layout design. Increases in performance were obtained by the use of boron-containing oxidation inhibitors. This was attributed to the oxidation of the boron forming a glass that sealed cracks in the matrix, thus minimizing the ingress of oxidation. The oxidation of graphite was the predominant mode of deterioration at low temperatures, but oxidation of SiC occurred at temperatures of 425°C. Verrilli et al. 1997 found similar results in their investigations of graphite-coated Nicalon™ fiber/SiC composites tested at 500–1300°C. Oxidation of the interfacial graphite occurred first and then oxidation of the SiC fibers occurred evidenced by the formation of surface pits and radius reduction (most severe between 700 and 800°C). Other investigators have reported the degradation of carbon-coated Nicalon™ fiber/SiC, Heredia et al. 1995, resulting from the oxidation of the fibers at intermediate temperatures (600–800°C).

The graphite oxidizes to CO and CO₂ (reactions 17.7 and 17.8). Then additional oxygen reacts with the SiC forming free Si, which then continues to react to form SiO₂, filling the space originally occupied by the graphite (reactions 17.9 and 17.10). These reactions are all temperature and oxygen partial pressure dependent as discussed in Chapter 2. This causes embrittlement and loss of toughness.

$$C + 1/2\,O_2 \rightarrow CO \tag{17.7}$$

$$CO + 1/2\,O_2 \rightarrow CO_2 \tag{17.8}$$

$$SiC + 1/2\,O_2 \rightarrow Si + CO \tag{17.9}$$

$$Si + 1/2\,O_2 \rightarrow SiO_{(g)} + 1/2\,O_2 \rightarrow SiO_2 \tag{17.10}$$

The latest preference for SiC/SiC composite is one with fibers of improved microstructure and chemistry called Sylramic™* incorporated into a matrix of melt-infiltrated SiC matrix. In addition, the interphase material of choice

---

* Sylramic™, Dow Corning Corp., Midland, MI.

has become BN, although its oxidation is essentially the same as carbon. Ogbuji 2003 attributed the problems with BN interphase to one involving a thin film of carbon that forms under the BN either from carbon-rich fibers or sizing char that oxidizes first exposing the BN interphase. PVA sizing and carbon-free fibers, although not completely removing this problem, at least tremendously decreased the severe pesting[*] that did occur.

In addition to the reactions described by equations 17.7 to 17.10, many others in the BN-coated SiC fiber/SiC composites produce gaseous phases, especially if moisture is present, More et al. 2003. These are shown below:

$$SiC + 3/2\ O_2 \rightarrow SiO_2 + CO_{(g)} \tag{17.11}$$

$$2BN + 3/2\ O_2 \rightarrow B_2O_3 + N_{2(g)} \tag{17.12}$$

$$Si + 2H_2O_{(g)} \rightarrow SiO_2 + 2H_{2(g)} \tag{17.13}$$

$$SiC + 3H_2O_{(g)} \rightarrow SiO_2 + CO_{(g)} + 3H_{2(g)} \tag{17.14}$$

$$SiO_2 + 2H_2O_{(g)} \rightarrow Si(OH)_{4(g)} \tag{17.15}$$

$$2BN + 3H_2O_{(g)} \rightarrow B_2O_{3(l)} + N_{2(g)} + 3H_{2(g)} \tag{17.16}$$

$$B_2O_{3(l)} + H_2O_{(g)} \rightarrow 2HBO_{2(g)} \tag{17.17}$$

These gases when liberated cause increased porosity, a porous scale, and possibly microcracks to develop. More et al. also concluded that a similar composite that had the SiC matrix infiltrated with molten silicon exhibited better oxidation resistance, presumably due to a lower porosity.

Moisture Attack

In the study of alumina composites reinforced with SiC whiskers, Kim and Moorhead 1991 found that the room temperature flexural strength after exposure to H_2/H_2O at 1300 and 1400°C was significantly affected by the pH_2O. Reductions in strength were observed when active oxidation of the SiC occurred at $pH_2O < 2 \times 10^{-5}$ MPa. Kim and Moorhead also reported that long-term exposures greater than 10 hours resulted in no additional loss in strength. At higher water vapor pressures, reductions in strength were less severe due to the formation of an aluminosilicate glass and mullite on the surface of the sample. For exposures at 1400°C for 10 hours above $pH_2O =$

---

[*] Pesting was originally used to describe the formation of a powder-like deposit on the surface of metallic silicides during oxidation; however, it is now used to describe a similar phenomenon on any material.

5×10^{-4} MPa strength increases were observed due to the healing of cracks caused by glass formation at the sample surface.

Even at high temperatures, moisture may attack silica-containing materials in the same fashion it does silicate glasses at ambient conditions. This is of concern for those materials like silicon nitride and carbide that form a protective layer of silica on their surfaces at high temperatures. Once the protective layer is broken, oxidation of the underlying material may take place. The protective layer does not even need to be broken for continued oxidation in moist environments. According to Williams 1965 the diffusion of oxygen through silica is an order of magnitude greater when moisture is present. This same enhanced diffusion of oxygen in moist environments was noted by Singhal 1976 and Narushima et al. 1990.

Other Gases

Hydrogen can react with SiC to form silicon or with carbon to form methane, Nelson 1993. This reaction is negligible below 1100°C in essentially dry environments (containing ≤ 100 ppm of moisture). With increasing amounts of moisture this temperature limit increases, reaching about 1300°C at moisture contents as high as 10%. The dissociation of molecular hydrogen occurs at temperatures above 1100°C or at lower temperatures by heterogeneous surface reactions. Once dissociated, hydrogen can become extremely reactive. The dissociation of hydrogen by heterogeneous surface reactions is much easier on metals than on carbon or ceramics. Hallum and Herbell 1988 reported a weight loss at 1000°C, grain boundary corrosion at 1100°C, and both grain and grain boundary corrosion at 1300°C for samples of SiC exposed to pure hydrogen. The effects of weight loss and corrosion were noted at times as low as 50 hrs. After 500 hrs at 1100 and 1300°C the room temperature MOR decreased by one-third.

Si_3N_4 Matrix Composites

A Si_3N_4 composite containing 30 wt% ZrO_2 (also containing 3 mol% Y_2O_3) when oxidized at 1200°C exhibited decomposition of the zirconia grains as reported by Falk and Rundgren 1992. The oxidation proceeded by first forming faceted cavities close to the zirconia grain boundaries due to release of nitrogen dissolved in the zirconia. Prolonged oxidation formed silica-rich films on the pore walls. Hot pressing at 1800°C apparently formed zirconia containing a variation in the amount of yttria that led to the formation of some monoclinic zirconia after oxidation for 20 min at 1200°C. At shorter times, only cubic and tetragonal zirconia were detected. Cristobalite formed in the oxide scale after 2 hrs of oxidation. Short-term oxidation was suggested as a means to enhance mechanical properties; however, long-term oxidation resulted in disintegration of the composite.

Klein et al. 2003 showed that in a Si_3N_4/TiN composite the oxidation of the TiN was more important below 1200°C and that above 1200°C the oxidation of the Si_3N_4 became more relevant. In a hot-pressed composite a multi-layered scale formed. The external layer was composed predominantly of TiO_2 and included some silica (cristobalite) and silicate phases. The internal porous layer was titanium depleted and rich in silica and SiN_xO_y. They also concluded that sintering additives enhance corrosion and therefore manu-facturing techniques should be used that minimize the amounts of these additives. All these results were confirmed by Medri et al. 2005. In a more detailed study Feldhoff et al. 2005 examined the oxidation of Si_3N_4/TiN *in situ* in an environmental scanning electron microscope in both dry pure oxy-gen and moist air at temperatures up to 1100°C. They found that oxidation of the TiN started at 650°C forming isolated nanocrystals of TiO_2 on the TiN particles exposed at the sample surface. Once the temperature was above the transition temperature[*] of the Si_3N_4 intergranular glassy phase (about 950°C), oxidation proceeded via dissolution and diffusional transport through the glassy phase. The isolated TiO_2 crystals continued to grow until they com-pletely covered the entire sample surface. No difference in oxidation kinetics was exhibited for dry oxygen or moist air. The same group of investigators reported on the isothermal oxidation of TiN at temperatures of 1000, 1200, and 1400°C for 4, 24, and 100 hrs in both dry and humid air, Mazerolles 2005. They found that a multilayered scale formed under the surface TiO_2 above 1000°C. This subsurface scale was composed of Si_3N_4, TiO_2, glass, and pores, similar to what Klein et al. 2003 found, as described above. They also reported that the humidity altered the glassy phase structure and thus the diffusion in the glassy phase. At 1000°C moist air enhanced the development of a more compact TiO_2 surface layer, presumably by enhancing the outward diffusion of Ti cations through the glassy phase. All these results lead one to conclude that the glassy phase contained in these materials is of critical importance to the oxidation characteristics.

In an attempt to prepare a Si_3N_4 composite with reduced wear resistance, the oxidation at 750°C for 10 hrs in nonflowing air of a Mo_5Si_3 particle rein-forced Si_3N_4 composite was investigated by IIzuka and Kita 2004a. They found the oxidation to be rather complex with variable oxidation of the Mo_5Si_3 that produced a Mo- and Si-containing dual-phase material, molybdenum oxide particles (both MoO_2 and MoO_3) surrounded by amorphous silica, and a porous amorphous silica phase. The porous silica phase was attributed to the vaporization of MoO_3. The MoO_3 is the desired phase for solid lubrica-tion characteristics. IIzuka and Kita concluded that oxidation must be kept below 750°C to ensure that the MoO_3 produced did not vaporize. A sample

---

[*] The transition temperature of a glass is when it changes from the low temperature elastic solid to the high-temperature viscoelastic state. This temperature is very dependent upon glass composition and thermal history.

oxidized at 700°C reduced the coefficient of friction by about 26% compared to a standard Si_3N_4. Iizuka and Kita 2004b extended the temperature of oxidation to 1400°C in another study of the same materials. At the higher temperatures the oxidation products were cristobalite and $Y_2Si_2O_7$ with scale thicknesses as great as 10–15 μm. At temperature the scale was presumed to be protective; however, upon cooling, many residual stress-induced cracks formed. Medri et al. 2005 reported that a composite reinforced with $MoSi_2$, if oxidized at temperatures of 1200 to 1500°C, formed a protective layer of cristobalite and Y-silicates. They attributed a rise in strength after treatment at temperatures >1200°C to the reaction of $MoSi_2$, O_2, and Si_3N_4 to form Mo_5Si_3 and additional Si_2N_2O and silica.

Vasques et al. 2005 reported on the oxidation of Si_3N_4-TiB_2 composite at temperatures up to 1400°C. Under non-isothermal conditions TiB_2 oxidized between 600 and 1000°C forming TiO_2 and B_2O_3, which readily vaporized. In addition a low-viscosity borosilicate glassy phase formed. They estimated that after 24 hrs at 1300°C 85 wt% of the B_2O_3 had vaporized.

Other Nonoxide Matrix Composites
The ternary compound Ti_3SiC_2 has recently been reported by Li et al. 2003a to have excellent properties. When reinforced with SiC (*in situ* formed) a weight gain of 7.9 mg/cm^2 was reported for heat treatment in air at 1200°C for 21 hrs. At all temperatures between 1000 and 1500°C TiO_2 formed on the surface. SiO_2 was present on the surface only at 1000°C. At higher temperatures the outer surface layer was porous allowing oxygen diffusion to the interior, which allowed the formation of silica along with TiO_2 in the interior layer. The porous outer layer caused the weight gain at 1300°C to be approximately 2.5 times greater than at 1200°C.

Tungsten carbide and titanium carbonitride matrix composites containing either cobalt or cobalt/nickel alloy, respectively, as binders were investigated by Pugsley and Sockel 2004. The WC composite was tested in tannic acid whereas the Ti(C,N) was tested in deionized water. Both materials experienced corrosion by selective dissolution of the binder phase. In addition both materials exhibited a decrease in strength by a factor of two after exposure for 168 hrs. The WC composite exhibited stress corrosion cracking, which was attributed to that material being susceptible to hydrogen embrittlement.

The multilayered scale that formed on AlN/SiC composites at temperature up to 1550°C was found to be $Al_{10}N_8O_2$ and cristobalite by Lavrenko et al. 1998. Intermediate layers contained either β-SiAlON or Al_2O_3 depending upon the AlN content. The outer layer was composed mostly of mullite. In an attempt to increase the oxidation resistance, Lavrenko et al. 2005 added TiB_2 to the AlN/SiC composite. At temperatures between 1200 to 1300°C rutile, alumina, and silica formed for short exposure times to air. Between 1350 to 1550°C for longer times mullite and β-Al_2TiO_5 formed. They concluded that the best oxidation resistance was obtained with a TiB_2 content of between 2 and 5%. Above 10% the formation of pores and volatile B_2O_3 caused loss of protection.

The oxidation at temperatures up to 1300°C of an AlN composite reinforced with 15% SiC and 30% ZrB_2 was studied by Brach et al. 2005. Thermogravimetric analysis was carried out in flowing synthetic air (30 mL/min). Oxidation was quite slow at low temperatures (between 600 and 1100°C) but increased rapidly above 1200°C. Due to the fact that this composite contained three oxidizable phases the data were fitted to be a multiple-law model as shown below:

$$\omega = K_{lin} \cdot t + K_{par} \cdot \sqrt{t} + K_{log} \cdot \log(t) \qquad (17.18)$$

where

ω = mass change per unit area,

t = time, and

K_{lin}, K_{par}, and K_{log} = kinetic parameters for linear, parabolic, and logarithmic behavior.

In the temperature range 700 to 900°C the oxidation was a combined parabolic/linear law with parabolic dominating, whereas at 1100°C all three terms applied with the logarithmic one dominating. Between 1200 and 1300°C the kinetics were again parabolic/linear with parabolic dominating. Between 700 and 900°C the oxidation was predominately that of ZrB_2 to ZrO_2 and B_2O_3. Equation 17.18 exhibited a negative linear term due presumably to the volatility of the B_2O_3. Between 1000 to 1100°C alumino borate formed either as a reaction of B_2O_3 with AlN or Al_2O_3 or both. Energetically the reaction with Al_2O_3 is the favored one. In addition, the SiC oxidized to form silica and aluminosilicate and borosilicate glasses. Between 1200 to 1300°C the stoichiometry of the alumino borate increased in aluminum and silica and alumina reacted to form mullite.

An AlN composite containing 15 vol% SiC and 30 vol% $MoSi_2$ was investigated by Medri et al. 2005 for oxidation resistance at temperatures between 800 and 1500°C. Even though $MoSi_2$ suffers from excessive oxidation and pesting between 500 and 800°C its resistance to oxidation above 1000°C is excellent due to the formation of a silica protective layer, Maruyama and Yanagihara 1997. The likely incorporation of Al into the lattice of $MoSi_2$ was thought by Medri et al. to be the reason for the elimination of pesting between 500 and 800°C. In addition at higher temperatures (1200–1500°C), a protective scale based upon mullite formed.

The high oxidation resistance at temperatures above 1350°C of AlN composites containing 10% of a solid solution of TiB_2 and $TiSi_2$ was attributed to the formation of a very dense surface scale of aluminum titanate (β-Al_2TiO_5—tialite) by Lavrenko et al. 2003. Other phases also formed, which included rutile, silica, and iron titanate. Mattia et al. 2005a prepared an AlN composite with 27 vol% HfB_2 for oxidation studies in flowing oxygen at temperatures up to 1425°C. Very low mass gain was exhibited below 1350°C after which the gain was quite rapid. At 1000°C HfO_2 appeared along with a glassy phase. At 1200°C an additional oxide phase of $Al_{18}B_4O_{33}$ appeared. The oxide scale was protective up to about 1100°C. At higher temperatures the greater vaporization of B_2O_3 prevented the glassy phase from being protective and also provided for easy ingress of oxygen. In the study of wetting of AlN/TiB_2 composite by various liquid metals, Mattia et al. 2005b found that in the case of Ni, corrosion was due to diffusion of Ni into the composite and dissolution of the aluminum-based phases. The corrosion in sessile drop experiments (performed at 1460°C for 30 min) exhibited two layers of reaction with the formation of various Al-N and Ti-Ni secondary phases. The size of the corroded layers increased with TiB_2 content (35, 55, and 65 mol%).

Nicalon™ fiber reinforced TiB_2 matrix composites have been tested for use in the production of aluminum. Exposure to molten aluminum for 24 hrs caused heavy attack of the TiB_2 and delamination of the composite. This deterioration was attributed by Lowden et al. 1995 to retain chlorine that caused premature dissolution of TiB_2 into aluminum. The retained chlorine was from the $TiCl_4$ precursor material and an infiltration temperature of 900°C. When an infiltration temperature of 1200°C was used the TiB_2 was unaffected even after 10 weeks of exposure to molten aluminum. These longer exposure times, however, resulted in attack of the reinforcement fibers.

Various ceramic matrix materials have been manufactured from the pyrolysis of polymers, called *preceramic polymer precursors*. Although these materials are supposedly nonoxides one must be careful as to the actual production route and source. According to French 1995 the polysilazane materials are moisture sensitive and therefore yield a ceramic high in oxygen. Decomposition reactions of the type:

$$polysilazane\ containing\ moisture \rightarrow SiO_g + CO_g \qquad (17.19)$$

may occur. French commented that reaction 17.19 was not appreciable below 1400°C.

One of the early investigations into improving the oxidation of carbides was that by Accountius et al. 1954. They attempted to form glassy protective layers on TiC, B_4C, and SiC.

Borides, although having excellent properties, suffer from the formation of volatile B_2O_3 at temperatures below about 1300°C. Various researchers have tried incorporating materials such as SiC that react with the B_2O_3 forming a protective borosilicate glass that improves the overall oxidation resistance of these composites. A HfB_2 matrix composite containing 19 vol% SiC was investigated for oxidation resistance up to 1600°C by Monteverde and Bellosi 2005. This composite also contained 5.8 vol% Si_3N_4 as a sintering aid. Even though the formation of volatile B_2O_3 at temperatures below 1350°C was detrimental, above 1400°C the reaction of the B_2O_3 with silica from the oxidation of SiC formed a protective borosilicate glass. In a companion study Monteverde 2005 reported that the oxidation of either a HfB_2 or a ZrB_2/HfB_2 matrix composite (both containing 3 vol% HfN as sintering aid) reinforced with 19.5 vol% SiC particles obeyed paralinear kinetics. The products of oxidation were HfO_2, ZrO_2, borosilicate glass, and volatile B_2O_3, with the borosilicate glass being protective above 1400°C. Monteverde 2007 subsequently studied ZrB_2 matrix composites containing either 15 vol% SiC or 15 vol% SiC and 10 vol% HfB_2. Oxidized in air at 1400°C for 20 hrs these composites formed a protective surface layer of silicate glass. The deviation from parabolic oxidation kinetics was attributed to the evolution of volatile products, the active oxidation of SiC, and the faster oxidation of the borides compared to SiC.

Karlsdottir and Halloran 2008 studied the oxide scale formation on ZrB_2–15 vol% SiC at 1550°C for times of 3 and 4 hrs. The outward flow of a boria-rich silicate liquid with subsequent vaporization of the B_2O_3 led to the formation of a silica-rich scale on the surface. This low-viscosity liquid also provided a path for the easy ingress of oxygen that allowed subscale oxidation to take place. The importance of this outward liquid flow (called convection cells) in the formation of the surface scale was discussed by Karlsdottir et al. 2007 and Karlsdottir and Halloran 2009. See Figure 17.1 for an SEM backscattered image of these convection cells. For a cross-sectional view of a similar sample heated to 1700°C for 15 min see Figure 17.2, which shows the various layers formed. The ZrO_2 layer that forms is a porous skeleton that acts as a substrate

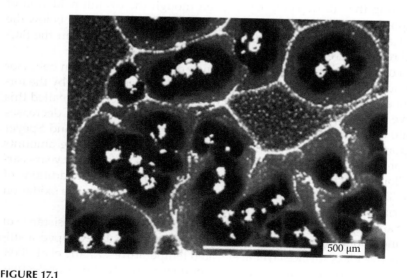

FIGURE 17.1
ZrB_2–15 vol% SiC composite oxidized at 1600°C for 30min. Convection cells showing ZrO_2 islands (white) surrounded by B_2O_3-rich areas (black) located in larger SiO_2 lagoons (gray). Scale bar at bottom right is 500 μm. (Karlsdottir and Halloran 2008, reprinted with permission of the American Ceramic Society/Wiley & Sons Copyright Clearance Center.)

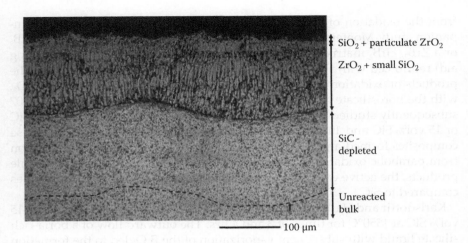

SiO₂ + particulate ZrO₂

ZrO₂ + small SiO₂

SiC - depleted

Unreacted bulk

— 100 μm

FIGURE 17.2
Cross section image of ZrB₂/15 vol% SiC heated to 1700°C for 15 min. Multilayer oxide scale consists of unaffected substrate, a SiC-depleted zone, a columnar ZrO₂ layer, and a thin layer of SiO₂ glass. (Karlsdottir and Halloran 2007, reprinted with permission of the American Ceramic Society/Wiley & Sons Copyright Clearance Center.)

for the borosilicate liquid,[*] Li et al. 2008. Since an oxygen chemical potential gradient develops across the diffusion barrier lower partial pressures should be sufficient to cause active oxidation of SiC at subsurface levels, Fahrenholtz 2007, forming the SiC-depleted layer. Even though the overall oxidation of the composite was passive, an active oxidation region was present below the surface. Calculated partial pressures of oxygen by Fahrenholtz in the SiC-depleted region were between 4.0×10^{-14} and 1.8×10^{-11} Pa.

The incorporation of SiC into ZrB₂ extends the passive oxidation behavior over a much greater temperature range (up to 1500 versus 1100°C) by the formation of a silicate glass protective layer. Some investigators have called this protective layer a borosilicate glass; however, the boron content decreases with time and temperature due to the volatility of B₂O₃. Peng and Speyer 2008 reported that oxidation resistance increased with increasing amounts of SiC from about 10 to 35 vol%. This same effect was reported 40 years earlier by Clougherty et al. 1968. Peng and Speyer also found that additions of TaB₂ and TaSi₂ to ZrB₂/SiC composites both increased resistance to oxidation by the formation of Ta₂O₅ or additional silicate glass.

All the above researchers attempted to improve the oxidation resistance of ZrB₂ by incorporating silicon-containing compounds that developed a silicate protective glass surface layer at elevated temperatures. Zhang et al. 2008, however, tried to develop a surface layer of dense ZrO₂ that was protective.

---

[*] This ZrO₂ skeleton layer embedded with a silicate glass is essentially identical to the surface layer formed on zircon refractories in the production of Li-Al-silicate glasses for pyrocerams. See Chapter 15.

This they thought could be accomplished by additions of WC. Since WC forms a solid solution with ZrB_2, during oxidation at 1500°C in air an intimately mixed scale formed containing both ZrO_2 and WO_3. The incorporation of WC enhanced the oxidation resistance as evidenced by both a lower mass gain (25 to 35%) and a thinner scale thickness. Although scale thickness increased with duration at temperature it increased at a slower rate when WC was present. The improved oxidation resistance of ZrB_2 with 10 mol% WC was attributed to the WO_3 (formed by oxidation) providing densification of the ZrO_2 surface scale through liquid phase sintering. Although the volume expansion of W to WO_3 is substantial (230%), apparently the 4 mol% WC was sufficiently low to not cause any mechanical problems.

Although the oxidation in air of ZrB_2-containing SiC particulates has been extensively studied, little has been reported on the oxidation at reduced oxygen pressures. Rezaie et al. 2006 examined the oxidation of a ZrB_2 composite containing 30 vol% SiC in a CO/CO_2 mixture that provided an oxygen partial pressure of 10^{-10} Pa at 1500°C. A thermodynamic model, which indicated that SiC should exhibit active oxidation under these conditions, accurately predicted the formation of a SiC-depleted zone. An external layer of porous ZrO_2 formed at a rate of 13.6 μm/hr, while the overall weight loss was 3.8 mg/cm² · hr.

LaB_6 matrix 21% ZrB_2 composite was shown to have excellent oxidation resistance below 1650°C, Chen 1997. However, at 800°C Gao et al. 2005 found that the LaB_6 was oxidized to $LaBO_3$ and B_2O_3, whereas the ZrB_2 was unaffected. At the higher temperature of 1100°C the $LaBO_3$ converted to $La(BO_2)_3$ and ZrO_2 formed. La_2O_3 first appeared at 1200°C and by 1300°C LaB_6 was completely oxidized.

The oxidation of a clay-bonded $MoSi_2$-based composite in oxygen and oxygen plus 10% H_2O at 600 and 700°C was examined by Hansson et al. 2005. MoO_3 and amorphous silica were the first products of oxidation to form. The next step was the formation of the volatile $(MoO_3)_3$ species in oxygen and the $MoO_2(OH)_2$ in the 10%H_2O environment. A protective scale was formed next by silica filling in the porosity left behind by the loss of Mo. The volatility of the species present depends not only upon their vapor pressure but also upon the water vapor content. Thus MoO_3 evaporated faster in the moist environment due to the additional formation of $MoO_2(OH)_2$. Since the vapor pressures of these species increase with temperature and water vapor content, the protective scale thickness was thinner for the 10%H_2O environment.

Metal Matrix Composites (MMC)

Although there isn't much on corrosion, the book by Taya and Arsenault 1989 contains a lot of information about the properties and behavior of MMC.

TABLE 17.3

Matrix Metals and Their Upper Limits of
Temperature Service

Metal	Upper Service Temperature (°C)
Aluminum	400
Titanium alloys	650
Superalloys (Ni)	980–1100
Refractory metals	1100–1650

Much of the corrosion of MMCs is related to corrosion of the metal matrix, a topic that will not be covered here.

Aluminum is probably the most common metal used for the production of metal matrix composites. This is due mostly to its low density, excellent mechanical properties, and low melting point that aids in fabrication. The addition of zinc creates an alloy with improved strength, wear resistance, and corrosion resistance. Since aluminum has a relatively low melting point, other metals have been used as the matrix when increased service temperatures are required. See Table 17.3 for a list of some of the metals used and their service temperature limits.

Reinforcements for aluminum can be of the insulating (Al_2O_3), conducting (carbon), or semiconducting (SiC) type. With the use of semiconducting reinforcements galvanic corrosion becomes a problem, Ding and Hihara 2009. Since SiC is a p-type semiconductor in aluminum MMC, galvanic corrosion is enhanced when the MMC is illuminated. Ding and Hihara proposed a photochemical-corrosion-diode model that described the galvanic corrosion of MMC containing semiconducting reinforcements. Their model correctly predicted the effects of illumination.

After learning that a 6092 Al composite reinforced with 20 vol% particulate Al_2O_3 underwent serious corrosion (Ding et al. 2009) and that localized corrosion sites formed that most likely were cathodic that caused the local pH to increase (Ding and Hihara 2005), Ding and Hihara 2010 examined this composite further and found that it contained a variety of Ti-containing particles (Ti_6O, Ti_3O, Ti_2O, TiO, TiO_2, and Ti-Zr-Al). The titanium most likely was present as an impurity in the alumina reinforcement particles. Ding and Hihara studied this composite in 0.5 M Na_2SO_4, 3.15 wt% NaCl,[*] and 1 mM $CuSO_4$ solutions. They found that corrosion was initiated from sites containing titanium and from Fe-Si-Al intermetallic particles. The electrically insulating particles of alumina did not substantially contribute to the overall corrosion of the composite. Ding et al. 2009 attributed the lower corrosion rate in

---

[*] A 3.15 wt% NaCl solution simulates seawater, both of which have a chloride ion concentration of 0.54 M. ASTM D-1141 Standard Practice for Preparation of Substitute Ocean Water includes inorganic salts in addition to NaCl.

seawater than in 3.15 wt% NaCl or 0.5 M Na_2SO_4 solutions to the formation of precipitates (Al-Mg-hydrotalc-like compounds) on the surface of the composite and to the pH buffering effect of the sea water.

The weight change in a 3.5 wt% NaCl solution for up to 28 dys of an Al matrix (containing 4 wt% Mg) SiC (60 vol%) particulate composite was studied by Candan and Bilgic 2004. They found that the formation of Mg_2Si produced a beneficial effect upon corrosion resistance. At treatment times greater than 4 days, Candan and Bilgic found the composite to exhibit greater corrosion resistance than a pure aluminum matrix composite.

The stress corrosion cracking in a 3.5% NaCl solution of a pure aluminum matrix composite reinforced by alumina borate whiskers ($Al_{18}B_4O_{33}$) was studied by Hu et al. 2002a. Crack propagation was predominantly along the whisker/matrix interface. This was attributed by Hu et al. to dissolution of the matrix surrounding the whiskers as evidenced by pits that formed at the crack tips. In a complementary study Hu et al. 2002b determined that the interfacial phases of spinel ($MgAl_2O_4$) and Al_2Cu^* formed during the NaCl solution treatment at 495°C. Hu et al. 2004 concluded that any interfacial reactions that occurred are critical to the electrochemical corrosion of the composite.

The electrochemical characteristics of a 6092 Al matrix reinforced with 20 vol% of particulate $B_4C^\dagger$ was investigated by Ding and Hihara 2011 in an air-exposed 3.15 wt% solution. Corrosion was initiated from carbon particles within the matrix, presumably introduced during processing. Results suggested that galvanic corrosion was enhanced under illumination but limited in the dark. The enhanced corrosion under illumination was a result of the photo-enhanced cathodic activity of the B_4C. The actual mechanisms are still controversial since B_4C has been described as a degenerate semiconductor or as a nondegenerate p-type semiconductor with a high density of gap states, Werheit 2007.

A SiC-coated graphite fiber/Al alloy matrix composite was developed to overcome the reaction of graphite with aluminum above 500°C forming Al_4C_3, Cornie et al. 1986. This phase formed as hexagonal plates with subsequent degradation of fiber strength. In addition Al_4C_3 is hygroscopic, causing further deterioration. A silicon enrichment on the outer surface of the fibers inhibited the formation of Al_4C_3; however, it allows improvements of only short duration.

Although increased amounts of albite ($NaAlSi_3O_8$) reinforcement increased the corrosion resistance of an aluminum alloy composite, the corrosion resistance of the composite was not as good as the unreinforced matrix alloy, Seah et al. 2002. Corrosion tests were performed in 1 M HCl at 20 to 80°C and from 24 to 96 hrs. A black surface film formed that supposedly retarded the corrosion rate. This black film has been observed by others, De Salazar et al. 1999 and Castle et al. 1994, although different identifications were given,

---

* The Cu came from the 4.4 wt% contained in the aluminum 2024 matrix alloy.
† The exact stoichiometry of boron carbide is slightly deficient in carbon.

hydrogen hydroxy chloride and aluminum hydroxide, respectively. The differences apparently were due to different reinforcement materials and possibly different aluminum alloys. McCafferty 1998 believed that the increased resistance was due to an increase in Al^{3+} ion concentration in the corroding solution. Seah et al. gave no experimental data that would allow one to determine if Al^{3+} ion concentration buildup was a problem.

The seawater corrosion of SiC/Al was found to be more resistant than graphite/Al by Aylor and Kain 1986. This was attributed to a lack of a galvanic driving force between the SiC and the aluminum matrix although both composites exhibited similar mechanisms of corrosion—essentially pitting of the metal matrix around the reinforcement material.

A 6063 aluminum composite reinforced with AlN particles was investigated for its durability against immersion in alkaline solution of $pH = 12$, Liu et al. 2006. After a rapid initial weight loss during the first 3 days, the weight loss tended to level off. A white precipitate of $Al(OH)_3$ formed on the surface. Aside from the dissolution of aluminum, the AlN particles hydrolyzed, contributing to the high initial rate of corrosion and to the formation of micro-crevices.

Galvanic corrosion (see the section on electrochemical corrosion in Chapter 1) between the matrix metal and the reinforcement fibers can occur, especially when in contact with aqueous solutions. This has been reported by Trzaskoma 1986 for magnesium alloy matrix–graphite fiber composites.

The reactions of hydrogen and SiC-reinforced titanium-based matrix composites have been described by Nelson 1993. Since these reactions are ones involving various metallic phases rather than the SiC fiber or the interface, the details will not be given here. However, as reported by Nelson, the solubility of hydrogen in metals is problematic, in some cases forming metal hydrides (generally at higher temperatures) that lead to mechanical degradation.

The high-temperature application of $(Ni,M)/YSZ^*$ as a solid electrolyte (for fuel cells) where M = Co, Cu, or Fe was investigated by Ringuedé et al. 2004. Evaluation of metal/YSZ/metal cells was done in wet H_2 or wet $N_2 + H_2$ atmospheres at temperatures of 20 to 56°C. The best performance was found for $Ni_{1-x}Co_x/YSZ$.

The stability of TiN toward reaction with various metals such as iron and nickel aluminides has been reported to be better than SiC by Nolan et al. 1991. Thus TiN would be the better choice as the reinforcement for these metal matrix composites.

Since the intermetallic Ni- and Ti-aluminides have service temperatures limited to about 1200°C, molybdenum disilicide has been investigated for applications where greater temperatures may be reached. One of the major drawbacks of $MoSi_2$ is its oxidation resistance. Cook et al. 1991 investigated the incorporation of 30 vol% TiB_2, ZrB_2 HfB_2, and SiC as a reinforcement in hopes of developing a composite of greater oxidation resistance than the base

---

* YSZ = yttrium-stabilized zirconia.

$MoSi_2$. Specimens were exposed to isothermal testing at 800, 1200, 1400, and 1500°C for 24 hrs in air, in addition to a thermal cycle consisting of 55 min at 1200 or 1500°C and then 5 min ambient cooling with subsequent reheating. All the boride-containing materials exhibited a greater deterioration than the silicon carbide-containing composite, although none exhibited a greater oxidation resistance than the base $MoSi_2$. See Chapter 15 on borides for a discussion of the oxidation of these materials.

Li et al. 2003b studied the oxidation at 700 and 800°C in air of a TiC particulate reinforced Ti_3Al intermetallic matrix composite. The oxide scale consisting of TiO_2 (rutile) and minor alumina crystals that formed on the samples heated to 700°C exhibited severe spallation. At the higher temperature of 800°C a thick but fragile multilayer oxide (mainly TiO_2) scale formed. An outer TiO_2 layer, an intermediate alumina enriched layer, and an inner porous TiO_2 layer were formed. Li et al. found that increased milling times of the starting materials increased oxidation mass gain. Increased milling times increases the density of grain boundaries, dislocations, and other short-circuit paths for an enhanced diffusion flux of reactants. Their results showed that the oxidation of the composite was lower than pure Ti_3Al at both temperatures.

An alumina particulate (42–50 vol%) reinforced TiAl intermetallic matrix composite was investigated by Li et al. 2004 for its oxidation resistance at temperatures between 800 and 900°C for up to 500 hrs. At 800°C the composite exhibited a mass gain that was 2 to 4 times lower than a TiAl intermetallic with no reinforcement. At 900°C the mass gain was 10 times lower. This improvement in oxidation resistance was attributed to a lower difference between the substrate coefficient of expansion and the oxide scale compared to that of the intermetallic with no reinforcement.

The high-temperature oxidation in air at 1873 K (1600°C) for up to 8 hrs of a multiphase intermetallic, $(Mo,Cr)(Si,Al)_2$, matrix reinforced with SiC particulates was investigated by Mitra et al. 2004. Oxidation led to the formation of cristobalite, Cr_2O_3, Al_2O_3, and Cr_2MoO_6 on the surface and Mo_5Si_3 below this surface scale. The preferential selective oxidation of chrome was attributed to the high diffusivity of the Cr^{3+} ions and the high volatility of CrO_3.[*]

Although not generally thought of as metal matrix composites, a relatively new class of materials called *fibrous monolithic ceramics*, Coblenz 1988, actually may contain a metal as the matrix that surrounds cells of a fibrous polycrystalline ceramic. One example of such a material investigated by Baskaran et al. 1994 contained fibrous polycrystalline alumina cells surrounded by nickel. The nickel cell boundary thickness varied from 1 to about 15 μm. Oxidation at 1200°C for 10 hrs initially formed NiO that subsequently reacted with the alumina, forming $NiAl_2O_4$. The formation of the aluminate was thought to provide protection toward additional oxidation.

---

[*] Above 1000°C Cr_2O_3 oxidizes to CrO_3.

Polymer Matrix Composites (PMC)

Two publications by ASTM discuss the environmental effects upon polymeric composites, Harris and Gates 1993 and Gates and Zureick 1997. The largest amount of composites produced is probably of this type reinforced with glass fibers, called glass-reinforced plastics, polymers, or polyesters (GRP). Degradation in aqueous environments generally occurs by fiber/matrix debonding. Since glass fibers are attacked by moisture, which drastically reduces their strength, glass fibers are given a protective coating.

Graphite/carbon fiber/epoxy composites (CFRP) have seen some recent use in marine environments. In many cases they are generally used in contact with metals. In a seawater environment the graphite fibers act as the cathode for accelerated galvanic corrosion of the metals.

Electrochemical impedance spectroscopy was used by Wall et al. 1993 to monitor the damage in graphite fiber/bismaleimide composites in contact with aluminum, steel, copper, and titanium immersed into aerated 3.5 wt% NaCl solution. Decomposition of the bismaleimide polymer was thought to occur by the action of hydroxyl ions, which break imide linkages. The production of hydroxyl ions occurred through the following reaction:

$$O_2 + 2H_2O + 4e^- \rightarrow 4OH^- \tag{17.20}$$

at the surface of the graphite fibers. They concluded that the corrosion concentrated at the fiber/matrix interface was caused by cathodic polarization and was dependent upon the overpotential and the cathodic reaction rate. Oxidation of the matrix and fibers was thought to be the cause of ablation of the composite.

Aylor 1993 reported increased galvanic action (i.e., initial current level) with increased amounts of fiber exposure for a graphite fiber/epoxy composite in contact with either HY80 steel or nickel aluminum bronze subjected to seawater at ambient temperature for 180 days. Even when no fibers were exposed to the environment galvanic corrosion occurred. This phenomenon was attributed by Aylor to the absorption of moisture through the epoxy to the fibers. The galvanic current determined during the tests was found to display several distinct regions. These have been identified by Aylor as follows:

Region I   activation of surface
Region II   film formation
Region III   reduction of active surface areas
Region IV   buildup of calcareous deposit on composite

These regions were attributed to localized differences in active anodic and cathodic areas, which could also be affected by the stability of the films formed on the surfaces of the metal and composite. The calcareous deposits on the surfaces of the graphite fibers were reported as the result of formation of hydroxyl ions at the cathode with an associated increase in pH and precipitation of $CaCO_3$ and $Mg(OH)_2$. Actual seawater galvanic corrosion rates would be significantly affected by the stability of the films formed in Region II, and most likely would be much greater than the rates found in the laboratory tests.

A mica-flake-filled polyester, when used as a lining material for the outlet duct of a coal-fired power plant, formed the compound jarosite, $KFe_3(SO_4)_2(OH)_6$, at the mica/polyester interface. Subsequent wedging* of these materials resulted in failure of the lining, Koch and Syrett 1985.

Additional Recommended Reading

Delmonte, J. 1989. History of Composites, *Reference Book for Composites Technology*, S. Lee (ed.), Technomics Publ. Co., Lancaster, PA.

Evans, A.G., He, M.Y. and Hutchinson, J. W. 1989. Interface Debonding and Fiber Cracking in Brittle Matrix Composites, *J. Am. Ceram. Soc.*, 72, 2300–303.

Kerans, R.J., Hay, R.S., Parthasarathy, T.A. and Cinibulk, M.K. 2002. Interface Design for Oxidation-Resistant Ceramic Composites, *J. Am. Ceram. Soc.*, 85(11), 2599–632.

Lewis III, D. 1995. Continuous Fiber-Reinforced Ceramic Matrix Composites: A Historical Overview, pp. 1–34 in *Handbook on Continuous Fiber-Reinforced Ceramic Matrix Composites*, Lehman, R.L., El-Rahalby, S.K. and Wachtman, J.B. Jr. (eds.), CIAC Purdue University, West Lafayette, IN, and Am. Ceram. Soc., Westerville, OH.

Logan, K.V. (ed.), 1995. *Advanced Synthesis and Processing of Composites and Advanced Ceramics, Ceramic Transactions Vol. 56*, Am. Ceram. Soc., Westerville, OH.

Lowden, R.A. 1991. Fiber Coatings and the Mechanical Properties of Fiber-Reinforced Ceramic Composites, *Ceram. Trans.*, 19, 619–30.

Taya, M. and Arsenault, R.J. 1989. *Metal Matrix Composite Thermomechanical Behavior*, Pergamon Press, New York, NY, p. 264.

Exercises, Questions, and Problems

1. Develop a definition for a composite material by listing the various characteristics and explain the reason for each. What is the advantage of using a composite over that of a single component material?

---

* Wedging is a procedure where ceramic bodies are prepared by hand kneading. This is done to uniformly disperse water and remove air pockets and laminations.

2. Discuss why the adhesion of matrix to reinforcement is the region of greatest importance during corrosion.

3. Discuss how a difference in thermal expansion between the matrix and the reinforcement is related to corrosion.

4. Why is the corrosion process of oxidation a problem for so many composites?

5. How does the thermal expansion mismatch between surface layers formed by corrosion and the underlying substrate materials affect corrosion?

6. Discuss how the manufacturing process of a particular reinforcement fiber may affect the corrosion of a composite.

7. What does the term *embrittlement* mean when related to the corrosion of composites?

8. Discuss the difference that occurs during the oxidation of a composite having a SiC matrix and SiC fibers with either a BN or carbon interphase.

9. Compare the oxidation of TiB_2 that is used as a reinforcement for Si_3N_4 and BN that is used as an interphase for SiC fiber–reinforced composites.

References

Accountius, O., Sisler, H., Sheblin, S. and Bole, G. 1954. Oxidation Resistances of Ternary Mixtures of the Carbides of Titanium, Silicon, and Boron, *J. Am. Ceram. Soc.*, 37(4), 173–77.

Arun, R., Subramanian, M. and Mehrotra, G.M. 1990. Oxidation Behavior of TiC, ZrC, and HfC Dispersed in Oxide Matrices, in *Ceramic Transactions Vol. 10: Corrosion and Corrosive Degradation of Ceramics*, Tressler, R. E. and McNallan, M. (eds.), Am. Ceram. Soc., Westerville, OH, pp. 211–23.

Aveston, J. and Kelly, A. 1973. Theory of Multiple Fracture of Fibrous Composites, *J. Mat. Sci.*, 8, 352–62.

Aylor, D.M. 1993. The Effect of a Seawater Environment on the Galvanic Corrosion Behavior of Graphite/Epoxy Composites Coupled to Metals, in *High Temperature and Environmental Effects on Polymeric Composites*, STP 1174, Harris, C.E. and Gates, T.S. (eds.), ASTM, Philadelphia, PA, pp. 81–94.

Aylor, D.M. and Kain, R.M. 1986. *Recent Advances in Composites in the United States and Japan, ASTM STP 864*, Vinson, J.R. and Taya, M. (eds.), pp. 718–29.

Baskaran, S., Nunn, S.D. and Halloran, J.W. 1994. Fibrous Monolithic Ceramics: IV, Mechanical Properties and Oxidation Behavior of the Alumina/Nickel system, *J. Am. Ceram. Soc.*, 77(5), 1256–62.

Baudin, C. and Moya, J.S. 1990. Oxidation of Mullite-Zirconia-Alumina-Silicon Carbide Composites, *J. Am. Ceram. Soc.*, 73(5), 1417–20.

Bender, B., Shadwell, D., Bulik, C., Incorvati, L. and Lewis III, D. 1986. Effect of Fiber Coatings and Composite Processing on Properties of Zirconia-Based Matrix SiC Fiber Composites, *Ceram. Bull.*, 65(2), 363–69.

Bentur, A., Ben-Bassat, M. and Schneider, D. 1985. Durability of Glass-Fiber-Reinforced Cements with Different Alkali-Resistant Glass Fibers, *J. Am. Ceram. Soc.*, 68(4), 203–8.

Bischoff, E., Ruhle, M., Sbaizero, O. and Evans, A.G. 1989. Microstructural Studies of the Interfacial Zone of a SiC-Fiber-Reinforced Lithium Aluminum Silicate Glass-Ceramic, *J. Am. Ceram. Soc.*, 72(5), 741–45.

Bonney, L.A. and Cooper, R.R. 1990. Reaction Layer Interfaces in SiC-Fiber-Reinforced Glass-Ceramics: A High-Resolution Scanning Transmission Electron Microscopy Analysis, *J. Am. Ceram. Soc.*, 73(10), 2916–21.

Borom, M.P., Bolon, R.B. and Brun, M.K. 1988a. Oxidation Mechanism of MoSi2 Particles in Mullite, *Adv. Ceram. Mat.*, 3(6), 607–11.

Borom, M.P., Brun M.K., and Szala, L.E. 1988b. Kinetics of Oxidation of Carbide and Silicide Dispersed Phases in Oxide Matrices, *Adv. Ceram. Mat.*, 3(5), 491–97.

Brach, M., Sciti, D., Balbo, A. and Bellosi, A. 2005. Short-Term Oxidation of a Ternary Composite in the System AlN-SiC-ZrB$_2$, *J. Euro. Ceram. Soc.*, 25, 1771–80.

Brennan, J.J. 1986. Interfacial Characterization of Glass and Glass-Ceramic Matrix/Nicalon SiC Composites, pp. 549–60 in *Tailoring Multiphase and Composite Ceramics*, Mater. Res. Soc. Symp. Proc., 20, Plenum Press, New York, NY.

Candan, S. and Bilgic, E. 2004. Corrosion Behavior of Al-60 Vol% SiC$_p$ Composites in NaCl Solution, *Materials Letters*, 58, 2787–90.

Caputo, J. Lackey, W.J. and Stinton, D.P. 1985. Development of a New, Faster Process for the Fabrication of Ceramic Fiber-Reinforced Ceramic Composites by Chemical Vapor Infiltration, *Ceram. Eng. Sci. Proc.*, 6(7–8), 694–706.

Castle, J.E., Sun, L. and Yan, H. 1994. The Use of Scanning Auger Microscopy to Locate Cathodic Centres in SiC$_p$/Al-6061 MMC and to Determine the Current Density at Which They Operate, *Corros. Sci.*, 36(6), 1093–1110.

Cawley, J.D., Ünal, Ö. and Eckel, A.J. 1993. Oxidation of Carbon in Continuous Fiber Reinforced Ceramic Matrix Composites, pp. 541–52 in *Ceramic Transaction, Vol. 38, Advances in Ceramic-Matrix Composites*, Bansal, N.P. (ed.), Am. Ceram. Soc., Westerville, OH.

Chaim, R. and Heuer, A.H. 1991. Carbon Interfacial Layers Formed by Oxidation of SiC in SiC/Ba-Stuffed Cordierite Glass-Ceramic Reaction Couples, *J. Am. Ceram. Soc.*, 76(7), 1666–67.

Chen, C.M. 1997. PhD dissertation, Northwestern Polytechnic University, Xian, China.

Clark, T.J., Arons, R.M. and Stamatoff, J.B. 1985. Thermal Degradation of Nicalon™ SiC Fibers, pp. 576–88 in *Ceramic Engineering and Science Proceedings*, Smothers, W.J. (ed.), Vol. 6, (7–8).

Clark, T.J., Jaffe, M., Rabe, J. and Langley, N.R. 1986. Thermal Stability Characterization of SiC Ceramic Fibers: I, Mechanical Property and Chemical Structure Effects, *Ceram. Eng. Sci. Proc.*, 7(7–8), 901–13.

Clougherty, E.V., Pober, R.L. and Kaufman, L. 1968. Synthesis of Oxidation Resistant Metal Diboride Composites, *Trans. Met. Soc.*, ASME, 242(6), 1077–82.

Coblenz, W.S. 1988. *Fibrous Monolithic Ceramic and Method for Production*, U.S. Patent No. 4,772,524, September 20.

Cook, J., Mahapatra, R., Lee, E.W., Khan, A. and Waldman, J. 1991. Oxidation Behavior of MoSi$_2$ Composites, pp. 1656–70 in *Ceramic Engineering and Science Proceedings*, Wachtman, J.B. (ed.), Vol. 12, (9–10).

Cooper, R.F. and Chyung, K. 1987. Structure and Chemistry of Fiber-Matrix Interfaces in Silicon Carbide Fiber-Reinforced Glass-Ceramic Composites: An Electron Microscopy Study, *J. Mater. Sci., 22*(9), 3148–60.

Cooper, R.F. and Hall, P.C. 1993. Reactions between Synthetic Mica and Simple Oxide Compounds with Application to Oxidation-Resistant Ceramic Composites, *J. Am. Ceram. Soc., 76*(5), 1265–73.

Cornie, J.A., Chiang, Y-M., Uhlmann, D.R., Mortensen, A. and Collins, J.M. 1986. Processing of Metal and Ceramic Matrix Composites, *Ceram. Bull., 65*(2), 293–304.

Courtright, E.L. 1993. Engineering Limitations of Ceramic Composites for High Performance and High Temperature Applications, pp. 21–32 in *Proc. 1993 Conf. on Processing, Fabrication and Applications of Advanced Composites*, Long Beach, CA, 9–11 Aug 1993, Upadhya, K. (ed.), ASM, OH.

Curtin, W.A. 1991. Theory of Mechanical Properties of Ceramic-Matrix Composites, *J Am. Ceram. Soc., 74*(11), 2837–45.

Curtin, W.A. 1994. *In Situ* Fiber Strengths in Ceramic-Matrix Composites from Fracture Mirrors, *J. Am. Ceram. Soc., 77*(4), 1075–78.

Davidge, R.W. and Briggs, A. 1989. The Tensile Failure of Brittle Matrix Composites Reinforced with Unidirectional Continuous Fibers, *J. Mat. Sci., 24*, 2815–19.

De Salazar, J.M.G., Urefia, A., Manzanedo, S., and Barrena, M.I. 1999. Corrosion Behavior of AA6061 and AA7005 Reinforced with Al$_2$O$_3$ Particulates in Aerated 3.5% Chloride Solution: Potentiodynamic Measurements and Microstructure Evaluation, *Corros. Sci., 41*, 529–45.

Ding, H., Hawthorn, G.A. and Hihara, L. 2009. Inhibitive Effect of Seawater on the Corrosion of Particulate-Reinforced Aluminum-Matrix Composites and Monolithic Aluminum Alloy, *J. Electrochem. Soc., 156*(10), C352–59.

Ding, H. and Hihara, L. 2005. Localized Corrosion Currents and pH Profile over B$_4$C, SiC, and Al$_2$O$_3$ Reinforced 6092 Aluminum Composites I. In 0.5 M Na$_2$SO$_4$ Solution, *J. Electrochem. Soc., 152*(4), B161–67.

Ding, H. and Hihara, L. 2009. Galvanic Corrosion in Metal-Matrix Composites Containing Semiconducting Constituents, *J. Electrochem. Soc., 156*(12), C422–27.

Ding, H. and Hihara, L. 2010. Effect of Embedded Titanium-Containing Particles on the Corrosion of Particulate Alumina Reinforced Aluminum-Matrix Composite, *J. Electrochem. Soc., 157*(2), C79–85.

Evans, A.G. and Marshall, D.B. 1990. The Mechanical Behavior of Ceramic Matrix Composites, pp. 1–39 in *Fiber Reinforced Ceramic Composites*, Mazdiyansni, K.S. (ed.), Noyes, Park Ridge, NJ.

Faber, K.T. 1997. Ceramic Composite Interfaces: Properties and Design, *Annu. Rev. Mater. Sci., 27*, 499–524.

Fahrenholtz, W.G. 2007. Thermodynamic Analysis of ZrB$_2$-SiC Oxidation: Formation of a SiC Depleted Region, *J. Am. Ceram. Soc., 90*(1), 143–48.

Falk, L.K.L. and Rundgren, K. 1992. Microstructure and Short-Term Oxidation of Hot-Pressed Si$_3$N$_4$/ZrO$_2$(+Y$_2$O$_3$) Ceramics, *J. Am. Ceram. Soc., 75*(1), 28–35.

Fareed, A.S., Schiroky, G.H. and Kennedy, C.R. 1993. Development of BN/SiC Duplex Fiber Coatings for Fiber-Reinforced Alumina Matrix Composites Fabricated by Directed Metal Oxidation, *Ceram. Eng. Sci. Proc., 14*(9–10), 794–801.

Feldhoff, A., Trichet, M.-F., Mazerolles, L., and Backhaus-Ricoult, M. 2005. Electron Microscopy Study on the High Temperature Oxidation of Si_3N_4-TiN Ceramics: In Situ and Ex Situ Investigations, *J. Euro. Ceram. Soc., 25*, 1733–42.

Filipuzzi, L., Camus, G., Naslain, R. and Thebault, J. 1994. Oxidation Mechanisms and Kinetics of 1D-SiC/C/SiC Composite Materials: I, An Experimental Approach, *J. Am. Ceram. Soc., 77*(2), 459–66.

Fitzer, E. 1990. Oxidation of Molybdenum Disilicide, in *Ceramic Transactions Vol. 10: Corrosion and Corrosive Degradation of Ceramics*, Tressler, R.E. and McNallan, M. (eds.), Am. Ceram. Soc., Westerville, OH, pp. 19–41.

French, J.E. 1995. Ceramic Matrix Composite Fabrication and Processing: Polymer Pyrolysis, pp. 269–99 in *Handbook on Continuous Fiber-Reinforced Ceramic Matrix Composites*, Lehman, R.L., El-Rahalby, S.K. and Wachtman Jr., J.B. (eds.), CIAC Purdue University, West Lafayette, IN, and Am. Ceram. Soc., Westerville, OH, 1995.

Galasso, F.S. 1989. *Advanced Fibers and Composites*, Gordon and Breach Science Publishers, New York, NY, pp. 178.

Gates, T.S. and Zureick, A-H. (eds.). 1997. *High Temperature and Environmental Effects on Polymeric Composites*, 2nd Vol., STP 1302, ASTM, Philadelphia, PA.

Hallum, G.W and Herbell, T.P. 1988. Effects of High Temperature Hydrogen Exposure on Sintered α-SiC, *Adv. Ceramic Mater., 3*, 171–75.

Harris, C.E. and Gates, T.S. (eds.). 1993. *High Temperature and Environmental Effects on Polymeric Composites*, STP 1174, ASTM, Philadelphia, PA.

Heredia, F.E., McNulty, J.C., Zok, F.W. and Evans, A.G. 1995. Oxidation Embrittlement Probe for Ceramic Matrix Composites, *J. Am. Ceram. Soc., 78*(8), 2097–100.

Hermes, E.E. and Kerans, R.J. 1988. Degradation of Non-Oxide Reinforcement and Oxide Matrix Composites, in *Materials Research Society Symposium Proceedings, Vol. 125: Materials Stability and Environmental Degradation*, Barkatt, A., Verink, Jr., E.D., Smith, L.R. (eds.), Mat. Res. Soc., Pittsburgh, PA, pp. 73–78.

Herron, M.A. and Risbud, S.H. 1985. Characterization of Oxynitride Glass-Ceramic Matrix SiC Fiber Composites, in *Ceramic Engineering and Science Proceedings*, W.J. Smothers (ed.), Vol. 6, (7–8), pp. 622–31.

Herron, M.A. and Risbud, S.H. 1986. Characterization of SiC-Fiber-Reinforced Ba-Si-Al-O-N Glass-Ceramic Composites, *Ceram. Bull., 65*(2), 342–46.

Holmes, M. and Just, D.J. 1983. *GRP in Structural Engineering*, p. 10, Applied Science Publishers, New York, NY, pp. 298.

Hu, J., Chen, C.S., Xu, L.X., Yao, C.K. and Zhao, L.C. 2002a. Effect of Whisker Orientation on the Stress Corrosion Cracking Behavior of Alumina Borate Whisker Reinforced Pure Al Composite, *Materials Letters, 56*, 642–46.

Hu, J., Xu, L.X., and Yao, C.K. 2002b. Location Corrosion of Alumina Borate Whisker Reinforced AA2024 T6 Composite in Aqueous 3.5% NaCl Solution, *Materials Chem. Phys., 76*, 290–94.

Hu, J., Chu, W.Y., Fei, W.D. and Zhao, L.C. 2004. Effect of Interfacial Reaction on Corrosion Behavior of Alumina Borate Whisker Reinforced 6061 Al Composite, *Materials Sci. and Engr., A374*, 153–59.

Ilzuka, T. and Kita, H. 2004a. Oxidation Mechanism of Mo_5Si_3 Particle in Si_3N_4 Matrix Composite at 750°C, *Materials Sci. and Engr., A366*, 10–16.

Ilzuka, T. and Kita, H. 2004b. Oxidation Behavior and Effect of Oxidation on Mechanical Properties of Mo_5Si_3 Particle-Reinforced Si_3N_4 Composites, *Materials Sci. and Engr., A374*, 115–21.

Karlsdottir, S.N. and Halloran, J.W. 2007. Rapid Oxidation Characterization of Ultra-High Temperature Ceramics, *J. Am. Ceram. Soc.*, *90*(10), 3233–38.

Karlsdottir, S.N. and Halloran, J.W. 2008. Formation of Oxide Scales on Zirconium Diboride-Silicon Carbide Composites during Oxidation: Relation of Subscale Recession to Liquid Oxide Flow, *J. Am. Ceram. Soc.*, *91*(11), 3652–58.

Karlsdottir, S.N. and Halloran, J.W. 2009. Formation of Oxide Films on ZrB_2-15 Vol% SiC Composites during Oxidation: Evolution with Time and Temperature, *J. Am. Ceram. Soc.*, *92*(6), 1328–32.

Karlsdottir, S.N., Halloran, J.W. and Henderson, C.E. 2007. Convections Patterns in Liquid Oxide Films on Zirconium Diboride-Silicon Carbide Composites Oxidized at High Temperatures, *J. Am. Ceram. Soc.*, *90*(9), 2863–67.

Kerans, R.J., Hay, R.S., Parthasarathy, T.A. and Cinibulk, M.K. 2002. Interface Design for Oxidation-Resistant Ceramic Composites, *J. Am. Ceram. Soc.*, *85*(11), 2599–2632.

Kim, H-E. and Moorhead, A.J. 1991. Corrosion and Strength of SiC-Whisker-Reinforced Alumina Exposed at High Temperatures to H_2-H_2O Atmospheres, *J. Am. Ceram. Soc.*, *74*(6), 1354–59.

Klein, R., Medri, V., Desmaison-Brut, M., Bellosi, A. and Desmaison, J. 2003. Influence of Additives Content on the High Temperature Oxidation of Silicon Nitride Based Composites, *J. Euro. Ceram. Soc.*, *23*, 603–11.

Koch, H. and Syrett, B.C. 1985. Progress in EPRI Research on Materials for Flue Gas Desulphurization Systems, pp. 105–24 in *Dewpoint Corrosion*, Holmes, D.R. (ed.), Ellis Horwood Ltd., Chichester, UK.

Labruquere, S., Blanchard, H., Pailler, R. and Naslain, R. 2002. Enhancement of the Oxidation Resistance of Interfacial Area in C/C Composites. Part II: Oxidation Resistance of B-C, Si-B-C and Si-C Coated Carbon Preforms Densified with Carbon, *J. European Ceram. Soc.*, *22*, 1011–21.

Labruquere, S., Gueguen, J.S., Pailler, R. and Naslain, R. 2002. Enhancement of the Oxidation Resistance of Interfacial Area in C/C Composites. Part III: The Effect of Oxidation in Dry or Wet Air on Mechanical Properties of C/C Composites with Internal Protections, *J. European Ceram. Soc.*, *22*, 1023–30.

Lavrenko, V.A., Desmaison-Brut, M., Panasyuk, A.D. and Desmaison, J. 1998. Features of Corrosion Resistance of AlN-SiC Ceramics in Air up to 1600°C, *J. Euro. Ceram. Soc.*, *18*, 2339–44.

Lavrenko, V.A., Desmaison, J., Panasyuk, A.D. and Desmaison-Brut, M. 2003. Oxidation Resistance of AlN-(TiB_2-$TiSi_2$) Ceramics in Air up to 1450°C, *J. Euro, Ceram. Soc.*, *23*, 357–69.

Lavrenko, V.A., Desmaison, J., Panasyuk, A.D., Desmaison-Brut, M. and Fenard, E. 2005. High-Temperature Oxidation of AlN-SiC-TiB_2 Ceramics in Air, *J. Euro. Ceram. Soc.*, *25*, 1781–87.

Li, S-B., Xie, J-X., Zhang, L-T and Cheng, L-F. 2003a. Mechanical Properties and Oxidation Resistance of Ti_3SiC_2/SiC Composite Synthesized by In Situ Displacement Reaction of Si and TiC, *Materials Letters*, *57*, 3048–56.

Li, Z., Gao, W. Liang, J. and Zhang, D.L. 2003b. Oxidation Behavior of Ti_3Al-TiC Composites, *Materials Letters*, *57*, 1970–76.

Li, Z.W., Gao, W., Zhang, D.L. and Cai, Z.H. 2004. High Temperature Oxidation Behavior of a TiAl-Al_2O_3 Intermetallic Matrix Composite, *Corrosion Sci.*, *46*, 1997–2007.

Li, J., Lenosky, T.J., Forst, C.J. and Yip, S. 2008. Thermodynamical and Mechanical Stabilities of the Oxide Scale of ZrB_2+SiC and Oxygen Transport Mechanisms, *J. Am. Ceram. Soc.*, *91*(5), 1475–80.

Lin, H-T. and Becher, P. F. 1997. Stress-Temperature-Lifetime Response of Nicalon Fiber-Reinforced Silicon Carbide Composites in Air, pp. 128–41 in *Thermal and Mechanical Test Methods and Behavior of Continuous-Fiber Ceramic Composites*, ASTM STP 1309, Jenkins, M.G., Gonczy, S.T., Lara-Curzio, E., Ashbaugh, N.E. and Zawada, L.P. (eds.), ASTM, West Conshohocken, PA.

Liu, Z., Huang, B. and Gu, M. 2006. Corrosion Behavior of Al/AlN_p Composite in Alkaline Solution, *Materials Letters*, *60*, 2024–28.

Lowden, R.A., Stinton, D.P. and Besmann, T.M. 1995. Ceramic Matrix Composite Fabrication and Processing: Chemical Vapor Infiltration, pp. 205–68 in *Handbook on Continuous Fiber-Reinforced Ceramic Matrix Composites*, Lehman, R.L., El-Rahalby, S.K. and Wachtman Jr., J.B. (eds.), CIAC Purdue Univ., West Lafayette, IN, and Am. Ceram. Soc., Westerville, OH.

Luthra, K.L. 1987. Oxidation of SiC-Containing Composites, Ceram. Eng. Sci. Proc., *8* (7–8) 649–53.

Luthra, K.L. and Park, H.D. 1990. Oxidation of Silicon Carbide-Reinforced Oxide-Matrix Composites at 1375 and 1575°C, *J. Am. Ceram. Soc.*, *73*(4), 1014–23.

Marshall, D.B. and Evans, A.G. 1985. Failure Mechanisms in Ceramic-Fiber/Ceramic-Matrix Composites, *J. Am. Ceram. Soc.*, *68*(5), 225–31.

Maruyama, T. and Yanagihara, K. 1997. High Temperature Oxidation and Pesting of $Mo(Si,Al)_2$, *Mater. Sci. Eng. A, 239/240*, 828–41.

Mattia, D., Desmaison-Brut, M., Dimovski, S., Gogotsi, Y. and Desmaison, J. 2005a. Oxidation Behavior of an Aluminium Nitride-Hafnium Diboride Ceramic Composite, *J. Euro. Ceram. Soc.*, *25*, 1789–96.

Mattia, D., Desmaison-Brut, M., Tetard, D. and Desmaison, J. 2005b. Wetting of HIP $AlN-TiB_2$ Ceramic Composites by Liquid Metals and Alloys, *J. Euro. Ceram. Soc.*, *25*, 1797–1803.

Mazerolles, L., Feldhoff, A., Trichet, M.-F. and Backhaus-Ricoult, M. 2005. Oxidation Behavior of Si_3N_4-TiN Ceramics under Dry and Humid Air at High Temperature, *J. Euro. Ceram. Soc.*, *25*, 1743–48.

McCafferty, E. 1998. Effect of Electrolyte Volume on the Acid Dissolution of Aluminium Alloy 7075, *Corrosion*, *54*(11), 862.

McCracken, W.J., Clark, D.E. and Hench, L.L. 1980. Surface Characterization of Ceramed Composites and Environmental Sensitivity, pp. 311–17 in *Ceramic Engineering and Science Proceedings*, Vol. 1, (7–8A).

Medri, V., Bracisiewicz, M., Krnel, K., Winterhalter, F. and Bellosi, A. 2005. Degradation of Mechanical and Electrical Properties after Long-Term Oxidation and Corrosion of Non-oxide Structural Ceramic Composites, *J. Euro. Ceram. Soc.*, *25*, 1723–31.

Mendelson, M.I. 1985. SiC/Glass Composite Interphases, pp. 612–21 in *Ceramic Engineering and Science Proceedings*, Smothers, W.J. (ed.), Vol. 6, (7–8).

Metcalfe, A.G. and Schmitz, G.K. 1972. Mechanism of Stress Corrosion in E Glass Filaments, *Glass Technol.*, *13*(1), 5–16.

Mukerji, J. and Biswas, S.K. 1990. Synthesis, Properties, and Oxidation of Alumina-Titanium Nitride Composites, *J. Am. Ceram. Soc.*, *73*(1), 142–45.

Mitra, R., Khanna, R. and Rao, V.V.R. 2004. Microstructure, Mechanical Properties and Oxidation Behavior of a Multiphase (Mo,Cr)(Si,Al)$_2$ Intermetallic Alloy–SiC Composite Processed by Reaction Hot Pressing, *Materials Sci. and Engr., A382*, 150–61.

Monteverde, F. 2005. The Thermal Stability in Air of Hot-Pressed Diboride Matrix Composites for Uses at Ultra-High Temperatures, *Corrosion Sci., 47*, 2020–33.

Monteverde, F. and Bellosi, A. 2005. The Resistance to Oxidation of an HfB$_2$-SiC Composite, *J. Euro. Ceram. Soc., 25*, 1025–31.

Monteverde, F. and Scatteia, L. 2007. Resistance to Thermal Shock and to Oxidation of Metal Diborides-SiC Ceramics for Aerospace Application, *J. Am. Ceram. Soc., 90*(4), 1130–38.

More, K.L., Tortorelli, P.F. and Walker, L.R. 2003. High-Temperature Stability of SiC-Based Composites in High-Water-Vapor-Pressure Environments, *J. Am. Ceram. Soc., 86*(8), 1272–81.

Munson, K.L. and Jenkins, M.G. 1997. Retained Tensile Properties and Performance of an Oxide-Matrix Continuous-Fiber Ceramic Composite after Elevated-Temperature Exposure in Ambient Air, pp. 176–89 in *Thermal and Mechanical Test Methods and Behavior of Continuous-Fiber Ceramic Composites*, ASTM STP 1309, Jenkins, M.G., Gonczy, S.T., Lara-Curzio, E., Ashbaugh, N.E. and Zawada, L.P. (eds.), ASTM, West Conshohocken, PA.

Narushima, T., Goto, T., Iguchi, Y. and Hirai, T. 1990. High-Temperature Oxidation of Chemically Vapor-Deposited Silicon Carbide in Wet Oxygen at 1823 to 1923 K, *J. Am. Ceram. Soc., 73*(12), 3580–84.

Nelson, H.G. 1993. A Challenge to Materials: Advanced Hypersonic Flight Hydrogen and High Temperature Materials, pp. 11–20, *Proc. 1993 Conf. on Processing, Fabrication and Applications of Advanced Composites*, Long Beach, CA, 9–11 Aug, Upadhya, K. (ed.), ASM, OH.

Nolan, T.A., Allard, L.F., Coffey, D.W., Hubbard, C.R. and Padgett, R.A. 1991. Microstructure and Crystallography of Titanium Nitride Whiskers Grown by a Vapor-Liquid-Solid Process, *J. Am. Ceram. Soc., 74*(11), 2769–75.

Ogbuji, L.U.J.T. 2003. Pest-Resistance in SiC/BN/SiC Composites, *J. Eur. Ceram. Soc., 23*, 613–17.

Otsuka, A., Waku, Y. and Tanaka, R. 2005. Corrosion of a Unidirectionally Solidified Al$_2$O$_3$/YAG Eutectic Composite in a Combustion Environment, *J. Euro. Ceram. Soc., 25*, 1269–74.

Panda, P.C. and Seydel, E.R. 1986. Near-Net-Shape Forming of Magnesia-Alumina Spinel/Silicon Carbide Fiber Composites, *Ceram. Bull., 65*(2), 338–41.

Pannhorst, W., Spallek, M., Brückner, R., Hegeler, H., Reich, C., Grathwohl G., Meier, B. and Spelmann, D. 1990. Fiber-Reinforced Glasses and Glass-Ceramics Fabricated by a Novel Process, *Ceram. Eng. Sci. Proc., 11*(7–8), 947–63.

Pantano, C.G., Spear, K.E., Qi, G. and Beall, D.M. 1993. Thermochemical Modeling of Interface Reactions in Glass Matrix Composites, pp. 173–98 in *Ceramic Transaction, Vol. 38, Advances in Ceramic-Matrix Composites*, Bansal, N.P. (ed.), Am. Ceram. Soc., Westerville, OH.

Peng, F. and Speyer, R.F. 2008. Oxidation Resistance of Fully Dense ZrB$_2$ with SiC, TaB$_2$, and TaSi$_2$ Additives, *J. Am. Ceram. Soc., 91*(5), 1489–94.

Pilling, N.B. and Bedworth, R.E. 1923. The oxidation of metals at high temperatures, *J. Inst. Met., 29*, 529–82.

Prewo, K.M., Brennan, J.J. and Layden, G.K. 1986. Fiber Reinforced Glasses and Glass-Ceramics for High Performance Applications, *Ceram. Bull.*, 65(2), 305–13.

Pugsley, V.A. and Sockel, H.-G. 2004. Corrosion Fatigue of Cemented Carbide Cutting Tool Materials, *Materials Sci. and Engr.*, A366, 87–95.

Ready, D.W. 1992. High Temperature Gas Corrosion of Ceramic Composites, *Ceram. Eng. Sci. Proc.*, 13(7–8), 301–18.

Revankar, V., Hexemer, R., Mroz, C., Bothwell, D., Goel, A., Bray, D. and Blakely, K. 1995. Novel Process for Titanium Nitride Whisker Synthesis and Their Use in Alumina Composites, pp. 135–46, in *Advanced Synthesis and Processing of Composites and Advanced Ceramics*, Logan, K.V. (ed.), *Ceramic Transactions*, Vol. 56, American Ceramic Society, Westerville, OH.

Rezaie, A., Fahrenholtz, W.G. and Hilmas, G.E. 2006. Oxidation of Zirconium Diboride-Silicon Carbide at 1500°C at Low Partial Pressure of Oxygen, *J. Am. Ceram. Soc.*, 89(10), 3240–45.

Rice, R.W. 1990. Toughening in Ceramic Particulate and Whisker Composites, *Ceram. Eng. Sci. Proc.*, 11(7–8), 667–94.

Ringuedé, A., Fagg, D.P. and Frade, J.R. 2004. Electrochemical Behavior and Degradation of (Ni,M)/YSZ Cermet Electrodes (M = Co, Cu, Fe) for High Temperature Applications of Solid Electrolytes, *J. Euro. Ceram. Soc.*, 24, 1355–58.

Sambell, R.A., Bowen, D. and Phillips, D.C. 1972. Carbon Fiber Composites with Ceramic and Glass Matrices, Part 1, Discontinuous Fibers, *J. Mat. Sci.*, 7, 663–75.

Sato, T. and Tokunaga, Y. 1988. Corrosion of Silicon Nitride Ceramics in Aqueous Hydrogen Chloride Solutions, *J. Am. Ceram. Soc.*, 71, 1074–79.

Sawyer, L.C., Chen, R.T., Haimbach IV, F., Harget, P.J., Prack, E.R. and Jaffe, M. 1986. Thermal Stability Characterization of SiC Ceramic Fibers: II, Fractography and Structure, *Ceram. Eng. Sci. Proc.*, 7(7–8), 914–30.

Seah, K.H.W., Krishna, M., Vijayalakshmi, V.T. and Uchil, J. 2002. Effects of Temperature and Reinforcement Content on Corrosion Characteristics of LM13/Albite Composites, *Corros. Sci.*, 44(4), 761–72.

Shin, H.H., Berta, Y. and Speyer, R.F. 1993. Fiber-Matrix Interaction in SiC Fiber Reinforced Ceramic and Intermetallic Composites, pp. 235–48 in *Ceramic Transaction, Vol. 38, Advances in Ceramic-Matrix Composites*, Bansal, N.P. (ed.), Am. Ceram. Soc., Westerville, OH.

Singh, R.N. and Brun, M.K. 1987. Effect of Boron Nitride Coating on Fiber-Matrix Interactions, *Ceram. Eng. Sci. Proc.*, 8(7–8), 636–43.

Singhal, S.C. 1976. Effect of Water Vapor on the Oxidation of Hot-Pressed Silicon Nitride, *J. Am. Ceram. Soc.*, 59(1–2), 81–82.

Strife, J.R. 1991. Fundamentals of Protective Coating Strategies for Carbon-Carbon Composites, pp. 121–27 in *Damage and Oxidation Protection in High Temperature Composites*, Vol. 1, Haritos, G.K. and Ochoa, O.O. (eds.), ASME, New York, NY.

Studt, T. 1991. Breaking Down the Barriers for Ceramic Matrix Composites, *R & D Magazine*, Aug, pp. 36–42.

Tampieri, A. and Bellosi, A. 1992. Oxidation Resistance of Alumina-Titanium Nitride and Alumina-Titanium Carbide Composites, *J. Am. Ceram. Soc.*, 75(6), 1688–90.

Taya, M. and Arsenault, R.J. 1989. *Metal Matrix Composites Thermomechanical Behavior*, Pergamon Press, New York, NY, p. 264.

Thomas, J.M. 1965. in *Chemistry and Physics of Carbon*, Vol. 1, Walker Jr., P.L. (ed.), Marcel Dekker, New York, NY, pp. 135–68.

Tian, J. and Shobu, K. 2003. Improvement in the Oxidation Resistance of Oxide-Matrix Silicon Carbide-Particulate Composites by Mullite Infiltration, *J. Am. Ceram. Soc.*, 86(10), 1806–8.

Trzaskoma, P.P. 1986. Corrosion Behavior of a Graphite Fiber/Magnesium Metal Matrix Composite in Aqueous Chloride Solution, *Corrosion*, 42(10), 609–13.

Vasques, B., Tixier, C., Klein, R., Hidalgo, H., Tristant, P. and Desmaison, J. 2005. Modification of the Oxidation Behavior of Si_3N_4-TiB_2 Composites by PECVD Alumina Coatings, *J. Euro. Ceram. Soc.*, 25, 1749–55.

Verrilli, M.J., Calomino, A.M. and Brewer, D.N. 1997. Creep-Rupture Behavior of a Nicalon/SiC Composite, pp. 158–75 in *Thermal and Mechanical Test Methods and Behavior of Continuous-Fiber Ceramic Composites*, ASTM STP 1309, Jenkins, M.G., Gonczy, S.T., Lara-Curzio, E., Ashbaugh, N.E. and Zawada, L.P. (eds.), ASTM, West Conshohocken, PA.

Wall, F.D., Taylor, S.R. and Cahen, G.L. 1993. The Simulation and Detection of Electrochemical Damage in BMI/Graphite Fiber Composites Using Electrochemical Impedance Spectroscopy, pp. 95–113 in *High Temperature and Environmental Effects on Polymeric Composites*, STP 1174, Harris, C.E. and Gates, T.S. (eds.), ASTM, Philadelphia, PA.

Wang, T.C., Chen, R.Z. and Tuan, W.H. 2003. Oxidation Resistance of Ni-Toughened Al_2O_3, *J. Eur. Ceram. Soc.*, 23, 927–34.

Wang, H.J., Gao, P.Z. and Jin, Z.H. 2005. Preparation and Oxidation Behavior of Three-Dimensional Braided Carbon Fiber Coated by SiC, *Materials Letters*, 59, 486–90.

Werheit, H. 2007. Are There Bipolarons in Icosahedral Boron-Rich Solids?, *J. Phys.: Condens. Matter*, 19(18), 186–207.

Wetherhold, R.C. and Zawada, L.P. 1991. Heat Treatments as a Method of Protection for a Ceramic Fiber-Glass Matrix Composite, *J. Am. Ceram. Soc.*, 74(8), 1997–2000.

Williams, E.L. 1965. Diffusion of Oxygen in Fused Silica, *J. Am. Ceram. Soc.*, 48(4), 190–94.

Wu, X., Holmes, J.W. and Hilmas, G.E. 1995. Environmental Properties of Ceramic Matrix Composites, pp. 431–71 in *Handbook on Continuous Fiber-Reinforced Ceramic Matrix Composites*, Lehman, R.L., El-Rahalby, S.K. and Wachtman Jr. J.B. (eds.), CIAC Purdue Univ., West Lafayette, IN, and Am. Ceram. Soc., Westerville, OH.

Xing, H., Cao, X., Hu, W., Zhao, L. and Zhang, J. 2005. Interfacial Reactions in 3D-SiC Network Reinforced Cu-Matrix Composites Prepared by Squeeze Casting, *Materials Letters*, 59, 1563–66.

Zawada, L.P., Staehler, J. and Steel, S. 2003. Consequence of Intermittent Exposure to Moisture and Salt Fog on the High-Temperature Fatigue Durability of Several Ceramic Matrix Composites, *J. Am. Ceram. Soc.*, 86(8), 1282–91.

Zhang, S.C., Hilmas, G.E. and Fahrenholtz, W.G. 2008. Improved Oxidation Resistance of Zirconium Diboride by Tungsten Carbide Additions, *J. Am. Ceram. Soc.*, 91(11), 3530–35.

Section IV

Properties and Corrosion

18

Mechanisms

... if anything is to be hoped, everything ought to be tried.

James Madison

Introduction

Probably the most important property that is affected by corrosion is that of mechanical strength. Other properties are also affected by corrosion; however, they generally do not lead to failure, as is often the case with changes in strength. Strength loss is not the only mechanical effect of corrosion, since there are also many cases where the effects of corrosion lead to increased strength. Increases in strength due to corrosion are the result of healing of cracks and flaws in the surface layers of a specimen due, quite often, to the diffusion of impurities from the bulk to the surface. This change in chemistry at the surface may lead to the formation of a compressive layer on the surface due to differential thermal expansion between the surface layer and the bulk. Compressive surface layers may also form due to surface alteration layers having a larger specific volume than the bulk.

Environmentally enhanced strength loss may arise through the following phenomena:

1. cracking of the surface alteration layers due to excessive mismatch in thermal expansion between the surface and the bulk,
2. melting of secondary phases at high temperature,
3. lowering of the viscosity of a glassy grain boundary phase at high temperature,
4. surface cracking caused by polymorphic transitions in the crystalline phases at the surface,
5. alteration that forms low-strength phases,
6. formation of voids and pits, especially true for corrosion by oxidation, and
7. crack growth.

The term used to describe these phenomena is *stress corrosion* or *stress corrosion cracking (SCC)*, which occurs when a material is subjected to a corrosive environment while being under the influence of an external mechanical load. Stress corrosion cracking implies that the pair of parameters, applied stress and corrosive environment, must both be active. Removal of either the applied stress or the corrosive environment will prevent cracking.

Oxidation often leads to compositional and structural alteration, especially of surface layers and grain boundary phases, of a ceramic that subsequently leads to considerable changes in the physical properties. Such alterations can lead to changes in density, thermal expansion, and thermal and electrical conductivity. The influence that these changes have upon mechanical properties can be deduced only through a thorough investigation of the mechanisms and kinetics of corrosion. For example, the oxidation of silicon-based ceramics has been shown to be either active or passive depending upon the partial pressure of oxygen present during exposure (see Chapter 15 for a discussion of the oxidation of SiC and Si_3N_4). When the pO_2 is low, gaseous SiO is formed, leading to rapid material loss and generally to a loss in strength. When the pO_2 is high, SiO_2 is formed that can lead to strength increases depending upon the actual temperature and time of exposure and whether the strength test is conducted at room or an elevated temperature. The investigator should be well aware that conducting mechanical property tests in air (which may also include moisture) at elevated temperatures constitutes exposure to a corrosive environment for many materials.

The failure of ceramics after long exposure to a constant applied load, well below the critical stress, is called *static fatigue* or *delayed failure*. If the load is applied under constant stress rate conditions, it is called *dynamic fatigue*. If the load is applied, removed, and then reapplied, the failure after longtime cycling is called *cyclic fatigue*. It is now well known that brittle fracture is quite often preceded by subcritical crack growth that leads to a time dependence of strength. It is the effect of the environment upon the subcritical crack growth that leads to the phenomenon termed stress corrosion cracking. Thus fatigue (or delayed failure) and stress corrosion cracking relate to the same phenomenon. In glassy materials, this delayed failure has been related to glass composition, temperature, and the environment (e.g., pH). Failure is due to the chemical reaction that takes place preferentially at strained bonds at the crack tip with the rate being stress sensitive. Some crystalline materials exhibit a delayed failure similar to that in glasses.

The experimental relationship between crack velocity and the applied stress (i.e., stress intensity factor K_I) is therefore of utmost importance. Attempts to fit various mathematical relationships to the experimental data have led to an assortment of equations of either the commonly used power law type or of some exponential form. The power law:

$$v = A(K_I/K_{IC})^n \qquad (18.1)$$

where A is a material constant (strong dependency upon environment, temperature, etc.), n is the stress corrosion susceptibility parameter (weak dependency upon environment), K_I is the applied stress intensity, and K_{IC} is the critical stress intensity factor, has been used most often due to its simplicity. It is the value of n (and also A) that determines a material's susceptibility to subcritical crack growth. Final lifetime predictions are very sensitive to the value of n. The power law, however, does not always lead to the best representation. Jakus et al. 1981 evaluated the prediction of static fatigue lifetimes from experimental dynamic fatigue data for four different materials and environments. These were hot-pressed silicon nitride at 1200°C in air, alumina in moist air, optical glass fiber in air, and soda-lime glass in water. They found that the exponential forms of the crack velocity equations allowed better predictions of lifetimes for the silicon nitride and optical glass fiber, but the power law form of the crack velocity equation allowed better predictions for alumina and soda-lime glass. Thus they concluded that one should collect data for several different loading conditions and then select the crack velocity equation that best represents all the data for making lifetime predictions. Matthewson 1993 has reported that one particular optical fiber material gave a best fit to the exponential form when tested in ambient air but gave a best fit to the power law when tested at 25°C in a $pH = 7$ buffer solution. Matthewson suggested that one kinetics model unique to all environments probably does not exist, and that since the power law yields the most optimistic lifetimes, it is unsatisfactory for design purposes.

Crack velocity can be evaluated by direct and indirect methods. In the direct methods, crack velocity is determined as a function of the applied stress. These involve testing by techniques such as the double cantilever beam method, the double torsion method, and the edge or center cracked specimen method. Indirect methods, which are normally performed on opaque samples, infer crack velocity data from strength measurements. A common indirect method is to determine the time-to-failure as a function of the applied load. In addition to the constant load technique, the constant strain technique has also been used. Other methods that have been used to evaluate the effects of corrosion upon the mechanical properties of ceramics include the following:

1. the percent loss in fracture strength after exposure to a corrosive environment (strength test conducted at room temperature),

2. the fracture strength at some elevated temperature during exposure to a corrosive environment,

3. the evaluation of creep resistance during exposure to a corrosive environment, and

4. the determination of the strength distribution (at room temperature) after exposure to a corrosive environment and a static load. Generally this type of evaluation indicates the dynamic nature of the flaw population.

Because silicate glasses are isotropic and homogeneous, most of the investigations into mechanisms have been done on these materials.

Mechanisms

Crystalline Materials

There have been several mechanisms described in the literature, some of which are due to variations in the environment. Probably the most important area where questions still arise is what actually is occurring at the crack tip. Although the mechanism described by Johnson et al. 1984, Evans and Blumenthal 1984, and Cao et al. 1987 involved the effects of an intrinsic, small quantity of a secondary amorphous phase, the overall effect should be very similar to the case when a solid is in contact with a corrosive environment that either directly supplies the amorphous phase to the crack tips or forms an amorphous phase at the crack tips through alteration. In essentially single-phase polycrystalline alumina, Johnson et al. 1984 attributed cracking to the penetration into the grain boundaries of amorphous phase that was contained at the crack tip of intrinsic cracks, which subsequently caused localized creep embrittlement. Crack blunting will occur if the amorphous phase becomes depleted at the crack tip.

Strength degradation at high temperatures according to Lange 1974 and 1979 is due to crack growth at stress levels below the critical applied stress required for fracture. This type of crack growth is called *subcritical crack growth* and is caused by cavitation of the glassy grain boundary phase located at grain junctions. The stress field surrounding the crack tip causes the glassy phase to cavitate facilitating grain boundary sliding, thus allowing cracks to propagate at stress levels less than critical. Surface and grain boundary self-diffusion were reported by Chuang 1982 to be the accepted controlling factors in cavity growth at high temperatures, although other factors such as grain sliding and dislocation slip may also be present.

A mechanism for stress corrosion cracking at high temperatures was believed by Cao et al. 1987 to be due to stress-enhanced diffusion through the corrosive amorphous phase from crack surfaces, causing accelerated crack propagation along grain boundaries. They made the following assumptions:

1. flat crack surfaces behind the crack tip,
2. principal flux toward the crack tip,
3. equilibrium concentration of the solid in the liquid,
4. reduced solid in liquid at the crack tip caused by crack surface curvature,

5. sufficiently slow crack tip velocity to allow viscous flow of liquid into the tip, and

6. chemical potential gradient normal to the crack plane was ignored.

Cao et al. pointed out that this mechanism was most likely to occur in materials where the amorphous phase was discontinuous. Systems that contained a small dihedral angle (see Chapter 8, Surface Effects, for a discussion of dihedral angles) at the grain boundary and contained low-viscosity amorphous phases were the ones that were the most susceptible to rapid crack propagation. Thus the wetting of the solid by the amorphous phase is of primary importance, since phases that wet well form small dihedral angles that induce sharp crack tips.

Glassy Materials

It is a well-known fact that silicate glasses can be strengthened by etching in hydrofluoric acid. This phenomenon has been explained by Hillig and Charles 1965 to be one that involved the increase in the radius of curvature of the tips of surface cracks caused by the uniform rate of attack, which depends upon the curvature, by the corrosive medium. This increase in radius of curvature or rounding of the crack tips increases the critical stress required for failure. Bando et al. 1984 gave direct TEM evidence of crack tip blunting in thin foils of silica glass, supporting the dissolution/precipitation theory of crack tip blunting suggested by Ito and Tomozawa 1982, although it was not clear that the precipitated material caused any significant strength increase. However, under the influence of an applied stress, Charles 1958 concluded that the corrosion reaction rate was stress sensitive, leading to an increased rate of attack at the crack tip and thus a decrease in the radius of curvature (i.e., a sharper crack tip) and a lower strength.

The fact that glass suffers from static fatigue has been known for many years and studies over the past few decades have elucidated the reasons for this behavior. It is now believed that the reaction between water vapor and the glass surface is stress dependent and leads to eventual failure when glass is subjected to static loading. As reported by Wiederhorn 1967, three regions of behavior are exhibited when crack velocity is plotted versus applied force (depicted in Figure 18.1). In the first region, the crack velocity (as low as 10^{-10} m/s) is dependent upon the applied force, with the exact position of the curve and its slope being dependent upon the humidity. At higher humidities the crack growth occurred more rapidly and at a lower force. In addition Wiederhorn and Bolz 1970 have shown that the slope and position of the curve in this first region was dependent upon the glass composition (stress corrosion resistance was in the order fused silica > aluminosilicate > borosilicate > soda-lime silicate > lead silicate) and Wiederhorn and Johnson 1973 have shown that it was dependent upon the pH. In the second region,

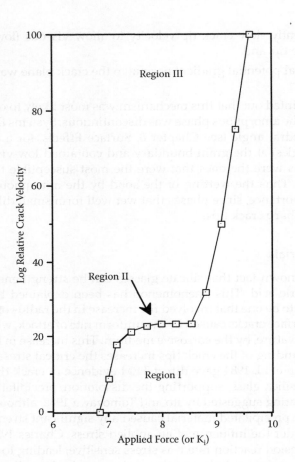

FIGURE 18.1
Crack velocity versus applied force (or K_i). (After Wiederhorn, 1967.)

the crack velocity is essentially independent of the applied force. At higher humidities, this portion of the curve shifts to higher velocities. In the third region, the crack velocity was again dependent upon the applied force but the slope is much steeper, indicative of a different mechanism for crack propagation. This third region was also independent of the humidity.

Wiederhorn 1972 has shown that the crack growth in glasses was dependent upon the pH of the environment at the crack tip and was controlled by the glass composition. Wiederhorn and Johnson 1973 clarified that even further by reporting that at high crack velocities the glass composition (for silica, borosilicate, and soda-lime glasses) controlled the pH at the crack tip and that at low crack velocities the electrolyte controlled the pH at the crack tip. They studied the crack velocity as a function of the applied stress intensity, which they determined by the following equation for a double cantilever beam specimen:

$$K_I = \frac{PL \ (3.467 + 2.315 \ t/L)}{(wa)^{1/2}t^{3/2}}$$ (18.2)

where:

P = applied load,
L = crack length,
w = total thickness,
a = web thickness, and
t = half width.

The actual shape of the velocity versus K_I curves is determined by a balance between the corrosion process, which tends to increase the crack tip radius, and the stress-corrosion process, which tends to decrease the crack tip radius, Simmons and Freiman 1981.

Wiederhorn et al. 1982 gave an equation of the following type for determining the crack velocity in aqueous media:

$$v = v_o a_{H_2O} \exp(-\Delta G^*/RT)$$ (18.3)

where:

v = crack velocity,
v_o = empirical constant,
a_{H_2O} = activity of water,
ΔG^* = free energy of activation,
R = gas constant, and
T = temperature.

They derived this equation from reaction rate theory, assuming that crack velocity was directly proportional to the reaction rate. In addition, they assumed that the reaction order was equal to one with respect to water in solution. This, it was pointed out, was reasonable at the high relative humidities of their work but was most likely incorrect at low relative humidities, where it is probably one-half based upon the work of Freiman 1974 in alcohols. The activity of water vapor over a solution is equal to the ratio of the actual vapor pressure to that of pure water. For water dissolved in a non-aqueous liquid, this ratio is equivalent to the relative humidity over the solution. This is why the crack velocity is dependent upon the relative humidity and not the concentration of the water, Freiman 1974. Thus it is important not to assume that a liquid is inert just because it has a low solubility for water. In the region of high crack velocities (i.e., region III), it is the chain length of the alcohol for N between 6 and 8 that determines crack velocity.

The pH at the crack tip was dependent upon the reaction of the solution at the crack tip with the glass composition and diffusion between the bulk solution and the solution at the crack tip. Ion exchange at the crack tip between protons from the solution and alkalies from the glass produced $(OH)^-$ ions and thus a basic pH at the crack tip. Ionization of the silicic acid and silanol groups at the glass surface produced an acid pH at the crack tip. Estimated crack tip pH ranged from about 4.5 for silica glass to about 12 for soda-lime glass. At high crack velocities reaction rates at the crack tip are fast and the glass composition controls the solution pH. At low velocities, diffusion depletes the solution at the crack tip, which is then similar to the bulk solution. Wiederhorn and Johnson 1973 concluded that silica exhibited the greatest resistance to static fatigue in neutral and basic solutions whereas borosilicate glass exhibited the greatest resistance in acid solutions.

Michalske and Bunker 1987 gave an equation that relates crack velocity of a silica glass to the applied stress intensity (K_I) for environments of ammonia, formamide, hydrazine, methanol, N-methylformamide, and water. This equation is given below:

$$V = V_o \exp(nK_I) \tag{18.4}$$

where:
V = crack velocity,
V_o = empirical constant,
K_I = applied stress intensity, and
n = slope of the exponential plot.

Crack velocity versus applied stress intensity plots (same as Figure 18.1) yielded region I behavior for ammonia, formamide, hydrazine, methanol, N-methylformamide, and water. A small amount of residual water contained in aniline, n-propylamine, and *tert*-butylamine yielded a behavior representative of regions I and II. Moist N_2 exhibited a behavior represented by all three regions. Michalske and Bunker 1987 interpreted the mechanism for each region based upon the representations given in Table 18.1.

All the chemicals that exhibited region I behavior only have at least one lone pair electron orbital close to a labile proton. Using the shift in the

TABLE 18.1

Representation of Crack Growth for Each Region of Figure 18.1

Region	Represented By
I	Velocity (<10^{-5} m/s) exponentially dependent upon the applied K_I. Crack growth controlled by corrosive action of water (ion exchange) that ruptures Si-O-Si bonds as shown in Figure 18.2.
II	Velocity ($10^{-5} < v < 10^{-3}$ m/s) independent of applied K_I. Crack growth controlled by transport rate of water to the crack tip.
III	Velocity (>10^{-3} m/s) exponentially dependent upon the applied K_I. Crack growth controlled purely by mechanical bond rupture.

vibrational frequencies of the OH groups on silica surfaces, Michalske and Bunker concluded that all nine of the chemicals tested acted as effective bases toward the silica surface silanol groups and thus one would expect a similar behavior based solely upon chemical activity.

Michalske and Bunker, therefore, developed a steric hindrance model to explain why aniline, *n*-propylamine, and *tert*-butylamine exhibited a bimodal behavior. These molecules were the largest of all those examined, and a critical diameter of <0.5 nm for molecular diffusion to the crack tip opening was suggested. They also noted that as the size of the corrosive environment molecule exhibiting region I behavior only increased, its effectiveness decreased. This whole area of the effects of steric hindrance and chemical activity upon stress corrosion fracture kinetics appears to be one of some importance, not only to glass but also to crystalline materials.

For environments to enhance the crack growth, they must be both electron and proton donors, Michalske and Freiman 1983. In soda-lime glass the modifier ions do not participate directly in the fracture process but may change the reactivity of the Si-O bridging bonds and affect the elastic properties of the network bonds, Freiman et al. 1985. Thus static fatigue is controlled by the stress-enhanced reaction rate between the Si-O bond and the environment at the crack tip.

Michalske and Freiman 1983 described a three-step mechanism for the reaction of water with strained Si-O bonds. These were as follows:

1. Water molecule aligns its oxygen lone electron pair orbitals toward the Si with hydrogen bonding to the oxygen of the silica (a strained Si-O bond enhances reaction at this site).

2. Electron transfer from oxygen of water to Si along with proton transfer to oxygen of silica.

3. Rupture of hydrogen bond to oxygen of water and the transferred hydrogen yielding Si-OH bonds on each fracture surface.

This mechanism is depicted in Figure 18.2. This mechanism appears to be a general one, at least for cations that are attracted to the oxygen's (of water) lone electron pair. Michalske et al. 1986 have shown that this dissociative chemisorption mechanism is the same for alumina, although the details differ. In alumina it is not necessary for the bonds to be strained for adsorption to occur.

White et al. 1987 reported that Li^+ ions in solution negated the enhanced rates noted in high-*pH* solutions where OH^- controls the rate of bond breaking by readily associating with the OH^-, and not allowing OH^- to react with the Si-O-Si bond at the crack tip. This type of reaction is not exhibited by other alkalies, since they do not readily react with OH^- ions.

A more recent review of this topic of stress corrosion in silicate glasses was given by Gy 2003.

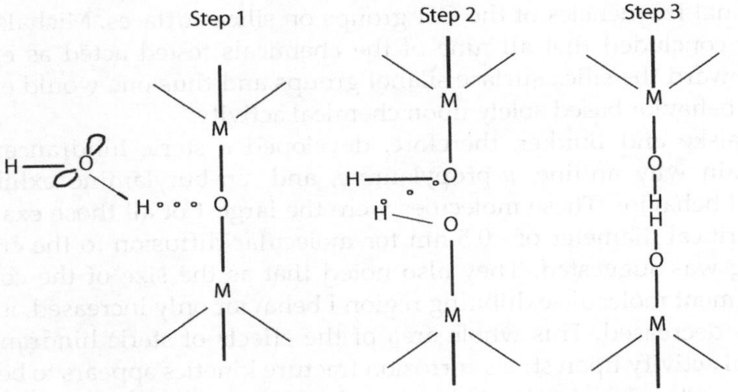

Step 1   Step 2   Step 3

FIGURE 18.2
Mechanism of bond rupture. (After Michalske and Freiman, 1983.)

Additional Recommended Reading

Evans, A.G. (ed.). *Fracture in Ceramic Materials*, Noyes Publications, Park Ridge, NJ, 1984, 420 pp.

Frechette, V.D. and Varner, J.R. (eds.) *Ceramic Transactions, Vol. 17, Fractography of Glasses and Ceramics II*, Amer. Ceram. Soc., Westerville, OH, 1991, 548 pp.

Varner, J.R. and Frechette, V.D. (eds.). *Advances in Ceramics, Vol. 22, Fractography of Glasses and Ceramics*, Amer. Ceram. Soc., Westerville, OH, 1988, 442 pp.

Exercises, Questions, and Problems

1. Environmentally enhanced strength loss may arise through seven different phenomena. Discuss three of these and their relationship to corrosion.

2. Describe *stress corrosion cracking* and the consequences that relate to engineering materials.

3. Describe the differences among *static, dynamic, delayed,* and *cyclic* fatigue.

4. How does *stress corrosion cracking* relate to the type of fatigue listed in question #3?

5. How does one determine whether to use the power law or the exponential form to represent best the static fatigue lifetimes?

6. Discuss how cracks may propagate at a stress level less than that of the critical one for crack growth.

7. Describe several methods that have been used to evaluate the effects of corrosion upon the mechanical properties of ceramics.
8. Why is it that silicate glasses can be strengthened by etching in acid?
9. Discuss the three regions of behavior related to crack velocity and applied force for glassy materials. What role does relative humidity play?
10. What is the applied stress intensity (K_I)?

References

Bando, Y., Ito, S. and Tomozawa, M. 1984. Direct Observation of Crack Tip Geometry of SiO_2 Glass by High-Resolution Electron Microscopy, *J. Am. Ceram. Soc.*, *67*(3), C36–37.

Cao, H.C., Dalgleish, B.J., Hsueh, C-H. and Evans, A.G. 1987. High-Temperature Stress Corrosion Cracking in Ceramics, *J. Am. Ceram. Soc.*, *70*(4), 257–64.

Charles, R.J. 1958. Static Fatigue of Glass. I, *J. Appl. Phys.*, *29*(11), 1549–54.

Chuang, T-J. 1982. A Diffusive Crack-Growth Model for Creep Fracture, *J. Am. Ceram. Soc.*, *65*(2), 93–103.

Evans, A.G. and Blumenthal, W. 1984. High Temperature Failure Mechanisms in Ceramic Polycrystals, in *Deformation of Ceramics II*, Tressler, R.E. and Bradt, R.C. (eds.), Plenum Publishing Co., New York, NY, pp. 487–505.

Freiman, S.W. 1974. Effect of Alcohols on Crack Propagation in Glass, *J. Am. Ceram. Soc.*, *57*(8), 350–53.

Freiman, S.W., White, G.S. and Fuller Jr., E.R. 1985. Environmentally Enhanced Crack Growth in Soda-Lime Glass, *J. Am. Ceram. Soc.*, *68*(3), 108–12.

Gy, R. 2003. Stress Corrosion of Silicate Glass: A Review, *J. Non-Cryst. Solids*, *316*, 1–11.

Hillig, W.B. and Charles, R.J. 1965. Surfaces, Stress-Dependent Surface Reactions, and Strength, Chapter 17 in *High Strength Materials*, Zackey, V.F. (ed.), Wiley & Sons, New York, NY, pp. 682–705.

Ito, S. and Tomozawa, M. 1982. Crack Blunting of High-Silica Glass, *J. Am. Ceram. Soc.*, *65*(8), 368–71.

Jakus, K., Ritter Jr., J.E. and Sullivan, J.M. 1981. Dependency of Fatigue Predictions on the Form of the Crack Velocity Equation, *J. Am. Ceram. Soc.*, *64*(6), 372–74.

Johnson, M.J., Dalgleish, B. J. and Evans, A.G. 1984. High Temperature Failure of Polycrystalline Alumina: III. Failure Times, *J. Am. Ceram. Soc.*, *67*(11), 759–63.

Lange, F.F. 1974. High-Temperature Strength Behavior of Hot-Pressed Si_3N_4: Evidence for Subcritical Crack Growth, *J. Am. Ceram. Soc.*, *57*(2), 84–87.

Lange, F.F. 1979. Evidence for Cavitation Crack Growth in Si_3N_4, *J. Am. Ceram. Soc.*, *62*(3–4), 222–23.

Matthewson, M.J. 1993. Models for Fiber Reliability, *Proc. Int. Symp. Fiber Optic Networks & Video Communications*, Berlin, Germany, Apr.

Michalske, T.A. and Bunker, B.C. 1987. Steric Effects in Stress Corrosion Fracture of Glass, *J. Am. Ceram. Soc.*, *70*(10), 780–84.

Michalske, T.A., Bunker, B.C. and Freiman, S.W. 1986. Stress Corrosion of Ionic and Mixed Ionic/Covalent Solids, *J. Am. Ceram. Soc.*, 69(10), 721–24.

Michalske, T.A. and Freiman, S.W. 1983. A Molecular Mechanism for Stress Corrosion in Vitreous Silica, *J. Am. Ceram. Soc.*, 66(4), 284–88.

Simmons, C.J. and Freiman, S.W. 1981. Effect of Corrosion Processes on Subcritical Crack Growth in Glass, *J. Am. Ceram. Soc.*, 64(11), 683–86.

White, G.S., Freiman, S.W., Wiederhorn, S.M. and Coyle, T.D. 1987. Effects of Counterions on Crack Growth in Vitreous Silica, *J. Am. Ceram. Soc.*, 70(12), 891–95.

Wiederhorn, S.M. 1967. Influence of Water Vapor on Crack Propagation in Soda-Lime Glass, *J. Am. Ceram. Soc.*, 50(8), 407–14.

Wiederhorn, S.M. 1972. A Chemical Interpretation of Static Fatigue, *J. Am. Ceram. Soc.*, 55(2), 81–85.

Wiederhorn, S.M. and Bolz, L.H. 1970. Stress Corrosion and Static Fatigue of Glass, *J. Am. Ceram. Soc.*, 53(10), 543–48.

Wiederhorn, S.M., Freiman, S.W., Fuller Jr., E.R. and Simmons, C.J. 1982. Effect of Water and Other Dielectrics on Crack Growth, *J. Mater. Sci.*, 17(12), 3460–78.

Wiederhorn, S.M. and Johnson, H., 1973. Effect of Electrolyte pH on Crack Propagation in Glass, *J. Am. Ceram. Soc.*, 56(4), 192–97.

19

Degradation by Specific Environments

Homogeneous bodies of materials—I was told—do not exist, homogeneous states of stress are not encountered.

Otto Mohr

Degradation by Aqueous Solutions

Bioactive Materials

Bioactive ceramics include those materials that rapidly react with human tissue to form direct chemical bonds across the interface. Poor bonding across this interface and a sensitivity to stress corrosion cracking has limited the use of some materials. Alumina is one material that has received a reasonable amount of study. Porous alumina has been shown to lose 35% of its strength *in vivo* after 12 weeks, Frakes et al. 1974. Seidelmann et al. 1982 have shown that alumina loses about 15% of its strength after exposure to deionized water or blood when subjected to a constant stress. They also concluded that the service life of a hip endoprosthesis was dependent upon the density of the alumina. Ritter et al. 1979 studied the effects of coating alumina with a bioactive glass that retarded the fatigue process.

Bioactive glasses, although bonding well to bone and soft tissue, generally lack good mechanical properties. Bioactive glasses are especially sensitive to stress corrosion cracking. Barry and Nicholson 1988 reported that a soda-lime phosphosilicate bioactive glass was unsuitable for prosthetic use at stresses above 15 MPa, thus limiting its use to tooth prostheses. This glass sustained a tensile stress of 17 MPa for only 10 years in a $pH = 7.4$ environment. Troczynski and Nicholson 1990 then studied the fatigue behavior of particulate- and fiber-reinforced bioactive glass of the same composition. The reinforcement materials were either minus 325 mesh silver powder or silicon carbide whiskers. These materials were mixed with powdered glass and hot-pressed at 700°C and 30 MPa for 30 minutes. The composite containing the silver particulates exhibited a decreased sensitivity to stress corrosion cracking, while the composite containing the silicon carbide whiskers exhibited a sensitivity similar to that of the pure glass. Comparing the 10-year lifetimes of the two composites indicated that the particulate-containing material survived a static stress of 22 MPa and the whisker-containing material survived a static stress of 34 MPa.

Fractography results indicated agglomerate-initiated failure for the composites as opposed to surface machining defects for the pure bioactive glass.

Glassy Materials

In their investigation of silica optical fibers, Dabbs and Lawn 1985 presented data that questioned the acceptance of the Griffith flaw concept, which assumed that the flaws were exclusively cracklike and were free of preexisting influences. The real problem lies in predicting fatigue parameters for ultrasmall flaws from macroscopic crack velocity data. Abrupt changes in lifetime characteristics can occur due to evolution of flaws long after their inception. To conduct experiments with well-defined flaws, many investigators are now using microindentation techniques. Lawn and Evans 1977 reported that the formation of radial cracks from indentations was dependent upon the applied load. There exists a threshold load below which no radial cracks are generated; however, radial cracks may spontaneously form at the corners of subthreshold indentations long after the initial indent has been implanted if the surface were exposed to water, Lawn et al. 1983. Dabbs and Lawn reported data for silica optical fibers showing an abrupt increase in strength under low-load conditions below the threshold for formation of radial cracks. They attributed this behavior to a transition from crack propagation-controlled failure to one of crack initiation-controlled failure. Although the subthreshold indents had no well-developed radial cracks, they were still the preferred site for fracture origin and, therefore, must overcome crack initiation first. This crack initiation step, being close to the sample's free surface, is thus sensitive to environmental interactions. This low-load region exhibited three general features when compared to the high-load region where failure is controlled by crack propagation: an increase in strength, an increase in fatigue susceptibility, and an increase in scatter of the data.

Matthewson and Kurkjian 1988, however, have suggested that dissolution of high-strength silica fibers, with the subsequent formation of surface pits, was the cause of enhanced fatigue at low stress levels and not the spontaneous crack "pop-in" as suggested by Dabbs and Lawn. Pop-in does occur for weaker fibers. Their dissolution theory of enhanced fatigue is supported by data of Krause 1980 who reported a two- to three-fold reduction in strengths after exposure to water under zero stress. Since the time-to-failure is essentially linear with pH over the entire pH range, Matthewson and Kurkjian stated that the link between fatigue and dissolution was unclear. Matthewson et al. 1992 showed that by incorporating colloidal silica into a polymer coating substantial improvements in static fatigue and zero stress aging behavior could be obtained. This essentially delayed the onset of the fatigue knee (discussed below), leading to greater times-to-failure. The abrupt change of slope (or change in the fatigue parameter, n) in plots of applied stress versus time-to-failure has been called the *fatigue knee* (see Figure 19.1). If one

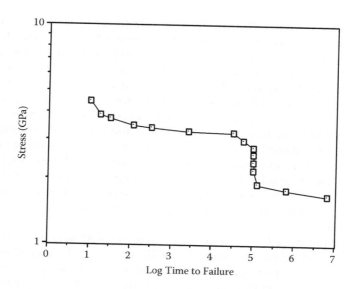

FIGURE 19.1
Applied stress vs. time-to-failure; the so-called fatigue knee.

were to extrapolate short-term data to longer times, a fatigue life very much shorter would be predicted. This fatigue knee, which has been well established for liquid environments, also has been recently established for vapor environments, Matthewson and Yuce 1994. Matthewson and Kurkjian 1992 have shown that the reduction in strength of silica fiber exposed to water under zero stress occurred at a time similar to that of the fatigue knee, and thus attributed both phenomena to the formation of surface pits by dissolution. These data all strongly suggested that enhanced fatigue at low stress levels was caused by the initiation of new surface flaws from dissolution pits and not by the propagation of cracks from preexisting defects. Thus it is best not to base lifetimes on extrapolated data but to study the behavior in the strength range of interest.

Ito and Tomozawa 1982 investigated the effects of exposure to water and to $Si(OH)_4$ aqueous solution at room temperature and 88°C upon the strength of high-silica glass rods. Mechanical strengths were determined at a constant stressing rate at room temperature in the two aqueous solutions and in liquid nitrogen. The room temperature strength after exposure to $Si(OH)_4$ at 88°C increased more rapidly during the first 250 hours than that of rods exposed to water. Strengths leveled off after 240 hours for the $Si(OH)_4$ exposure, whereas strengths for rods exposed to water increased gradually throughout the entire range of exposure times approaching those of the $Si(OH)_4$ exposed rods after 360 hours. Maximum obtained strengths were about 30% higher than for unexposed samples. The strengths of samples exposed to room temperature solutions were essentially unchanged. The weight loss at 88°C in water was much higher than in $Si(OH)_4$ by a factor of about 10. Since the strength increase was observed only when an observable weight loss was recorded, Ito and Tomozawa attributed the strength increase to a mechanism involving glass dissolution that increased the crack tip radii (i.e., crack blunting). If dissolution were the only phenomenon involved, strengths for water exposed samples should be higher than those for $Si(OH)_4$-exposed samples, since the dissolution was greater for samples exposed to water. Since solubility is a function of surface curvature and if solubility and dissolution were proportional, the dissolution rate would decrease with decreasing crack tip radius. This leads to a variation in dissolution rate around the crack tip leading to diffusion of dissolved glass and the combined effect of dissolution and precipitation, Iler 1979. Ito and Tomozawa, therefore, attributed the strength increasing mechanism to one of crack tip blunting caused by dissolution and precipitation.

Crack tip blunting by a different mechanism was suggested by Hirao and Tomozawa 1987 for soda-lime, borosilicate, and high-silica glasses that had been annealed at or near their transition temperatures for 1 hour in air or a vacuum. Diffusion of water vapor into the glasses as they were being annealed in air was confirmed by infrared spectroscopy. The more rapid strength increases for glasses annealed in air compared to those annealed in a vacuum were attributed to the faster rate of viscous flow (causing more rapid crack tip blunting) in the less viscous water-containing glasses, indicating that the release of residual stresses by annealing was not the cause for the strength increase as suggested by Marshall and Lawn 1981. Hirao and Tomozawa thus suggested that the conventional idea of glass fatigue caused by crack propagation alone is not sufficient and must include a crack-sharpening step.

Environmentally enhanced crack growth was shown to be dependent upon composition in zirconia and barium fluoride glasses by Freiman and Baker 1988. They observed extended crack growth after 15 minutes in several different liquids and found them to increase in the order dry oil, heptane, acetonitrile, and water. The fact that crack growth in acetonitrile was greater than in heptane suggested that it was not the presence of dissolved water in the liquids but the acetonitrile molecule that led to the enhanced crack growth.

It should be obvious that stress corrosion cracking is a rather complex phenomenon and that its evaluation is not as straightforward as it might first appear. Exactly how crack tip blunting increases strength is still unclear. Decreases in strength generally are attributed to bond rupture at the crack tip caused by the presence of water molecules; however, it has been shown that other molecules (i.e., acetonitrile) act in a similar manner. Lifetime predictions are based upon selection of the proper crack velocity equation, and it has been shown that it is best to use an equation that represents the data of several loading conditions. In addition, the equation selected most likely will not be unique to all environments.

Nitrides

In the evaluation of several hot isostatically pressed silicon nitrides, Sato et al. 1988 found that the dissolution in HCl of the sintering aids (Y_2O_3 and Al_2O_3) from the grain boundaries decreased the three-point flexural strength. Their test variables included acid concentration, temperature, duration of dissolution, and crystallinity of the grain boundary phase. In general, the flexural strength decreased with increasing dissolution of Y^{3+} and Al^{3+} cations. Strengths were decreased by at least 50% after being exposed to 1 M HCl solution for 240 hours at 70°C. As expected, the grain boundary phase having the highest degree of crystallinity exhibited the highest strength (i.e., it is easier to leach cations from a glass than from a crystal). A control composition containing no sintering aids exhibited little, if any, strength degradation after the HCl treatment, although the strengths were considerably below those materials containing sintering aids (initially 240 vs. 600 MPa).

Lifetimes that are predicted from different fatigue tests will vary. Slow crack growth has been reported by Kawakubo and Komeya 1987 to accelerate under cyclic conditions, especially of the tension-compression type cycle at room temperature for sintered silicon nitride. They also reported a plateau at about 70 to 90% of the stress intensity factor, when crack velocity was plotted versus K_I. Three regions in the data were observed, very similar to that reported for glasses as shown in Figure 18.1. Since the materials studied had a glassy grain boundary phase, the fatigue mechanism was assumed to be the same as that reported for glassy materials, Wiederhorn 1967 (i.e., stress corrosion cracking due to moisture in the air). Fett et al. 1986 reported that at 1200°C, the lifetimes for cyclic loads were higher than for static loads. Tajima et al. 1989 reported that a gas pressure sintered silicon nitride was resistant to slow crack growth up to 900°C but then was susceptible to slow crack growth at 1000°C due to the softening of the glassy grain boundary phase. A higher fatigue resistance was reported for higher frequencies of the load cycle due to the viscoelastic nature of the glassy grain boundary phase.

Degradation by Molten Metals

The strength degradation of sintered α-silicon carbide was evaluated in both an as-received and as-ground (600 grit) condition after exposure to molten lithium by Cree and Amateau 1987. Transgranular fracture was exhibited for all samples when treated at temperatures below 600°C. At temperatures above 600°C, both transgranular and intergranular fracture occurred. The transgranular fracture strengths were generally greater than 200 MPa, whereas the intergranular strengths were less than 200 MPa. The low-strength intergranular failure was attributed to lithium penetration along grain boundaries beyond the depth of the uniform surface layer that formed on all samples. Grain boundary degradation was caused by the formation of Li_2SiO_3, from the reaction of oxidized lithium and silica. The formation of lithium silicate was accompanied by an increase in volume by as much as 25%, depending upon the temperature of exposure. The localized stresses caused by this expansion promoted intergranular crack propagation.

Degradation by Molten Salts

Carbides and Nitrides

The strength loss of α-SiC and siliconized-SiC tubes exposed to a combustion flame into which a sodium silicate/water solution was injected was evaluated by Butt and Mecholsky 1989. The corrosive exposure was for times up to 373 hours, at temperatures from 900 to 1050°C, with an oxygen partial pressure of about 4 kPa. Strength losses exceeded 50% for the α-SiC and were 25 to 45% for the siliconized-SiC. Strength tests were conducted on C-ring samples after most of the reaction products were removed. Those samples for which the reaction products were not removed prior to strength testing exhibited no significant loss of strength, although an increase in scatter of the data was reported. Surface or corrosion pits were identified as the fracture origin for both types of SiC. In addition the α-SiC exhibited grain boundary attack, whereas the siliconized-SiC exhibited oxidation of the silicon matrix and attack of the large SiC grains.

In a study of the effects of molten salt upon the mechanical properties of silicon nitride, Bourne and Tressler 1980 reported that hot-pressed silicon nitride exhibited a more severe degradation in flexural fracture strength than did reaction sintered silicon nitride, even though the weight loss of the hot-pressed material was less than that of the sintered one as reported by Tressler et al. 1976 in a previous study. Their strength data are shown in Figure 19.2. The exposure to a eutectic mixture of NaCl and Na_2SO_4 was more severe than to molten NaCl alone for the hot-pressed material, whereas for the reaction-sintered material the effect was about the same. The differences between these two materials were attributed to the diffusion of contaminants along grain boundaries in the hot-pressed material and penetration of contaminants into pores of the reaction-sintered material. This was based upon the observation that the grain boundaries of the hot-pressed material were more severely affected than those of the reaction-sintered material, which did not contain an oxide grain boundary phase. The lowered fracture strengths resulted from an increase in the critical flaw size and a decrease in the critical stress intensity factor. The slight increase in fracture strengths at 1200°C was due to a slight increase in the critical stress intensity factor. The

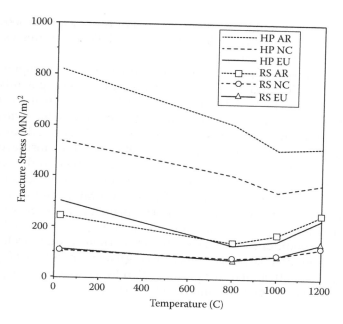

FIGURE 19.2
Fracture strength of Si_3N_4 vs. molten salt composition. (HP = hot pressed; RS = reaction sintered; AR = as-received; NC = NaCl; and EU = eutectic mixture of NaCl + Na_2SO_4). (After Bourne and Tressler 1980.)

$NaCl/Na_2SO_4$ eutectic mixture, being more oxidizing than the NaCl melt, caused a greater increase in the critical flaw size.

In the application of ceramics to turbine engines, the static fatigue life is of prime importance. Compared to the other types of mechanical testing in corrosive environments, little work has been reported on the longtime exposure effects to static fatigue life. Swab and Leatherman 1992 reported that at stresses between 300 and 500 MPa, there was a significant decrease in the time-to-failure for Si_3N_4 containing magnesia exposed to Na_2SO_4 at 1000°C. At stresses above 500 MPa and below 300 MPa, little change in the time-to-failure was noted. Since the molten salt was not replenished during the test, corrosion pits were unable to grow to a size sufficient to decrease the time-to-failure at stress levels between 300 and 500 MPa. Although the decrease in room temperature strength for a yttria-containing silicon nitride after exposure to sodium sulfate was about 35%, it retained a greater strength than the magnesia-containing material (549 MPa for the Y-containing material vs. 300 MPa for the Mg-containing material), Letterman et al. 1993. Fox and Smialek 1990 tested sintered silicon nitride in a simulated gas turbine rig where the corrosive environment was continued throughout the 1000°C/40 hrs of the test. Room-temperature MOR fracture origins were located at pits in 17 of 22 samples. Pit formation was attributed to gas evolution during the oxidation of the silicon nitride and subsequent reaction of the silica with sodium sulfate forming a low-viscosity sodium silicate liquid. Fracture stresses were on the order of 300 MPa after exposure.

Boron- and carbon-doped injected molded sintered α-SiC sprayed with thin films of Na_2SO_4 and Na_2CO_3 were exposed to several gas mixtures at 1000°C for 48 hours by Smialek and Jacobson 1986. The gas mixtures used were 0.1%SO_2 in oxygen and 0.1%CO_2 in oxygen in combination with the sulfate or carbonate thin films, respectively. The sulfate-covered sample was also exposed to pure air. The strength degradation was the most severe in the sulfate/SO_2 exposure (49% loss in strength), intermediate in the sulfate/air exposure (38% loss in strength), and the least in the carbonate/CO_2 exposure. The latter exposure caused a statistically insignificant decrease in strength when analyzed by the Student's t-test. The primary mode of degradation was the formation of pits that varied in size and frequency depending upon the corrosion conditions. The size of the pits correlated quite well with the strength degradation (i.e., larger pits caused greater strength loss). Jacobson and Smialek 1986 attributed this pit formation to the disruption of the silica scale by the evolution of gases and bubble formation.

Zirconia-Containing Materials

Although a considerable amount of scatter existed in the data of Swab and Leatherman 1992, they concluded that Ce-TZP survived 500 hours at 1000°C in contact with Na_2SO_4 at stress levels below 200 MPa. At stress levels greater than 250 MPa, failure occurred upon loading the samples. Swab and Leatherman also reported a 30% decrease in the room-temperature strength of Y-TZP after 500 hours at 1000°C in the presence of Na_2SO_4. This lowered strength for Y-TZP was probably due to leaching of the yttria from the surface, which caused transformation of the tetragonal phase to the monoclinic phase.

Degradation by Oxidation

Carbides and Nitrides

When evaluating the effects of corrosion one must be alert to the changes that occur if samples are corroded and then cooled to room temperature for mechanical testing. McCullum et al. 1991 found that the room-temperature flexural strength of SiC increased with exposure time to an air environment at 1300°C, whereas it decreased for Si_3N_4. They attributed this increased strength for SiC to the formation of a thin silica surface layer that healed surface flaws. The decreased strength for Si_3N_4 was attributed to the formation of a much thicker silica surface layer that cracked upon cooling. This cracking of the surface oxide layer was caused by stresses arising from the volume difference between the nitride and the oxide and to the polymorphic transformation of either cristobalite or tridymite. Exposure times beyond 100 hours did not yield continued lower strengths for the Si_3N_4, since the layer thickness remained essentially constant for exposure times greater than 100 hours. Hench et al. 1980 reported that if Si_3N_4 contained 8 wt% ZrO_2 as a hot-pressing additive, an amorphous silicate film formed ($\leq 18\ \mu m$ thick) that provided an effective diffusion barrier to oxygen when heated to about 1430°C in air for 100 hrs. The room-temperature MOR of this material decreased about 40% after oxidation, whereas the MOR of a material containing 3 wt%MgO decreased by more than 60%. The difference was attributed to the oxidized layer formed on the zirconia-containing material being essentially amorphous and that of the magnesia-containing material being totally crystalline and much thicker (~ 3x).

When tested at temperature, McCullum et al. 1991 found the flexural strength for SiC remained constant with increasing exposure times to oxidation, with values less than when tested at room temperature. This lower strength obtained when tested at temperature when compared to room-temperature strength was attributed to the formation of a compressive layer on the surface when cooled. In contrast, the Si_3N_4 exhibited a slight increase in strength with exposure time when tested at temperature with values greater than the room-temperature values. This was attributed to the integrity of the surface layer at temperature. These data are generalized in Figure 19.3.

McCullum et al. 1991 gave the following equation to evaluate the dynamic fatigue of several different SiC and Si_3N_4 samples:

$$s_f = A's^{(1/n+1)} \qquad (19.1)$$

where:

s_f = fracture strength,
A' = material constant,
s = loading rate, and
n = stress corrosion susceptibility constant.

Although they showed no strength versus loading rate plots, one can obtain the value of n from the slopes of these plots. The values for the stress corrosion susceptibility constant, obtained in this manner for several different materials, ranged from infinity to about eight, over a range of temperatures from 20 to 1400°C, respectively.

Due to a larger quantity of sintering aids for the Si_3N_4 as compared to the SiC samples, the nitride samples were more susceptible to strength loss due to increased temperatures and decreased loading rates than the carbide. Variations in the amount and chemistry of the sintering aids in Si_3N_4 causes a variation in the mechanical behavior. Tensile test results followed the same

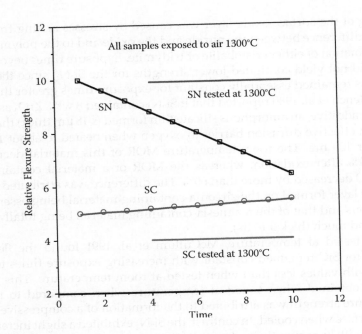

FIGURE 19.3
Fracture strength vs. time for Si_3N_4 and SiC. (After McCullum et al. 1991.)

general trends as flexural test results. In general the tensile strength values were lower than the flexural values.

In the evaluation of a sintered and a hot-pressed Si_3N_4 under pure oxidation and oxidation under a static load, Easler et al. 1982 found that, when exposed to air at 1370°C for times ranging from 0.5 to 50 hours, fracture data indicated that the range in flaw sizes decreased, whereas it increased after exposure for 1 hour under a static load. For the sintered material, strengths increased for oxidation under a static load of 23 or 45 MPa; however, the higher load condition resulted in a wider range of flaw sizes. In contrast, the hot-pressed material exhibited lower strengths for static loads of 45 and 160 MPa during oxidation. Under pure oxidation, the strengths of both materials increased for short (0.5 hr) oxidation times and then decreased at longer times. The increased strengths were attributed to flaw tip blunting. The lowered strengths under static loading conditions were attributed to subcritical crack growth. Easler et al. 1982 concluded that the strength-controlling mechanisms, at least for silicon nitride, were dynamic in nature and very material-specific.

Rapid oxidation in air of Y-doped, sintered reaction bonded Si_3N_4 at 1000°C was reported by Govila et al. 1985 to lower the strength and cause early failure. The fracture origins were determined to be β–Si_3N_4 needles. An excessive weight gain was reported to occur at 1000°C that was attributed to oxygen and nitrogen absorption of the matrix and secondary phases, one of which was reported to be YSiO$_2$N. The oxidation of YSiO$_2$N to Y$_2$Si$_2$O$_7$ is accompanied by a 12% molar volume change. This anomalously high weight gain was accompanied by a 15% loss in the room-temperature strength. Stress rupture tests indicated the presence of stress-enhanced oxidation at 1000°C with failure times ranging from 19 to 93 hours at an applied load of 138 MPa, and 14 to 31 hours at an applied load of 276 MPa. Losses in strength at temperatures greater than 1200°C were attributed to the softening of the glassy grain boundary phase, which leads to creep by grain boundary sliding. Samples exposed to oxidation at 1200°C at an applied load of 344 MPa did not fail, even after 260 hours, although some slight deformation had occurred.

In an effort to determine the effects of oxidation upon the flexural strength of Si$_3$N$_4$, Kim and Moorhead 1990 evaluated the room-temperature four-point bend strength of HIP-SN (with 6 wt% Y$_2$O$_3$ and 1.5 wt% Al$_2$O$_3$) after exposure in either H$_2$/H$_2$O or Ar/O$_2$ at 1400°C for 10 hours. In both atmospheres, the strength was dependent upon the amount of oxidant present. The actual variation in strength was, however, different depending upon the alteration of the surface layers formed and their characteristics. In the H$_2$/H$_2$O atmosphere at low pH$_2$O a nonprotective, not well attached glasslike layer containing crystalline Y$_2$Si$_2$O$_7$ formed. Since this layer was relatively uniform with no new strength-limiting flaws being formed (although some large bubbles were found at the surface/substrate interface), the maximum reduction in strength was limited to about 20% at a pH$_2$O of 2×10^{-5} MPa. A significant strength increase occurred as the pH$_2$O was increased, which the authors attributed to blunting of preexisting cracks by the interfacial silicate phase. This silicate phase was a continuous dense layer of Y$_2$Si$_2$O$_7$ containing small isolated bubbles believed to be formed by nitrogen generation during oxidation of the Si$_3$N$_4$. In the Ar/O$_2$ atmosphere, a similar reduction and subsequent increase in strength was not found. Instead, at low pO$_2$ an increase in strength occurred with increasing pO$_2$. The maximum strength occurred at the pO$_2$ (10^{-5} MPa) that yielded the greatest weight loss. Even at low pO$_2$, a surface reaction product of Y$_2$Si$_2$O$_7$ formed in isolated pockets at grain junctions, presumably by the reaction of Y$_2$O$_3$ solid with SiO gas. Kim and Moorhead attributed the increased strengths observed to the formation of more Y$_2$Si$_2$O$_7$ as the pO$_2$ increased. At approximately a pO$_2$ of 10^{-5} MPa where the maximum strength was observed, the Y$_2$Si$_2$O$_7$ layer became interconnected and, although not continuous, blunted strength-limiting flaws. At higher pO$_2$, where weight gains were observed and a continuous layer containing Y$_2$Si$_2$O$_7$ and cristobalite formed, the increase in strength was not as significant. In this region, competition between crack blunting and formation of new flaws (cracks and bubbles) was suggested as the reason for the slightly lower strengths. This particular study by Kim and Moorhead points out very well the effects that the surface layer characteristics have upon the mechanical properties. Similar strength increases were found by Wang et al. 1989 for two silicon nitride materials, one containing 13.9% Y$_2$O$_3$ plus 4.5% Al$_2$O$_3$ and the other containing 15% Y$_2$O$_3$ plus 5% Al$_2$O$_3$ when exposed to air at 1200°C for 1000 hours prior to strength testing at 1300°C. Strength increases as high as 87% were reported when compared to the unoxidized 1300°C strength, although the preoxidized 1300°C strength was slightly less than the unoxidized room-temperature strength. Wang et al. attributed these strength increases to healing of surface flaws and crack blunting during oxidation along with purification of the grain boundaries that raised the viscosity of the glassy boundary phase. These beneficial effects were not present when oxidation was conducted at 900°C.

Lange and Davis 1979 have suggested that oxidation can lead to surface compressive stresses that, if optimum, may lead to increased apparent strengths. If the compressive stresses become too severe, then spalling may occur leading to lowered strengths. They demonstrated this concept with Si_3N_4 doped with 15 and 20% CeO_2 exposed to oxidation in air, at temperatures ranging from 400 to 900°C. The apparent critical stress intensity factor (K_a) increased for short exposure times at 400, 500, and 600°C. This increase in K_a was attributed to oxidation of the Ce-apatite secondary phase and subsequent development of a surface compressive layer. At longer times (~8 hrs) and the two higher temperatures, surface spalling caused a decrease in K_a. At higher temperatures (i.e., 1000°C), the compressive stresses that may cause spalling are relieved by extrusion of the oxide product from the interior of the material. Thus prolonged oxidation at 1000°C does not degrade this material.

Oxynitrides

In a study of β' and O' SiAlON solutions, O'Brien et al. 1993 found that the oxygen (or nitrogen) content significantly affected the performance of these materials. The grain boundary glassy phase viscosity increased as the nitrogen content increased, which subsequently slowed the healing of flaws (see Chapter 1, section on bulk glasses and Chapter 16, section on silicate glasses for a discussion of the effects of nitrogen on durability). The higher-viscosity glassy phase also trapped evolving gases more easily, creating additional flaws. In general, the mean retained flexural strengths after oxidation at 1273 K for 24 hours of the SiAlON solutions was higher than that of several silicon nitrides, with the strengths being generally proportional to the oxidation resistance. O'Brien et al. concluded that the retained strengths after oxidation were dependent upon the characteristics of the surface oxide layer that formed. At higher temperatures, the potential for flaw healing was dependent upon the amount and composition of the glassy phase formed.

A zirconium oxynitride with the stoichiometry $ZrO_{2-2x}N_{4x/3}$ has been reported by Claussen et al. 1977 to form as a secondary phase in hot-pressed ZrO_2-Si_3N_4. This phase readily oxidized to monoclinic ZrO_2 at temperatures greater than 500°C. Lange 1980 used the volume change (about 4–5%) associated with this oxidation to evaluate the formation of a surface compression layer on silicon nitride compositions containing 5 to 30 vol% zirconia. To develop the correct stress distribution for formation of the surface compressive layer, the secondary phase that oxidizes must be uniformly distributed throughout the matrix. When oxidized at 700°C for 5 hours, a material containing 20 vol% ZrO_2 exhibited an increase in strength from 683 to 862 MPa. Lange attributed this increase in strength to the oxidation-induced phase change of the zirconium oxynitride to monoclinic zirconia.

Degradation by Other Atmospheres

Carbides and Nitrides

Clark 1988 reported that Nicalon SiC fibers when aged in nitrogen or humid air at 1200°C for 2 hours lost about one-half their tensile strength. A more gradual strength decrease was observed for fibers that were exposed to hot argon. Although the time dependence of strength loss for the different aging environments was similar, the mechanisms causing strength loss were quite different. For exposure to nitrogen, Clark attributed the strength loss to crack propagation from existing flaws; for exposure to argon he attributed the loss to grain growth and porosity; and for exposure to humid air he attributed the strength loss to fiber coalescence at the silica surface, to poor adherence of the surface silica layer, to a cracked crystalline silica surface layer, and to bubbles at the silica/fiber interface. Clark also pointed out that thermal stability should not be based solely upon weight change data, since for this fiber the weight gain produced by oxidation to silica was offset by weight loss due to CO evolution.

Siliconized, boron-doped, and aluminum-doped SiC samples were exposed to gaseous environments containing mixtures of predominantly N_2, H_2, and CO, representative of metallurgical heat-treatment atmospheres at 1300°C for up to 1000 hours by Butt et al. 1992. They reported significant strength losses for all three materials for times less than 100 hours when exposed to a gas mixture containing about 40% nitrogen. At longer exposure times, no additional strength loss occurred. The aluminum-doped SiC, unlike the other two, exhibited a slight strength increase after 1000 hours when exposed to a gas mixture containing 98.2% nitrogen. The strength losses were attributed primarily to pitting that was related to the presence of transition metal impurities.

It has been shown by Li and Langley 1988 that ceramic fibers composed of Si-C-N-O experience various degrees of strength degradation when aged in atmospheres of various hot gases. The rate of strength loss experienced by fibers aged in these hot gases was related to the rate of diffusion of the gases formed by decomposition. The gases of decomposition (N_2, CO, and SiO) diffuse through the fiber porosity and any surface boundary layers present.

TABLE 19.1

Effects of Aging in Various Atmospheres upon Strength of Si-C-N-O Fibers

Unaged Fiber	Aged at 1400°C for 0.5 hr in	
	N_2	Ar
1517 MPa	717 MPa	276 MPa

Source: Li, C-T. and Langley, N.R. 1988. Development of a Fractographic Method for the Study of High-Temperature Failure of Ceramic Fibers, in *Advances in Ceramics Vol. 22, Fractography of Glasses and Ceramics*, Varner, J.R. and Frechette, V.D. (eds.), Am. Ceram. Soc., Westerville, OH, pp. 177–184.

The diffusion of these product gases can be controlled by aging the fibers in atmospheres of these gases. Thus, greater strength loss was exhibited when fibers were aged in argon compared to aging in nitrogen. This effect can be seen by examining the data of Table 19.1.

Zirconia-Containing Materials

Brinkman et al. 1989 studied the effects of a diesel engine environment upon the strength of two commercial zirconias partially stabilized with magnesia. The combustion environment at temperatures between 500 and 900°C contained Fe, Zn, Ca, Mg, and P contaminants from the fuel along with water vapor. The average flexural strength decreased by about 32% after exposure for 100 hours for the material rated as thermally shock resistant, and decreased by only 9% for the one rated as maximum strength. When the surface reaction products were removed from the thermally shock-resistant material before strength testing, the strength decreased 22%. Both materials when exposed to air for 100 hours at 700 and 750°C exhibited decreases in strength of 6 to 8%, indicating a much more significant effect of the actual diesel engine environment. They found that the strength decreased as the amount of monoclinic zirconia increased. Thus the primary mechanism of degradation was attributed to localized increases in the monoclinic content.

Additional Recommended Reading

Evans, A.G. (ed.). *Fracture in Ceramic Materials*, Noyes Publications, Park Ridge, NJ, 1984, 420 pp.

Frechette, V.D. and Varner, J.R. (eds.). *Ceramic Transactions, Vol. 17, Fractography of Glasses and Ceramics II*, Amer. Ceram. Soc., Westerville, OH, 1991, 548 pp.

Varner, J.R. and Frechette, V.D. (eds.). *Advances in Ceramics, Vol. 22, Fractography of Glasses and Ceramics*, Amer. Ceram. Soc., Westerville, OH, 1988, 442 pp.

Exercises, Questions, and Problems

1. Explain how Si_3N_4 may decrease in strength (room temperature) and SiC increase in strength (room temperature) after being exposed to air at 1300°C.

2. Discuss the differences that one may find when determining strengths at temperature versus at room temperature and why this difference occurs.

3. Is it possible for an oxidative corrosion reaction to produce zero weight gain or loss? Explain.

4. Discuss the problems that one may encounter when extrapolating data to extended lifetimes.

References

Barry, C. and Nicholson, P.S. 1988. Stress Corrosion Cracking of a Bioactive Glass, *Ad. Ceram. Mat., 3*(2), 127–30.

Bourne, W.C. and Tressler, R.E. 1980. Molten Salt Degradation of Si_3N_4 Ceramics, *Ceram. Bull., 59*(4), 443–46, 452.

Brinkman, C.R., Begun, G.M., Cavin, O.B., Foster, B.E., Graves, R.L., Kahl, W.K., Liu, K.C. and Simpson, W.A. 1989. Influence of Diesel Engine Combustion on the Rupture Strength of Partially Stabilized Zirconia, in *Proceedings of the 3rd International Symposium on Ceramic Materials and Components for Engines*, Tennery, V.J. (ed.), Am. Ceram. Soc., Westerville, OH, pp. 549–58.

Butt, D.P. and Mecholsky, J.J. 1989. Effects of Sodium Silicate Exposure at High Temperature on Sintered α-Silicon Carbide and Siliconized Silicon Carbide, *J. Am. Ceram. Soc., 72*(9), 1628–35.

Butt, D.P., Tressler, R.E. and Spear, K.E. 1992. Silicon Carbide Materials in Metallurgical Heat-Treatment Environments, *Am. Ceram. Soc. Bull., 71*(11), 1683–90.

Clark, T.J. 1988. Fracture Properties of Thermally Aged Ceramic Fiber Produced by Polymer Pyrolysis, pp. 279–93 in *Advances in Ceramics Vol. 22, Fractography of Glasses and Ceramics*, Varner, J.R. and Frechette, V.D. (eds.), Am. Ceram. Soc., Westerville, OH.

Claussen, N., Wagner, R., Gauckler, L.J. and Petzow, G. 1977. Nitride-Stabilized Cubic ZrO2, presented at the 79th Annual Mtg. Am. Ceram. Soc., Chicago, Apr. 23, 1977, paper no. 136-B-77, abstract in *Am. Ceram. Soc. Bull.*, *56*(3), 301.

Cree, J.W. and Amateau, M.F. 1987. Degradation of Silicon Carbide by Molten Lithium, *J. Am. Ceram. Soc.*, *70*(11), C318–21.

Dabbs, T.P. and Lawn, B.R. 1985. Strength and Fatigue Properties of Optical Glass Fibers Containing Microindentation Flaws, *J. Am. Ceram. Soc.*, *68*(11), 563–69.

Easler, T.E., Bradt, R.C. and Tressler, R.E. 1982. Effects of Oxidation and Oxidation under Load on Strength Distributions of Si_3N_4, *J. Am. Ceram. Soc.*, *65*(6), 317–20.

Fett, T., Himsolt, G. and Munz, D. 1986. Cyclic Fatigue of Hot-Pressed Si3N4 at High Temperatures, *Adv. Ceram. Mat.*, *1*(2), 179–84.

Fox, D.S. and Smialek, J.L. 1990. Burner Rig Hot Corrosion of Silicon Carbide and Silicon Nitride, *J. Am. Ceram. Soc.*, *73*(2), 303–11.

Frakes, J.T., Brown, S.D. and Kenner, G.H. 1974. Delayed Failure and Aging of Porous Alumina in Water and Physiological Media, *Am. Ceram. Soc. Bull.*, *53*(2), 183–87.

Freiman, S.W. and Baker, T.L. 1988. Effects of Composition and Environment on the Fracture of Fluoride Glasses, *J. Am. Ceram. Soc.*, *71*(4), C214–16.

Govila, R.K., Mangels, J.A. and Baer, J.R. 1985. Fracture of Yttria-Doped, Sintered Reaction-Bonded Silicon Nitride, *J. Am. Ceram. Soc.*, *68*(7), 413–18.

Hench, L.L., Ohuchi, F., Freiman, S.W., Wu, C.C. and McKinney, K.R. 1980. Infrared Reflection Analysis of Si_4N_4 Oxidation, pp. 318–30 in *Ceramic Engineering and Science Proceedings*, Vol. 1 (7-8A), Am. Ceram. Soc., Westerville, OH.

Hirao, K. and Tomozawa, M. 1987. Kinetics of Crack Tip Blunting of Glasses, *J. A. Ceram. Soc.*, *70*(1), 43–48.

Iler, R.K. 1979. *The Chemistry of Silica*, Wiley & Sons, New York, NY, pp. 3–115.

Ito, S. and Tomozawa, M. 1982. Crack Blunting of High-Silica Glass, *J. Am. Ceram. Soc.*, *65*(8), 368–71.

Jacobson, N.S. and Smialek, J.L. 1986. Corrosion Pitting of SiC by Molten Salts, *J. Electrochem. Soc.*, *133*(12), 2615–21.

Kawakubo, T. and Komeya, K. 1987. Static and Cyclic Fatigue Behavior of a Sintered Silicon Nitride at Room Temperature, *J. Am. Ceram. Soc.*, *70*(6), 400–405.

Kim, H-E. and Moorhead, A.J. 1990. High-Temperature Gaseous Corrosion of Si3N4 in H_2-H_2O and Ar-O_2 Environments, *J. Am. Ceram. Soc.*, *73*(10), 3007–14.

Krause, J.T. 1980. Zero Stress Strength Reduction and Transitions in Static Fatigue of Fused Silica Fiber Lightguides, *J. Non-Cryst. Solids*, *38–39*, 497–502.

Lange, F.F. 1980. Compressive Surface Stresses Developed in Ceramics by an Oxidation-Induced Phase Change, *J. Am. Ceram. Soc.*, *63*(1–2), 38–40.

Lange, F.F. and Davis, B.I. 1979. Development of Surface Stresses during the Oxidation of Several Si_3N_4/CeO_2 Materials, *J. Am. Ceram. Soc.*, *62*(11–12), 629–30.

Lawn, B.R., Dabbs, T.P. and Fairbanks, C.J. 1983. Kinetics of Shear-Activated Indentation Crack Initiation in Soda-Lime Glass, *J. Mater. Sci.*, *18*(9), 2785–97.

Lawn, B.R. and Evans, A.G. 1977. A Model for Crack Initiation in Elastic/Plastic Indentation Fields, *J. Mater. Sci.*, *12*(11), 2195–99.

Leatherman, G.L., Katz, R.N., Bartowski, G., Chadwick, T. and King, D. 1993. The Effect of Sodium Sulfate on the Room Temperature Strength of a Yttria Containing Silicon Nitride, *Ceram. Engr. & Sci. Proceedings*, *14*(7–8), 341–49.

Li, C-T. and Langley, N.R. 1988. Development of a Fractographic Method for the Study of High-Temperature Failure of Ceramic Fibers, in *Advances in Ceramics Vol. 22, Fractography of Glasses and Ceramics*, Varner, J.R. and Frechette, V.D. (eds.), Am. Ceram. Soc., Westerville, OH, pp. 177–84.

Marshall, D.B. and Lawn, B.R. 1981. Residual Stresses in Dynamic Fatigue of Abraded Glass, *J. Am. Ceram. Soc.*, 64(1), C6–7.

Matthewson, M.J. and Kurkjian, C.R. 1988. Environmental Effects on the Static Fatigue of Silica Optical Fiber, *J. Am. Ceram. Soc.*, 71(3), 177–83.

Matthewson, M.J., Rondinella, V.V. and Kurkjian, C.R. 1992. The Influence of Solubility on the Reliability of Optical Fiber, *Proc. SPIE, 1791*, Optical Materials Reliability and Testing, 52–60.

Matthewson, M.J. and Yuce, H.H. 1994. Kinetics of Degradation during Aging and Fatigue of Fused Silica Optical Fiber, *Proc. SPIE, 2290*, 204–10.

McCullum, D.E., Hecht, N.L., Chuck, L. and Goodrich, S.M. 1991. Summary of Results of the Effects of Environments on Mechanical Behavior of High-Performance Ceramics, *Ceram. Eng. Sci. Proc.*, 12(9,10), 1886–1913.

O'Brien, M.H., Huang, C.M. and Coon, D.N. 1993. Oxidation and Retained Strength of *In-Situ* O'-β' SiAlON Composites, *Ceram. Eng. Sci. Proc.*, 14(7–8), Part 1, 350–57.

Ritter Jr., J.E., Greenspan, D.C., Palmer, R.A. and Hench, L.L. 1979. Use of Fracture Mechanics Theory in Lifetime Predictions for Alumina and Bioglass-Coated Alumina, *J. Biomed. Mater. Res.*, 13, 251–63.

Sato, T., Tokunaga, Y., Endo, T., Shimada, M., Komeya, K., Komatsu, M. and Kameda, T. 1988. Corrosion of Silicon Nitride Ceramics in Aqueous Hydrogen Chloride Solutions, *J. Am. Ceram. Soc.*, 71(12), 1074–79.

Seidelmann, U., Richter, H. and Sotesz, U. 1982. Failure of Ceramic Hip Endoprostheses by Slow Crack Growth-Lifetime Prediction, *J. Biomed. Mater. Res.*, 16, 705–13.

Smialek, J.L. and Jacobson, N.S. 1986. Mechanism of Strength Degradation for Hot Corrosion of α–SiC, *J. Am. Ceram. Soc.*, 69(10), 741–52.

Swab, J.J. and Leatherman, G.L. 1992. Static-Fatigue Life of Ce-TZP and Si_3N_4 in a Corrosive Environment, *J. Am. Ceram. Soc.*, 75(3), 719–21.

Tajima, Y., Urashima, K., Watanabe, M. and Matsuo, Y. 1989. Static, Cyclic and Dynamic Fatigue Behavior of Silicon Nitride, in *Proceedings of the 3rd International Symposium on Ceramic Materials and Components for Engines*, Tennery, V.J. (ed.), Am. Ceram. Soc., Westerville, OH, pp. 719–28.

Tressler, R.E., Meiser, M.D. and Yonushonis, T. 1976. Molten Salt Corrosion of SiC and Si_3N_4 Ceramics, *J. Am. Ceram. Soc.*, 59(5–6), 278–79.

Troczynski, T.B. and Nicholson, P.S. 1990. Stress Corrosion Cracking of Bioactive Glass Composites, *J. Am. Ceram. Soc.*, 73(1), 164–66.

Wang, L., He, C. and Wu, J.G. 1989. Oxidation of Sintered Silicon Nitride Materials, in *Proceedings of the 3rd International Symposium on Ceramic Materials and Components for Engines*, Tennery, V.J. (ed.), Am. Ceram. Soc., Westerville, OH, pp. 604–11.

Wiederhorn, S.M. 1967. Influence of Water Vapor on Crack Propagation in Soda-Lime Glass, *J. Am. Ceram. Soc.*, 50(8), 407–14.

Section V

Minimization of Corrosion

20

Methods to Minimize Corrosion

Failure is only the opportunity to begin again more intelligently.

Henry Ford

Introduction

The control of the chemical reactivity of ceramics with their environment is one of the most important problems facing the ceramics industry today. Through the study of corrosion phenomena, one can learn how best to provide the control of the chemical reactivity that will provide a maximum service life expectancy at a minimum cost. Most methods used to minimize corrosion have generally been methods that slow the overall reaction rates. Once a complete understanding is available, however, one can attempt to change the reaction mechanism to something less harmful, in addition to slowing the rate. Corrosion reactivity is affected by the following items (not necessarily listed in the order of importance):

1. heat transfer,
2. mass transfer,
3. diffusion-limited processes,
4. contact area,
5. mechanism,
6. surface-to-volume ratio,
7. temperature, and
8. time.

The following discussion will address some of these items and how they may be used to minimize the effects of corrosion by discussing various examples. It should be pointed out that quite often in the development of materials for a particular application no consideration is given to the potential for corrosion. *It should be remembered that no engineering material is truly inert. Although they may appear inert, it's only a matter of time before some degradation becomes apparent.*

Crystalline Materials—Oxides

The most obvious method of providing better corrosion resistance is to change materials; however, this can only be done to a certain extent. There will ultimately be only one material that does the job best. Once this material has been found, additional corrosion resistance can be obtained only by property improvement or, in some cases, by altering the environment. Different parts of an industrial furnace generally involve variations in the corrosive environment, necessitating the use of different materials with the best properties for a particular location within the furnace. Furnace designers have thus for a long time used a technique called *zoning* to maximize overall service life by using different materials in different parts of the furnace.

Property Optimization

Since exposed surface area is a prime concern in corrosion, an obvious property to improve is the porosity. Much work has been done in finding ways to make polycrystalline materials less porous or denser. The most obvious is to fire the material during manufacture to a higher temperature. Other methods of densification have also been used. These involve various sintering or densification techniques: liquid-phase sintering, hot pressing, and others. If additives are used to cause liquid-phase sintering, care must be exercised that not too much secondary phase forms, which may have a lower corrosion resistance, even though porosity may be reduced.

Alterations in major component chemistry may aid in increasing corrosion resistance, but this is actually a form of finding a new or different material, especially if major changes are made.

The history of glass-contact refractories is a good example of corrosion resistance improvement in a polycrystalline material. Porous clay refractories were used originally. Changes in chemistry by adding more alumina were made first to provide a material less soluble in the glass. The first major improvement was the use of fusion-cast aluminosilicate refractories. These provided a material of essentially zero porosity. The next step was the incorporation of zirconia into the chemistry. Zirconia is less soluble than alumina or silica in most glasses. Because of the destructive polymorphic transformation of zirconia, a glassy phase had to be incorporated into these refractories. This glassy phase added a less corrosion-resistant secondary phase to the refractory. Thus the higher resistance of the zirconia was somewhat compromised by the lower resistance of the glassy phase. The final product, however, still had a corrosion resistance greater than the old product without any zirconia. Today several grades of ZrO_2-Al_2O_3-SiO_2 fusion-cast refractories are available. Those with the highest amount of zirconia and the lowest amount of glassy phase have the greatest corrosion resistance.

Another example from the glass industry is the development of furnace regenerator refractories through the optimization of materials made of fireclay, by using higher-purity raw materials and then increased firing temperatures. Changes in chemistry were then made by switching from the fireclay products to magnesia-based products. Again, improvements were made by using higher-purity raw materials and then increased firing temperatures. Minor changes in chemistry were also made during the process of property improvement. Changes in processing involving prereaction of raw materials have also been done. The evolution of regenerator refractories for the flat glass industry up to the mid-1970s has been described by McCauley 1978. A more recent development in regenerator refractories has been the use of fusion cast alumina-zirconia-silica cruciform products. These are in the shape of a cross and are stacked in interlocking columns. This represents not only a change in chemistry but also a change in the shape of the product, both of which lead to better overall performance.

A part of the concept of improvement through chemistry changes is that of improving resistance to corrosion of the bonding phases. Bonding phases normally have a lower melting point and lower corrosion resistance than does the bulk of the material. The development of high-alumina refractories is a good example of improvement based on the bonding phase. The best conventional high-alumina refractories are bonded by mullite or by alumina itself. To change this bond to a more corrosion-resistant material compatible with alumina, knowledge of phase equilibria played an important role. Alumina forms a complete series of crystalline solutions with chromia, with the intermediate compositions having melting points between the two end members. Thus, a bonding phase formed by adding chromia to alumina would be a solution of chromia in alumina with a higher melting point than the bulk alumina and thus a higher corrosion resistance. In addition to the more resistant bonding phase, these materials exhibit a much higher hot modulus of rupture (more than twice mullite or alumina-bonded alumina). Nothing is ever gained, however, without the expense of some other property. In this case the crystalline-solution-bonded alumina has a slightly lower thermal shock resistance than does the mullite-bonded alumina. Owing to the excellent resistance of these materials to iron oxide and acid slags, they have found applications in the steel industry.

The development of tar-bonded and tar-impregnated basic refractories to withstand the environment of the basic oxygen process of making steel is yet another example of a way to improve the corrosion resistance of a material. Tar-bonded products are manufactured by adding tar to the refractory grain before pressing into shape. In this way, each and every grain is coated with tar. When the material is heated during service, the volatiles burn off, leaving carbon behind to fill the pores. An impregnated product is manufactured by impregnating a finished brick with hot tar. This product, once in service, will similarly end up with carbon in the pores. Impregnated products do

FIGURE 20.1
Carbon-containing basic refractory showing carbon oxidation at the surface (white band) when in contact with molten steel/slag.

not have as uniform a carbon distribution as do the bonded types. Newer products incorporate graphite into the raw material mix. The carbon that remains within the refractory increases the corrosion resistance to molten iron and slags by physically filling the pores, by providing a nonwetting surface, and by aiding in keeping iron in the reduced state, which then does not react with the oxides of the refractory. Any oxygen that diffuses into the interior of the refractory causes carbon oxidation that slightly increases the pore pressure and thus minimizes slag and metal penetration. A thin layer on the hot face (1 to 2 mm) does lose its carbon to oxidation and various slag components penetrate and react within this layer (see Figure 20.1). This corrosion, however, is much slower than with a product that contains no carbon.

An additional improvement upon the carbon-containing magnesia refractories has been the incorporation of magnesium metal, as reported by Brezny and Semler 1984. Upon magnesium volatilization and diffusion toward the hot face, oxidation and precipitation enhances the formation of the dense magnesia-rich zone that forms behind the hot face and thus minimizes slag and metal penetration in addition to oxygen diffusion to the interior of the refractory (see Chapter 15, for a discussion on the formation of this magnesia-rich dense zone).

Others, Siljan et al. 2002, have tried to take advantage of the nonwetting enhancement caused by antiwetting additives; however, their success has been questionable.

The automotive industry, in their efforts to develop a gas turbine engine, has conducted a considerable amount of research upon low-expansion lithium alumino silicates (LAS) and magnesium alumino silicates (MAS) for a rotary wheel heat exchanger. The LAS materials are based on solid solutions of the high-temperature polymorphs of two different compounds: eucryptite ($Li_2O \cdot Al_2O_3 \cdot SiO_2$) and spodumene ($Li_2O \cdot Al_2O_3 \cdot 4SiO_2$). Both of these materials have an upper use temperature of about 1200°C. Both have a very low thermal expansion (eucryptite being slightly negative), which gives them excellent thermal shock resistance. These materials, however, suffer from corrosion problems when used in dirty environments. To overcome these corrosion problems, an alumino silicate (AS) material was developed by the acid leaching of lithium from LAS prior to application. This material had acceptable thermal expansion, although not as low as LAS, but did not distort or crack as much.

Since corrosion of ceramics quite often involves the diffusion of various cations and anions through an interfacial reaction layer, changes to the chemistry that would either provide a layer through which diffusion is more difficult or provide species that would form a reaction layer immune to continued corrosion should be investigated. This would undoubtedly involve considerable research into the diffusion of various cations and anions through various materials. Only then will it be possible to tailor a composition to provide minimum corrosion.

External Methods of Improvement

In chapters on fundamentals, the importance of temperature was stressed several times. Various techniques have been used to lower the temperature of the interface or hot face of the material (lower hot face temperatures mean less corrosion). Many applications of a ceramic material subject the material to a thermal gradient. By altering the material or providing a means to increase the heat flow through the material, the hot-face temperature can be lowered significantly, or more accurately the slope of the thermal gradient is changed. One means of doing this is by forcibly cooling the cold face. This provides faster heat removal and thus lowers the hot-face temperature. Most industrial furnaces use some means of forced cooling on the cold face by cooling-air systems or water-cooled piping. In a few cases, water has actually been sprayed onto the cold face of the refractory using the heat of vaporization of the water to extract heat from the refractory. If the thermal gradient through the material becomes too steep, failure may occur (this depends upon the thermal expansion characteristics of the material). Pilkington Bros. Ltd. placed water-cooled pipes into the glass inside the furnace at the glass line next to the refractory lining, essentially stopping corrosion at that location. This definitely minimizes corrosion of the refractories but causes a major penalty in thermal efficiency (i.e., higher fuel cost).

Another method that has been used to lower the hot-face temperature is to place metal plates either within individual bricks or between them. A large portion of the heat is thus conducted through the metal plate. A similar technique has been used by manufacturing a product containing oriented graphite particles.

Another way to take advantage of increased cooling is to use a thinner material in the beginning. This will automatically cause a thinner reaction layer to form on the surface. In general, glass furnace basin wall linings should not be greater than 10 to 12 inches thick. Anything greater than about 12 inches does not normally increase overall life but adds an economic penalty in refractory cost per campaign. The thickness at the flux-line generally is nine inches so that effective air cooling can be used. In fact, most linings could probably be less than 10 inches; however, the thermal-mechanical environment will determine the ultimate thickness that should be used.

If a refractory lining is insulated, a greater portion of the refractory will be at a higher temperature and corrosion will proceed at a faster rate. In these cases, a balance must be obtained between service life and energy conservation. Because of the potential for increased corrosion of insulated linings, the properties of the lining material must be carefully evaluated before insulation is installed. In many cases the engineer may want to upgrade the lining material if it is to be insulated. *In no case should an operating furnace be insulated due to the rebalancing of thermal gradients that may cause movement, cracking, chipping, and even failure.*

Historically it has been recommended that flue gas temperatures be 20–30°C higher than the dew point, Meadowcroft and Cox 1985. However, because of thermal efficiencies and the related cost, this has been lowered to 5–10°C higher than the dew point. Once condensation has occurred, re-evaporation of the water or other volatile can concentrate corrosive species, causing a more severe corrosion problem; thus condensates should be removed as rapidly as possible.

When added, redox couples in photoelectrochemical corrosion of electronically conductive materials in acids act on the environment to minimize corrosion. An example is the addition of cobalt as the redox couple to scavenge SO_4^- that is formed by the reaction of a positive hole with the sulfate ion, Harris et al. 1977. The positive hole is photogenerated in the valence band of an illuminated titania semiconductor. The reactions listed below act to minimize corrosion:

$$SO_4^= + p^+ \rightarrow SO_4^- \tag{20.1}$$

$$Co^{2+} + SO_4^- \rightarrow Co^{3+} + SO_4^{2-} \tag{20.2}$$

Crystalline Materials—Nonoxides

Property Improvement

Most of the items discussed earlier can also be applied to these materials. The one property improvement that should be discussed a little further is that of porosity. For example, Si_3N_4 is predominantly covalent and does not densify on heating as do conventional ionic ceramics. In applications such as turbine blades, a theoretically dense material is desired. Only through special densification procedures can theoretically dense materials be obtained. In the past, this could be accomplished for Si_3N_4 only through hot pressing with large amounts (up to 10 wt%) of additives at very high temperatures and pressures. SiC, in contrast, could be prepared in the fully dense state with only a few percent of additives. Newer techniques have recently been developed using gas pressure sintering and much lower amounts of additives that allow the production of materials that are fully dense. The additives in these processes cause a liquid phase to form at high temperatures, and therefore densification can proceed through liquid-phase sintering. This liquid either crystallizes or forms a glass phase upon cooling. Much work has been done in attempting to obtain either crystalline phases with higher melting points or glassy compositions with higher viscosities to improve the high-temperature properties. The densification processes using lower amounts of additives (generally < 2 wt%) help to maximize the high-temperature properties.

Improved corrosion resistance of porous materials can be obtained by impregnating with either a material of the same composition as the bulk or with a material that, in the case of SiC or Si_3N_4, is later exposed to a carbiding or nitriding treatment. Other pore-filling materials can also be used, such as nitrates or oxychlorides. Decomposition reactions then produce pore-filling oxides. Impregnation with organosilicon compounds will yield SiC as the pore filler.

Corrosion resistance can sometimes be improved by changing the processing method. Chemical vapor deposition (CVD) is one of the most attractive methods to produce high-purity dense materials, because the sintering process is not required if a bulk material can be obtained directly from the raw vapors or gases. Microstructures of CVD products are strongly dependent upon the deposition temperature and total gas pressure. CVD can produce materials with no grain boundary phases but which are highly oriented. It is a well-known fact that CVD materials contain residual internal stresses. At present, the effect of these stresses on high-temperature strength and corrosion are not well known.

Preoxidation under some conditions can form a protective oxide layer that will minimize or possibly eliminate continued corrosion, Lange et al. 1983. In addition, impurities present, generally in the form of sintering aids, may migrate toward the surface and become part of the protective oxide layer. This layer can then be removed, resulting in a purer material with subsequent improvement in mechanical properties.

The development of nitride-based materials today has progressed to the point of studying materials in $Si_aM_bO_cN_d$ systems, where M has been confined mostly to trivalent cations. Most work has been in systems where M = Al, Y, and/or Be. These materials form secondary grain boundary phases that are highly oxidation resistant and thus provide a better material than conventional Si_3N_4 materials.

Cemented carbide cutting tools made from WC wear rapidly due to local welding of the tool to the steel piece being cut. To overcome this welding, additions of TiC were made to the WC to form a TiO_2 surface layer that protected the tool from rapid wear. WO_3 also formed, but it was volatile and produced no protective layer. In addition, small amounts of TaC and NbC were added to increase the overall oxidation resistance by increasing the melting temperature of the carbide solution formed. Incorporation of second phases to improve corrosion resistance leads one into the whole area of composites. Many examples similar to the ones above exist. One example is the addition of aluminum to TiN forming a (TiAL)N composite that performs better than TiN in aqueous salt solutions.

External Methods of Improvement

One method of minimizing corrosion that is becoming more prevalent is that of coating the ceramic with a layer of more resistant material. These coatings have recently been called environmental barrier coatings or EBCs (see Table 20.1). Probably the best method to coat the ceramic is by CVD, Davies et al. 1988, or plasma-sprayed material of the same composition as the substrate, Gogotsi and Lavrenko 1992. CVD, in general, provides a better coating than plasma-sprayed coatings, since it is difficult to form pore-free coatings with uniform thickness using plasma spraying. This provides

TABLE 20.1

Environmental Barrier Coatings on Nonoxides

Substrate	Coating	Environment	Reference
Graphite	SiC/Si-MoSi$_2$	Air (1400°C)	Zhao et al. 2006
Carbon nanotubes	SiC	Air (650°C)	Morisada et al. 2004a
Diamond	SiC	Air (950°C)	Morisada et al. 2004b
C/C composite	MoSi$_2$/WSi$_2$	Air (1400°C)	Fu et al. 2009
C/C composite	SiC/Y$_2$Si$_2$O$_7$	Air (1500°C)	Huang et al. 2006
C/C composite	SiC/ZrB$_2$	Air (1700–2600°C)	Corral and Loehman 2008
C/C-SiC comp.	SiC-B$_4$C-SiC/Cordierite	Air (600–1300°C)	Schulte-Fischedick et al. 2004
C/C-SiC comp.	Mullite	Air (1300–1550°C)	Damjanović et al. 2005
SiC or Si$_3$N$_4$	Mullite, mullite/YSZ, Si/mullite/YSZ	Air or water vapor (1300–1500°C)	Lee et al. 2003
SiC/SiC comp or Si$_3$N$_4$	RE$_2$Si$_2$O$_7$, Si/mullite/BSAS	H$_2$O/O$_2$ (up to 1400°C)	Lee et al. 2005
SiC fiber	LaPO$_4$	Air (1000–1200°C)	Mogilevsky et al. 2006
Si$_3$N$_4$	LuSiO$_5$/Lu$_2$Si$_2$O$_7$	Steam (1300°C)	Ueno et al. 2006
Si$_3$N$_4$	HfSiO$_4$, ZrSiO$_4$, TiSiO$_4$	Water vapor (1500°C)	Ueno et al. 2005
Si$_3$N$_4$/TiB$_2$	Al$_2$O$_3$	O$_2$ (up to 1400°C)	Vasques et al. 2005

a well-attached, pure, nonporous layer that has a good thermal expansion match with the substrate. Coating conditions can be varied to produce layers of amorphous material covered by crystalline material of the same composition. This sometimes provides a more complex diffusion path that minimizes oxidation.

Although plasma or flame spraying can be used to deposit most materials, control of the spraying parameters confines the coating to mainly oxides. Other methods investigated have been cathode sputtering, Davies et al. 1988, Gogotsi and Lavrenko 1992, and Gregory and Richman 1984, glow-discharge cathode sputtering, electron beam evaporation, and detonation deposition. These methods are not necessarily confined to the coating of nonoxides; oxides can also be coated. An example of the use of plasma spraying to form protective coatings is that reported by Huang et al. 2006 for the coating of carbon/carbon composites with yttrium silicate. The carbon/carbon composite exhibited excellent resistance to oxidation at 1500°C for 73 hrs. Lee et al. 2003 discussed the use of plasma spraying to form multilayer coatings on SiC that consisted of a silicon bond layer, a mullite intermediate layer, and an outer layer of barium aluminum silicate (essentially celsian with some strontium). Another method (called the reaction method) that works well in many cases is the application of a slurry of the ingredients for the EBC onto the substrate and then sintering. This method was used by Fu et al. 2009 to form a multilayer of SiC inner layer and a Si-Mo-W ($MoSi_2$ and WSi_2) outer layer on a carbon/carbon composite. The excellent oxidation resistance in air of these coatings was attributed by Fu et al. to the formation of a silica film on the coating surface at temperatures between 1400 and 1500°C. Other multilayer coatings have been investigated to protect carbon-based materials: SiC/ Al_2O_3/mullite (Huang et al. 2003), SiC/mullite (Fritze et al. 1998), and SiC/ borosilicate glass (Li and Zeng 2004).

Even though the discussion of this chapter is about minimizing the corrosion of ceramics, the whole area of thermal and environmental barrier coatings that are applied to metals should not be ignored. Ceramic coatings of oxides and nonoxides, in either single or multiple layers, have been used to protect steels and some other metals in very harsh environments. Included in this area are the ceramic glazes applied to metals (normally called enamels). These enamels would fall under the title of environmental barrier coatings in a modern classification since any elevated temperatures are very low.

Wittmer and Temuri 1991, in their work on oxidation of carbon-carbon composites, have described a method of protection by coating first with a well-adhering solid oxygen barrier and then coating with a glass-forming material to seal any cracks that may develop from thermal expansion mismatch. The carbon-carbon composite coating system used for the Space Shuttle nose cap is composed of an inner SiC layer covered by a silicate glaze. This is probably the most successful example of the use of oxygen barrier coatings to protect carbon-carbon composites, Strife 1991.

Glassy Materials

Property Optimization

The development of more resistant glasses has been predominantly through optimization of compositions. Historically, small amounts of alumina have been added to the basic soda-lime-silicate composition to improve durability. In general, lowering the alkali content increases the durability. This, however, has practical limits based on melting temperatures, viscosities, softening points, and working ranges. Borosilicate glasses are, in general, more resistant than soda-lime silicate glasses. In general, silicate glasses are less resistant to alkali solutions than they are to acid solutions. Table 16.1 lists the corrosion resistance of many glasses of varying compositions.

One technique of composition variation to improve durability that has not received much attention is that of incorporation of nitrogen into the glass structure. Frischat and Sebastian 1985 have shown that soda-lime-silica glasses containing 1.1 wt% nitrogen exhibit considerable improvement toward leaching by water at 60°C over compositions containing no nitrogen. This improvement was attributed to a denser structure for the nitrogen-containing glass.

Small changes in the chemistry of the glass can cause a significant change in the dissolution mechanism as shown by Lehman and Greenhut 1982. They reported that 1 mol% P_2O_5 addition to a lead silicate glass caused the formation of lead phosphosilicate crystals on the glass surface when exposed to 1% acetic acid at 22°C. They attributed the reduction in dissolution to the reduction of the apparent average interdiffusion coefficient of lead by a factor of 11.3. This is an example of changing the material chemistry to form an interface reaction product that reduces the diffusion rate of the species being leached.

External Methods of Improvement

The development of coating technology has provided a means to improve corrosion resistance, abrasion resistance, and strength. Combinations of coatings applied while the glass is hot and after it has cooled have been developed that form a permanent bond to the glass. These coatings are not removed by cooking or washing.

The most commonly used metallic hot-end coatings are tin and titanium. As the piece goes through the annealing lehr, the metal oxidizes, forming a highly protective ceramic coating. Tin is easier to work with since a thicker coating can be applied before problems of iridescence occur. These hot-end metallic coatings give the glass a high glass-to-glass sliding friction and thus a cold-end coating must be applied over these metallic coatings. The cold-end coatings usually have a polyethylene or fatty acid base.

Another type of coating is one that reacts with the surface of the glass to form a surface layer that is more corrosion resistant than the bulk composition.

Chemically inert containers are needed to contain various beverages and pharmaceuticals. To provide increased corrosion resistance, these containers are coated internally to tie up the leachable components. Internal treatment with a fluoride gas provides a new surface that is more corrosion resistant than the original and is more economical than the older sulfur treatment.

Although not a true coating technique, the manufacturers of flat glass have for many years treated the surface of their glass with SO_2 gas just prior to the glass being annealed to increase the weatherability of their products. This surface treatment allows the sodium in the surface layers to react with the SO_2, forming sodium sulfate. The sulfate deposit that forms on the surface due to this reaction is then washed off prior to inspection and packing. The first step in weathering is then diminished due to the low alkali content of the surface.

It has been shown by Harvey and Litke 1984, that matrix dissolution of an aluminosilicate glass apparently does not occur if the leaching solution is saturated first with solution products of the same glass composition. This technique is an example of how dissolution can be minimized by decreasing the driving force for corrosion by lowering the concentration gradient between the material and leachant, thus minimizing or eliminating the diffusion of cations and anions across the interfacial boundary. Using a different approach to minimize dissolution of a predominantly soda-borosilicate glass, Buckwalter and Pederson 1982 have shown that the sorption of metal ions onto the glass surface and/or the buffering of the leachate solution caused by the corrosion of metal containers significantly lowered the rate of aqueous corrosion.

Exercises, Questions, and Problems

1. Describe three different ways to minimize corrosion. What are the advantages and disadvantages of each?

2. What parameters affect corrosion reactivity?

3. What properties of a material could be altered to improve its corrosion resistance?

4. What are environmental barrier coatings and how do they work?

References

Brezny, R. and Semler, C.E. 1984. Oxidation and Diffusion in Selected Pitch-Bonded Magnesia Refractories, *J. Am. Ceram. Soc.*, 67(7), 480–83.

Buckwalter, C.Q. and Pederson, L.R. 1982. Inhibition of Nuclear Waste Glass Leaching by Chemisorption, *J. Am. Ceram. Soc.,* 65(9), 431–36.

Corral, E. and Loehman, R.E. 2008. Ultra-High-Temperature Ceramic Coatings for Oxidation Protection of Carbon-Carbon Composites, *J. Am. Ceram. Soc.,* 91(5), 1495–1502.

Damjanović, T., Argirusis, C., Borchardt, G., Leipner, H., Herbig, R., Tomandl, G. and Weiss, R. 2005. Oxidation Protection of C/C-SiC Composites by an Electrophoretically Deposited Mullite Precursor, *J. European Ceram. Soc.,* 25, 577–87.

Davies, G.B., Holmes, T.M. and Gregory, O.J. 1988. Hot Corrosion Behavior of Coated Covalent Ceramics, *Adv. Ceram. Mat.,* 3(6), 542–47.

Frischat, G.H. and Sebastian, K. 1985. Leach Resistance of Nitrogen-Containing Na_2O-CaO-SiO_2 Glasses, *J. Am. Ceram. Soc.,* 68(11), C305–7.

Fritze, H., Jojie, J., Witke, T., Ruscher, C. Weber, S. Scherrer, S., et al. 1998. Mullite Based Oxidation Protection for SiC-C/C Composites in Air at Temperatures up to 1900 K, *J. European Ceram. Soc.,* 18(16), 2351–64.

Fu, Q-G., Li, H-J., Li, K-Z. and Tong, K. 2009. A Si-Mo-W Coating to Protect SiC-Coated Carbon/Carbon Composites Against Oxidation, *J. Am. Ceram. Soc.,* 92(9), 2132–35.

Gogotsi, Yu. G. and Lavrenko, V.A. 1992. Corrosion Protection and Development of Corrosion-Resistant Ceramics, Chp. 7 in *Corrosion of High-Performance Ceramics,* Springer-Verlag, Berlin, pp. 151–62.

Gregory, O.J. and Richman, M.H. 1984. Thermal Oxidation of Sputter-Coated Reaction-Bonded Silicon Nitride, *J. Am. Ceram. Soc.,* 67(5), 335–40.

Harris, L.A., Cross, D.R. and Gerstner, M.E. 1977. Corrosion Suppression on Rutile Anodes by High Energy Redox Reactions, *J. Electrochem. Soc.,* 124(6), 839–44.

Harvey, K.B. and Litke, C.D. 1984. Model for Leaching Behavior of Aluminosilicate Glasses Developed as Matrices for Immobilizing High-Level Wastes, *J. Am. Ceram. Soc.,* 67(8), 553–56.

Huang, J.F., Zeng, X.R. and Li, H.J. 2003. Mullite-Al_2O_3-SiC Oxidation Protective Coating for C/C Composites, *Carbon,* 41(14), 2825–29.

Huang, J-F., Li, H-J., Zeng, X-R. and Li, K-Z. 2006. Yttrium Silicate Oxidation Protective Coating for SiC Coated Carbon/Carbon Composites, *Ceram. International,* 32, 417–21.

Lange, F.F., Davis, B.I. and Metcalf, M.G. 1983. Strengthening of Polyphase Si3N4 Materials through Oxidation, *J. Mater. Sci.,* 18(5), 1497–505.

Lee, K.N., Fox, D.S., Eldridge, J.I., Zhu, D., Robinson, R.C., Bansal, N.P. and Miller, R.A. 2003. Upper Temperature Limit of Environmental Barrier Coatings Based on Mullite and BSAS, *J. Am. Ceram. Soc.,* 86(8), 1299–306.

Lee, K.N., Fox, D.S. and Bansal, N.P. 2005. Rare Earth Silicate Environmental Barrier Coatings for SiC/SiC Composites and Si_3N_4 Ceramics, *J. European Ceram. Soc.,* 25, 1705–15.

Lehman, R.L. and Greenhut, V.A. 1982. Surface Crystal Formation during Acid Corrosion of Phosphate-Doped Lead Silicate Glass, *J. Am. Ceram. Soc.,* 65(9), 410–14.

Li, H.J. and Zeng, X.R. 2004. A New SiC/yttrium Silicate/Glass Multi-layer Oxidation Protective Coating for Carbon/Carbon Composites, *Carbon,* 42(11), 2356–59.

McCauley, R.A. 1978. Evolution of Flat Glass Furnace Regenerators, *Glass Ind.,* 59(10), 26–28, 34.

Meadowcroft, D.B. and Cox, W.M. 1985. Dewpoint Corrosion: Mechanisms and Solutions, Chapter 2 in *Dewpoint Corrosion*, Holmes (ed.), Ellis Horwood Ltd., Chichester, UK.

Mogilevsky, P., Boakye, E.E., Hay, R.S., Welter, J. and Kerans, R.J. 2006. Monazite Coatings on SiC Fibers II: Oxidation Protection, *J. Am. Ceram. Soc., 89*(11), 3481–90.

Morisada, Y., Maeda, M., Shibayanagi, T. and Miyamoto, Y. 2004a. Oxidation Resistance of Multiwalled Carbon Nanotubes Coated with Silicon Carbide, *J. Am. Ceram. Soc., 87*(5), 804–8.

Morisada, Y., Miyamoto, Y., Moriguchi, H., Tsuduki, K. and Ikegaya, A. 2004b. Growth Mechanism of Nanometer-Sized SiC and Oxidation Resistance of SiC-Coated Diamond Particles, *J. Am. Ceram. Soc., 87*(5), 809–13.

Schulte-Fischedick, J., Schmidt, J., Tamme, R., Kröner, U., Arnold, J. and Zeiffer, B. 2004. Oxidation Behavior of C/C-SiC Coated with SiC-B$_4$C-SiC-Cordierite Oxidation Protection System, *Materials Sci. and Engr. A, 386*, 428–34.

Siljan, O-J., Rian, G., Pettersen, D.T., Solheim, A. and Schøning, C. 2002. Refractories for Molten Aluminum Contact Part I: Thermodynamics and Kinetics, *Refractories Applications and News, 7*(6), 17–25.

Strife, J.R. 1991. Fundamentals of Protective Coating Strategies for Carbon-Carbon Composites, pp. 121–27 in *Damage and Oxidation Protection in High Temperature Composites*, Vol. 1, Haritos, G.K. and Ochoa, O.O. (eds.), ASME, New York, NY.

Ueno, S., Jayaseelan, D.D., Ohji, T. and Lin, H-T. 2005. Corrosion and Oxidation Behavior of ASiO$_4$ (A = Ti, Zr, and Hf) and Silicon Nitride with HfSiO4 Environmental Barrier Coating, *J. Ceram. Processing Res., 6*(1), 81–84.

Ueno, S., Ohji, T. and Lin, H.-T. 2006. Comparison of Hot-Corrosion Behavior of Al$_3$O$_3$, Lu$_2$O$_3$ and Their Silicates, *J. Ceram. Processing Res., 7*(3), 201–5.

Vasques, B., Tixier, C., Klein, R., Hidalgo, H., Tristant, P. and Desmaisson, J. 2005. Modification of the oxidation behavior of Si$_3$N$_4$ – TiB$_2$ composites by PECVD coatings, *J. Euro. Ceram. Soc., 25*(10), 1743–8.

Wittmer, D.E. and Temuri, M.Z. 1991. Thermochemical Studies in Selected Metal-Carbon-Oxygen Systems, *J. Am. Ceram. Soc., 74*(5), 973–82.

Zhao, J., Liu, L., Guo, Q., Shi, J. and Zhai, G. 2006. Oxidation Protective Behavior of SiC/Si-MoSi$_2$ Coating for Different Graphite Matrix, *Materials Letters, 60*, 1964–67.

Epilogue

And one more thing.

Steve Jobs

The literature and data available on the corrosion of ceramics indicate that corrosion occurs by either one of several possible mechanisms or a combination of these mechanisms. Many similarities exist between the corrosion of crystalline and glassy ceramics, although in general glass corrodes more rapidly under identical environmental conditions.

Corrosion in either crystalline or glassy ceramics can occur by a direct process where the ceramic congruently dissolves into the corroding medium. Reaction rates are generally linear, being proportional to the duration of the test. One way to minimize this type is to saturate the corroding medium with the same chemical species that are dissolving from the ceramic. Another way to minimize this type of corrosion is to add something to the ceramic that will diffuse to the surface and react with the corroding medium, forming a protective interface layer.

In another type, apparently the more common type of corrosion process, indirect, in either crystalline or glassy ceramics, species from both the corroding medium and the ceramic counterdiffuse and react at the interface, forming a glassy, a crystalline, or a gaseous interface reaction product. If the interface reaction product is solid, continued corrosion can occur only by continued diffusion through the interface. In some cases, the interface reaction product may be multilayered. The reaction layer thickness may vary from a few nanometers to several hundred micrometers. Reaction rates are generally parabolic, being proportional to the square root of time. One way to minimize this type of corrosion is to pre-react the ceramic to form an initial interface reaction layer that, if protective, will slow continued reaction. Another way is to add something to the ceramic that will form a layer through which diffusion will be more difficult.

In the first case discussed above, the corroding medium can be either a liquid or a gas; however, in gaseous corrosion one may not consider the *dissolution* to be congruent if the products are two different gases, as in the active oxidation of SiC to SiO and CO_2. In the second case above, the medium can again be either liquid or gaseous with either all or part of the ceramic forming the layer. In most cases, only part of the ceramic forms the layer (i.e., selective dissolution). In corrosion by liquids, the mechanisms are different if the corroding medium is a glass/slag versus water. In water, the first step is usually ion exchange, whereas in glass/slag attack, the first step is counterdiffusion, not quite the same as ion exchange, although ion exchange may take place in glass/slag attack.

Multicomponent ceramics generally corrode by a mixed mechanism with each step exhibiting a different and unique reaction rate. In these cases, the overall reaction rate will exhibit a mixed rate law, being neither linear nor parabolic.

Extended duration tests have indicated that the mechanism of corrosion may change after some extended time. This is especially true for oxide layers formed on nonoxide ceramics during gaseous corrosion. This change in mechanism is due to one or more of the following changes: crystallization of amorphous layers, alteration of crystalline phases as diffusion continues, cracking due to crystallization and alteration, and spalling. The few studies that have shown these changes indicate that one must be careful in making lifetime predictions based upon data from short-time laboratory tests.

In all cases, an increase in temperature increases the rate of corrosion. The mechanism of corrosion, however, may change as temperature is increased due to crystallization of amorphous reaction layers, polymorphic transitions, melting of crystalline layers, vaporization of various species in the layer, cracking, etc.

One method of minimizing corrosion that requires more emphasis appears to be the various coating methods. These could be used to advantage in composites where the initial step is, for example, oxidation of SiC fibers. By coating the fibers before incorporation into the matrix, oxidation may be slowed or even eliminated. The object is to find a material through which the diffusion of oxygen is a minimum and then use this material to coat the fibers. The technique of electrostatic attraction in an aqueous dispersion appears attractive as a coating method for materials such as fibers.

Although the above discussion may be an oversimplification of the corrosion processes that occur in ceramics, it is a step in the direction of simplifying and unifying the whole area. All of the data and discussion about corrosion point toward the need for more in-depth diffusion and solubility studies of the various species in the different corroding media encountered in practice.

Corrosion, being an interfacial process, requires a thorough understanding of the surface structure of the materials being corroded. Surface energy is the greatest for the least densely packed planes. The most compact planes are also the most thermodynamically stable. Thus the study of single crystals is the best method to determine the fundamentals of corrosion mechanisms. Although the crystal surface characteristics determine short-term corrosion behavior, they may not be as important for long-term corrosion. Single crystals do lend themselves to the evaluation of the effects that various dopants have upon leaching kinetics. In addition, various types of defects (e.g., vacancies, dislocations, etc.) could be incorporated into the lattice during production of the single crystals.

A large amount of published data on the corrosion of crystalline and glassy ceramics points toward the fact that more compact structures are more durable. It has been reported that synthesis or processing steps of nonoxides that require higher temperatures result in more ordered structures that correlate with increased oxidation resistance. In the study of glasses, references are made to corrosion being a function of glass structures, which are related to parameters such as composition, the number of nonbridging oxygens, the amount of cross-linking of the network structure, the degree of network packing, the density, the strength of the bonding, and the amount of covalent bonding. References have also been made to compact, strongly bonded glass structures being those with low thermal expansion and high softening points. Thus a technique that would determine the structural tightness may be sufficient to rank the durability of various materials, at least in the various compositional classes and to a specific environment. In addition to thermal expansion and softening point determinations, the determination of hardness may also yield information related to durability. Hardness is a measurement, however, that must be performed with some care, since hardness varies with the applied load and cracking and friction may interfere with the measurements. There has been no systematic study reported in the literature of the corrosion of ceramics related to properties such as expansion, hardness, or softening point.

Only through a thorough understanding of all the parameters involved can the engineer make an intelligent selection of the material that will best resist corrosion for a particular application. Only through intelligent materials selection can the cost of corrosion be minimized. Since the application of ceramics requires the optimization of properties other than corrosion resistance, a compromise between corrosion resistance, properties, and cost is generally needed.

If we knew what we were doing, it would not be called research, would it?

Albert Einstein

Glossary of Terms

Cor · rode, v.t. to eat into or wear away gradually, as by rusting or by the action of chemicals.

Webster's New World Dictionary

Alteration: The change or modification of a material through interaction with its environment, generally by the formation of a new phase. This reaction need not be deleterious.

Associative decay: The type of corrosion when a third medium is the originator of the corrosion between two materials in intimate contact. The third medium reacts with the first material forming a product that reacts with the second material.

Atmospheric corrosion: The degradation of materials by natural atmospheric environments. *Atmospheric corrosion* is a term often used by the metallurgist, whereas an equivalent term used by ceramists is *weathering.*

Biodegradation: Has been used to describe the beneficial effects caused by the activities of organisms upon materials, although the actual degradation could be essentially the same as any deterioration.

Biodeterioration: Any undesirable change in the properties of a material caused by the vital activities of organisms.

Biodurability: The ability of a ceramic to withstand the chemical action of a biological environment (i.e., to resist chemical dissolution).

Biopersistence: The ability of a mineral to resist the physiological processes of the human body (includes biodurability and the various clearance mechanisms).

Biosolubility: The ability of body fluids to dissolve a mineral (i.e., ceramic).

Bioreactivity: The extent to which a mineral will react with body fluids. There is a difference between acute and chronic reactivity. Acute reactivity causes rapid changes in body fluids and possible tissue damage, whereas chronic reactivity is a very slow (being essentially biodurable) reaction releasing chemicals into the body fluids over a very long time.

Condensation corrosion: Equivalent to dew point corrosion.

Corrosion: The irreversible chemical interaction of a ceramic with its environment, generally producing a deleterious effect. This chemical reaction can, in some cases, be put to beneficial use.

Dealkalization: The corrosion of a ceramic through the selective solution of the alkalies into the corroding medium. Generally used to describe the removal of alkalies from glasses.

Dew point corrosion: The deterioration of a solid ceramic material caused by the condensation of a corrosive liquid from a saturated gas when the temperature is lowered below the point (the dew point) where the liquid will condense. A form of *atmospheric corrosion* and equivalent to *condensation corrosion*.

Dissolution—anisotropic: When the ceramic material dissolves at different rates parallel to different crystallographic or geometric directions.

Dissolution—congruent, direct, or homogeneous: When the total ceramic chemistry dissolves simultaneously into the environment.

Dissolution corrosion: The corrosion of a ceramic through the solution of its various components into the corroding medium (generally a liquid).

Dissolution—incongruent, indirect, or heterogeneous: When the ceramic dissolves in such a way as to leave behind a material chemically different than the original ceramic as an interface between the ceramic and the corroding medium. These terms generally imply that the dissolution is selective.

Dissolution—isotropic: When the ceramic material dissolves at the same rate for all crystallographic or geometric directions.

Dissolution—selective: The corrosion of a ceramic through the selective solution of one or more (but not all) species into the corroding medium.

Durability: The ability of a ceramic to withstand the action of its environment.

Degradation: A general decrease or lowering of the quality of a ceramic, often through corrosive action.

Electrochemical corrosion: The corrosion that takes place when the reaction occurring involves electronic charge transfer. Generally this type occurs when ceramics are in contact with aqueous media, but may also occur in other media.

Galvanic corrosion: The corrosion that takes place when two chemically dissimilar ceramics are in contact with one another, both of which are in contact with the same electrolyte. Reaction occurs only when current flows in an external circuit (not through the electrolyte). A type of electrochemical corrosion.

Hot corrosion: Normally used to designate high-temperature oxidation of ceramics in contact with molten salt deposits. This definition should probably not be used, since the term *hot corrosion* is nonspecific and could apply to any type of corrosion at an elevated temperature.

Intergranular or grain boundary corrosion: The corrosion through any mechanism that takes place preferentially along grain boundaries or between grains.

Leaching: To remove through dissolution a portion of a ceramic material.

Leaching—selective: Removes one species in preference to another. The use of the word *selective* in this case is superfluous.

Microbiologically influenced corrosion (MIC): Corrosion resulting from the presence and activities of microorganisms within biofilms on a material surface.

Oxidation embrittlement: The embrittlement of a composite caused by the oxidation of the interface between the reinforcement material and the matrix resulting in a strong bond between the two.

Pesting: The formation of a powder-like deposit on the exposed surface of metallic silicides (i.e., $MoSi_2$) during oxidation.

Photoelectrochemical corrosion: Electrochemical corrosion that takes place when the charge transfer involves the positive holes formed by photon illumination. Also called *photodissolution*.

Stress corrosion: Corrosion by any mechanism that is enhanced by the presence of either a residual or applied stress.

Thermo-oxidative stability: The resistance to oxidation at elevated temperatures. Generally used in the discussion of composite materials.

Tribocorrosion: The combined processes of chemical corrosion and mechanical erosion. A very important process in the grinding and polishing of samples, human hip joint implants, and dental implants.

Weathering: This term describes the atmospheric effects upon materials of the construction industry, mostly structural clay products, sandstone, limestone, marble, and glass, and is essentially the attack by water vapor, CO_2, and SO_2.

Index